PUMPING MACHINERY THEORY AND PRACTICE

PUMPING MACHINERY THEORY AND PRACTICE

Hassan M. Badr
Wael H. Ahmed
King Fahd University of Petroleum and Minerals
Saudi Arabia

Registered Office
John Wiley & Sons Ltd, The Atrium, Southern Gate, Chichester, West Sussex, PO19 8SQ, United Kingdom

For details of our global editorial offices, for customer services and for information about how to apply for
permission to reuse the copyright material in this book please see our website at www.wiley.com.

Library of Congress Cataloging-in-Publication Data

Badr, Hassan M.
Pumping machinery theory and practice / Hassan M. Badr, Wael H. Ahmed.
   pages   cm
   Includes bibliographical references and index.
   ISBN 978-1-118-93208-7 (hardback)
1. Pumping machinery.   I. Ahmed, Wael H.   II. Title.
   TJ900.B193 2015
   621.6–dc23

                                        2014031896

A catalogue record for this book is available from the British Library.

Set in 10/12pt Times by SPi Publisher Services, Pondicherry, India

1   2015

To my parents, my dear wife and my children

Contents

Preface

Energy consumption in pumping systems accounts for approximately 20% of the world's electrical energy demand. Moreover, the operational cost of pumping machinery far outweighs their capital cost. Accordingly, engineers strive for optimum equipment performance for achieving economic operation. A thorough understanding of the components and principles of operation of these machines will provide an opportunity to dramatically reduce energy, operational and maintenance costs. Reducing energy consumption will also complement the current thrust towards protecting our environment.

This book is intended to be a basic reference on theoretical foundation and applications of various types of pumping machinery. In view of the great importance of pumps and compressors in almost every engineering system, this book presents the fundamental concepts underlying the flow processes taking place in these machines and the transformation of mechanical energy into fluid power. Special emphasis is given to basic theoretical formulation and design considerations of pumps and compressors in addition to improving problem-solving skills. This is achieved through the presentation of solved examples of applied nature using analytical means and/or basic engineering practices.

The book consists of ten chapters covering two main themes: the first smoothly introduces the essential terminology, basic principles, design considerations, and operational-type problems in pumping machinery. This part is supported by a good number of worked examples plus problems at the end of each chapter for the benefit of senior undergraduate students and junior engineers. This is considered a key feature of this book because other books in this area rarely provide enough worked problems and exercises. The second theme focuses on advanced topic such as two-phase flow pumping systems targeting practicing field engineers and introductory research scientists.

The authors wish to acknowledge their students' encouragement to write this book. The idea was initiated by the first author after searching for a good textbook for an undergraduate course in pumping machinery that he has been teaching for over 20 years. The absence of a suitable textbook demanded the preparation of a set of course notes to help the students to a better understanding of the subject. The support received from King Fahd University of Petroleum & Minerals under Grant # IN111025 for the preparation of this textbook is greatly appreciated.

Nomenclature

A	area
b	vane width
BP	brake power
C	speed of sound
C_c	contraction coefficient
C_H	head coefficient
C_P	power coefficient
C_Q	flow coefficient
D	impeller diameter
f	friction coefficient
g	gravitational acceleration
h	enthalpy
h_{ss}	static suction head
h_{sd}	static delivery head
h_s	suction head
h_d	delivery head
H	pump total head
$\underline{H}$	angular momentum
I	rotational enthalpy
k	specific heat ratio
K	loss coefficient for pipe fittings
L	length of connecting rod
$\dot{m}$	mass flow rate
M	Mach number
$\underline{M}$	moment
n_s	specific speed in SI
N_s	specific speed in the American system
$NPSH$	net positive suction head

N	speed of rotation in rpm
p	pressure
p_o	stagnation pressure
p_v	vapor pressure
P	power
Q	volume flow rate
r	crank radius
R	gas constant
S	suction specific speed
T	torque or temperature
T_u	unbalanced thrust
u	tangential velocity
v	flow velocity
V	whirl velocity component
V_r	relative velocity
x	coordinate
y	coordinate
Y	radial velocity component
z	elevation (measured from selected datum)

Greek Symbols

α	flow angle
β	vane angle
γ	specific weight of fluid
η	efficiency
λ	degree of reaction
μ	fluid viscosity
ν	fluid kinematic viscosity
θ	crank angle
ρ	fluid density
σ	Thoma's cavitation factor
ω	angular velocity

Subscripts

atm.	atmospheric
crit.	critical
d	discharge
e	Euler
f	friction
hyd.	hydraulic
L	leakage

mech.	mechanical
o	overall
r	relative
p	pressure
s	suction
sd	static delivery
sn	suction nozzle
ss	static suction
st	total static
u	unbalanced
v	vane
vol.	volumetric/volute
V	velocity

1

Essentials of Fluid Mechanics

The basic fundamentals of fluid mechanics are essential for understanding the fluid dynamics of pumping machinery. This chapter aims to provide a quick revision of the definitions and basic laws of fluid dynamics that are important for a thorough understanding of the material presented in this book. Of particular interest are the kinematics of fluid flow; the three conservation principles of mass, momentum, and energy; relevant dimensionless parameters; laminar and turbulent flows; and friction losses in piping systems. Some applications of relevance to pumping machinery are also considered.

1.1 Kinematics of Fluid Flow

To fully describe the fluid motion in a flow field it is necessary to know the flow velocity and acceleration of fluid particles at every point in the field. This may be a simple task in laminar flows but may be difficult in turbulent flows. If we use the Eulerian method and utilize Cartesian coordinates, the velocity vector at any point in a flow field can be expressed as

$$\underline{V} = u\underline{i} + v\underline{j} + w\underline{k} \tag{1.1}$$

where $\underline{V}$ is the velocity vector; u, v, and w are the velocity components in the x, y, and z directions; and $\underline{i}$, $\underline{j}$, and $\underline{k}$ are unit vectors in the respective directions. In general, each of the velocity components can be a function of position and time, and accordingly we can write

$$u = u(x, y, z, t), \ v = v(x, y, z, t), \ w = w(x, y, z, t) \tag{1.2}$$

The components of acceleration in the three directions can be expressed as

Pumping Machinery Theory and Practice, First Edition. Hassan M. Badr and Wael H. Ahmed.
© 2015 John Wiley & Sons, Ltd. Published 2015 by John Wiley & Sons, Ltd.

$$a_x = \frac{\partial u}{\partial t} + u\frac{\partial u}{\partial x} + v\frac{\partial u}{\partial y} + w\frac{\partial u}{\partial z} \tag{1.3a}$$

$$a_y = \frac{\partial v}{\partial t} + u\frac{\partial v}{\partial x} + v\frac{\partial v}{\partial y} + w\frac{\partial v}{\partial z} \tag{1.3b}$$

$$a_z = \frac{\partial w}{\partial t} + u\frac{\partial w}{\partial x} + v\frac{\partial w}{\partial y} + w\frac{\partial w}{\partial z} \tag{1.3c}$$

The acceleration vector becomes

$$\underline{a} = a_x\underline{i} + a_y\underline{j} + a_z\underline{k} \tag{1.4}$$

This vector can be split into two components, the local component, $\underline{a}_{local}$, and the convective component, $\underline{a}_{conv.}$, that can be expressed as

$$\underline{a}_{local} = \frac{\partial u}{\partial t}\underline{i} + \frac{\partial v}{\partial t}\underline{j} + \frac{\partial w}{\partial t}\underline{k} \tag{1.5a}$$

$$\underline{a}_{conv.} = \left(u\frac{\partial u}{\partial x} + v\frac{\partial u}{\partial y} + w\frac{\partial u}{\partial z}\right)\underline{i} + \left(u\frac{\partial v}{\partial x} + v\frac{\partial v}{\partial y} + w\frac{\partial v}{\partial z}\right)\underline{j} + \left(u\frac{\partial w}{\partial x} + v\frac{\partial w}{\partial y} + w\frac{\partial w}{\partial z}\right)\underline{k} \tag{1.5b}$$

1.1.1   Types of Flows

The flow field can be described as *steady* or *unsteady*, *uniform* or *non-uniform*, *compressible* or *incompressible*, *rotational* or *irrotational*, *one-*, *two-*, or *three-dimensional*, and can also be described as *laminar* or *turbulent*. The flow is said to be steady if the velocity vector at any point in the flow field does not change with time.

Accordingly, the local component of acceleration ($\underline{a}_{local}$) vanishes if the flow is steady. The flow can also be described as uniform if the velocity vector does not change in the streamwise direction. For example, the pipe flow shown in Figure 1.1 is uniform since the velocity vector does not change downstream, but the flow in the bend shown in Figure 1.2 is non-uniform.

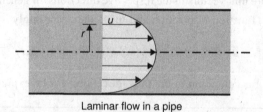

Laminar flow in a pipe

Figure 1.1   Laminar flow in a pipe as an example of uniform flow

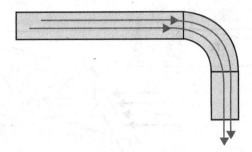

Figure 1.2   Flow in a 90° bend as an example of non-uniform flow

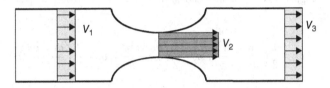

Figure 1.3   One-dimensional flow in a pipe with constriction

The flow is described as incompressible if the density change within the flow field does not exceed 5%. Accordingly, most of the flows in engineering applications are incompressible as, for example, flow of different liquids in pipelines and flow of air over a building. However, compressible flows occur in various applications such as flow in the nozzles of gas and steam turbines and in high speed flow in centrifugal and axial compressors. In general, the flow becomes compressible if the flow velocity is comparable to the local speed of sound. For example, the flow of air in any flow field can be assumed incompressible up to a Mach number of 0.3.

The flow is called one-dimensional (1-D) if the flow parameters are the same throughout any cross-section. These parameters (such as the velocity) may change from one section to another. As an approximation, we may call pipe or nozzle flows 1-D if we are interested in describing the average velocity and its variation along the flow passage. Figure 1.3 shows an example of 1-D flow in a pipe with constriction. On the other hand, the flow is called 2-D if it is not 1-D and is identical in parallel planes. For example, the viscous flow between the two diverging plates shown in Figure 1.4 is two-dimensional. In this case, two coordinates are needed to describe the velocity field.

If the flow is not 1-D or 2-D, it is then three-dimensional. For example, flow of exhaust gases out of a smoke stack is three-dimensional. Also, air flow over a car or over an airplane is three-dimensional.

1.1.2   *Fluid Rotation and Vorticity*

The rate of rotation of a fluid element represents the time rate of the angular displacement with respect to a given axis. The relationship between the velocity components and the rate of rotation can be expressed as

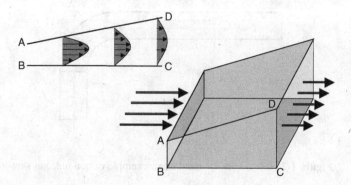

Figure 1.4 Two-dimensional flow between two diverging plates

$$\omega_x = \frac{1}{2}\left(\frac{\partial w}{\partial y} - \frac{\partial v}{\partial z}\right), \ \omega_y = \frac{1}{2}\left(\frac{\partial u}{\partial z} - \frac{\partial w}{\partial x}\right), \ \omega_z = \frac{1}{2}\left(\frac{\partial v}{\partial x} - \frac{\partial u}{\partial y}\right) \tag{1.6}$$

where ω_x, ω_y, ω_z represent the rate of rotation around the x, y, and z axes.

The vorticity ζ is defined as twice the rate of rotation. Accordingly, the vorticity vector $\underline{\zeta}$ can be expressed as

$$\underline{\zeta} = \zeta_x \underline{i} + \zeta_y \underline{j} + \zeta_z \underline{k} = 2\omega_x \underline{i} + 2\omega_y \underline{j} + 2\omega_z \underline{k} \tag{1.7}$$

The flow is called irrotational when the rate of rotation around the three axes is zero. In this case, we must have $\zeta_x = \zeta_y = \zeta_z = 0$ for irrotational flow. The components of the vorticity vector in cylindrical coordinates can be written as

$$\zeta_r = \frac{1}{r}\frac{\partial v_z}{\partial \theta} - \frac{\partial v_\theta}{\partial z} \tag{1.8a}$$

$$\zeta_\theta = \frac{\partial v_r}{\partial z} - \frac{1}{r}\frac{\partial}{\partial r}(rv_z) \tag{1.8b}$$

$$\zeta_z = \frac{1}{r}\frac{\partial}{\partial r}(rv_\theta) - \frac{1}{r}\frac{\partial v_r}{\partial \theta} \tag{1.8c}$$

1.2 Conservation Principles

1.2.1 Conservation of Mass

Considering the general case of a compressible flow through the control volume (c.v.) shown in Figure 1.5 and assuming that n is a unit vector normal to the elementary surface area dA and v is the flow velocity through this area, then the conservation of mass equation takes the form

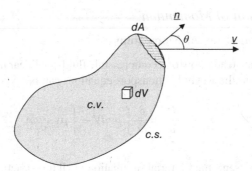

Figure 1.5 A schematic of an arbitrary control volume showing the flow velocity through a small elementary surface area

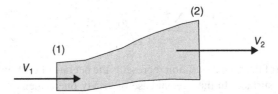

Figure 1.6 One-dimensional flow in a diverging flow passage

$$\frac{\partial}{\partial t}\int_{c.v.} \rho \, dV + \int_{c.s.} \rho \underline{v}.\underline{n} \, dA = 0 \qquad (1.9)$$

where ρ is the fluid density, $\underline{v}$ is the fluid velocity, dV is an elementary volume, and t is the time. When the control volume tends to a point, the equation tends to the differential form,

$$\frac{\partial \rho}{\partial t} + \frac{\partial(\rho u)}{\partial x} + \frac{\partial(\rho v)}{\partial y} + \frac{\partial(\rho w)}{\partial z} = 0 \qquad (1.10)$$

where u, v, and w are the velocity components in the x, y, and z directions. If the flow is incompressible, the above equation can be reduced to

$$\frac{\partial u}{\partial x} + \frac{\partial v}{\partial y} + \frac{\partial w}{\partial z} = 0 \qquad (1.11)$$

In the special case of 1-D steady flow in a control volume with one inlet and one exit (Figure 1.6), the conservation of mass equation takes the simple form,

$$\dot{m} = \rho_1 A_1 V_1 = \rho_2 A_2 V_2 = Const. \qquad (1.12a)$$

where $\dot{m}$ is the mass flow rate, V is the flow velocity, and A is the cross-sectional area.

1.2.2　Conservation of Momentum

1.2.2.1　Conservation of Linear Momentum

In the general case of unsteady flow of a compressible fluid, the linear momentum conservation equation (deduced from the Reynolds transport equation) can be expressed as

$$\sum \underline{F} = \frac{d\underline{M}}{dt} = \frac{\partial}{\partial t} \int_{c.v.} \rho \underline{v} dV + \int_{c.s.} \rho \underline{v}.\underline{v}.dA \tag{1.12b}$$

where the term $\Sigma \underline{F}$ represents the vectorial summation of all forces acting on the fluid body and $\underline{M}$ is its linear momentum.

　In case of steady flow, the first term on the right-hand side of Eq. (1.12) vanishes and the equation is reduced to

$$\sum \underline{F} = \int_{c.s.} \rho \underline{v}.\underline{v}.dA \tag{1.13}$$

The right-hand side of the above equation represents the net rate of outflow of linear momentum through the control surface. In the special case of steady one-dimensional flow, the equation can be written in the form

$$\sum \underline{F} = \sum \left(\dot{m} \underline{V} \right)_{out} - \sum \left(\dot{m} \underline{V} \right)_{in} \tag{1.14}$$

When the control volume is very small (tends to a point), the momentum equation tends to the following differential form (known as the Navier–Stokes equation):

$$\rho \left(\frac{\partial \underline{V}}{\partial t} + \left(\underline{V}.\underline{\nabla} \right) \underline{V} \right) = \underline{F} - \nabla p + \mu \nabla^2 \underline{V} \tag{1.15a}$$

If the flow is frictionless ($\mu = 0$), the diffusion term, $\mu \nabla^2 \underline{V}$, vanishes and the equation becomes

$$\rho \left(\frac{\partial \underline{V}}{\partial t} + \left(\underline{V}.\underline{\nabla} \right) \underline{V} \right) = \underline{F} - \nabla p \tag{1.15b}$$

The above equation is well-known as Euler's equation. The equation can be applied along a streamline to yield the following 1-D Euler equation

$$\frac{\partial}{\partial s} (p + \gamma z) = -\rho a_s \tag{1.16}$$

where a_s is the acceleration in the streamwise direction. If the above equation is further simplified for the case of steady, incompressible, frictionless flow, it results in Bernoulli's equation, which can be written as

$$\frac{p}{\gamma} + \frac{V^2}{2g} + z = Const. \tag{1.17}$$

The application of the momentum equation normal to the streamline results in an equation similar to (1.16) and can be written as

$$\frac{\partial}{\partial n}(p + \gamma z) = -\rho \, a_n \tag{1.18}$$

where n is a coordinate normal to the streamline.

1.2.2.2   Conservation of Angular Momentum

In the general case of unsteady flow of a compressible fluid, the angular momentum conservation equation (deduced from the Reynolds transport equation) can be expressed as:

$$\sum \underline{M} = \frac{\partial}{\partial t} \int_{c.v.} \rho(\underline{r} \times \underline{v}) \, dV + \int_{c.s.} \rho(\underline{r} \times \underline{v}) \, \underline{v} . dA \tag{1.19}$$

where the term $\sum \underline{M}$ represents the vectorial summation of all moment acting on the fluid body within the control volume, $\underline{v}$ is the velocity vector and dV is the elementary volume.

In the special case of steady one-dimensional flow, the first term in the right-hand side of Eq. (1.19) will vanish and the equation can be written in the form

$$\sum \underline{M} = \sum \left(\dot{m} \underline{V} \times \underline{r} \right)_{out} - \sum \left(\dot{m} \underline{V} \times \underline{r} \right)_{in} \tag{1.20}$$

The terms $\sum \left(\dot{m} \underline{V} \times \underline{r} \right)_{in}$ and $\sum \left(\dot{m} \underline{V} \times \underline{r} \right)_{out}$ represent the rates of inflow and outflow of angular momentum, respectively.

1.2.3   Conservation of Energy

Considering the case of steady 1-D flow, the application of the first law of thermodynamics for a control volume (Figure 1.7) results in a simplified form of the energy conservation equation that can be expressed as

$$\dot{Q}_{c.v.} + \sum \dot{m}_i \left(h_i + \frac{V_i^2}{2} + g z_i \right) = \sum \dot{m}_e \left(h_e + \frac{V_e^2}{2} + g z_e \right) + \dot{W}_{c.v.} \tag{1.21}$$

where h is the enthalpy, $\dot{Q}_{c.v.}$ is the rate of heat transfer to the c.v., and $\dot{W}_{c.v.}$ is the rate of doing work by the c.v. In the special case of one inlet and one exit, the above equation can be expressed in the form

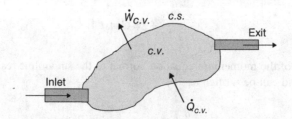

Figure 1.7 One-dimensional flow through a control volume

$$q + h_i + \frac{V_i^2}{2} + g z_i = h_e + \frac{V_e^2}{2} + g z_e + w \tag{1.22}$$

where q is the heat transfer per unit mass and w is the work done per unit mass. The case of steady incompressible flow with no heat transfer has many applications in fluid mechanics. Now, by writing $h = u + pv = u + p/\rho$, where u is the specific internal energy, Eq. (1.22) can be expressed in the form

$$\frac{p_1}{\gamma} + \frac{V_1^2}{2g} + z_1 = \frac{p_2}{\gamma} + \frac{V_2^2}{2g} + z_2 + w + h_L \tag{1.23}$$

where the term h_L represents the amount of heat generation due to fluid friction per unit weight of fluid. This term is normally referred to as the friction head loss. Every term in the above equation has a unit of length and represents energy per unit weight of fluid. Equation (1.23) can be simplified to

$$H_1 = H_2 + w + h_L \tag{1.24}$$

where $H = \frac{p}{\gamma} + \frac{V^2}{2g} + z$ and is called the total head, and the terms w and h_L are redefined in Eqs. (1.23) and (1.24) to represent the work done and the energy loss per unit weight of fluid, respectively.

1.3 Some Important Applications

a. In the case of a pump, the work is done by the prime mover, and the total head developed by the pump can be obtained by applying Eq. (1.24) between the inlet (1) and exit (2) sections shown in Figure 1.8 as follows:

$$H_1 + h_p = H_2 + h_L \tag{1.25}$$

where h_p is the head developed by the pump $(h_p = -w)$ and h_L is the friction head loss between sections 1 and 2.

The rate of doing work by the pump on the fluid, P_f, can be obtained from

$$P_f = \gamma Q h_p \tag{1.26}$$

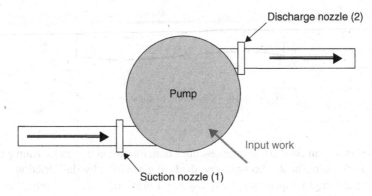

Figure 1.8   Schematic of a pump, showing the suction and discharge nozzles

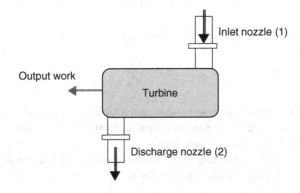

Figure 1.9   Schematic of a hydraulic turbine

b. The application of Eq. (1.24) to the case of a hydraulic turbine (Figure 1.9) results in

$$H_1 = H_2 + h_t + h_L \tag{1.27}$$

where h_t is the work produced by the turbine per unit weight of fluid $(h_t = w)$ and h_L is the friction head loss between sections 1 and 2. The power extracted from the fluid by the turbine will be

$$P_f = \gamma Q h_t \tag{1.28}$$

c. The application of Eq. (1.24) to the case of flow in a pipe (Figure 1.10) gives

$$h_L = H_1 - H_2 \tag{1.29}$$

The power loss in pipe friction can be obtained from

$$P_{friction} = \gamma Q h_L \tag{1.30}$$

Inflow section (1)   Outflow section (2)

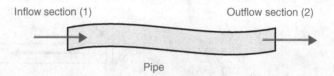

Pipe

Figure 1.10   Schematic of flow in a pipe

The power loss in fluid friction represents a transformation of energy from a useful form to a less useful form (heat). Accordingly, the heat generation by fluid friction will cause an increase in the fluid temperature. If we assume a thermally insulated pipe, the increase in fluid temperature (ΔT) can be obtained from the energy balance as follows:

$$P_{friction} = \gamma\,Q\,h_L = \dot{m}\,C_p\,\Delta T = \rho Q C_p\,\Delta T$$

This equation can be simplified to obtain

$$\Delta T = \frac{g\,h_L}{C_p} \tag{1.31}$$

The above temperature increase is only appreciable in long pipelines. In long crude oil pipelines, heat exchangers are used at intermediate stations for cooling the pumped fluid especially during the summer in hot areas.

d. The pressure variation in a rotating fluid (assuming solid body rotation or forced vortex) can be obtained by applying the Euler's equation as follows:

$$\text{Equation (1.18)} \rightarrow \frac{\partial}{\partial n}(p + \gamma z) = -\rho\,a_n$$

If we apply the above equation, considering the case of the rotating fluid shown in Figure 1.11, we obtain

$$\frac{p}{\gamma} - \frac{\omega^2 r^2}{2g} + z = Const. \tag{1.32}$$

The pressure variation in section A-A can be obtained from Eq. (1.32) by equating z to a constant and the pressure at any radius r can be expressed as

$$p_r - p_o = \frac{\gamma \omega^2}{2g} r^2 \tag{1.33}$$

where p_o is the pressure at the center O.

e. The hydraulic and energy gradient lines (HGL and EGL) are used for graphical representation of the variation of piezometric head $\left(\frac{p}{\gamma} + z\right)$ and the total head $\left(\frac{p}{\gamma} + \frac{V^2}{2g} + z\right)$ along the pipe respectively.

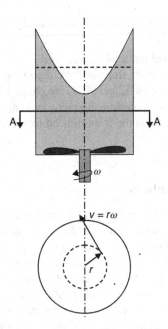

Figure 1.11   Forced vortical motion in a cylindrical tank

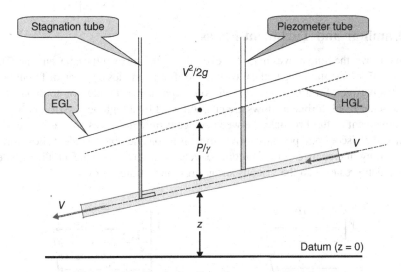

Figure 1.12   The hydraulic and energy gradient lines

As shown in Figure 1.12, the piezometric head at a point (or section) is the head that will be reached if a piezometer tube is attached to that point (or section). The energy gradient line is above the hydraulic gradient line by a distance equal to the velocity head ($V^2/2\,g$) and represents the variation of the total head (H) along the pipe.

Note:

1. The EGL is sometimes called the total energy line.
2. The EGL has always a downward slope in the direction of flow because of friction losses.
3. If the velocity is constant, the EGL and HGL are parallel lines.
4. If the HGL is above the pipe centerline the pressure is above atmospheric and vice versa.
5. The slope of the EGL represents the friction head loss/unit length.

1.4 Dimensionless Numbers

The important dimensionless numbers in fluid mechanics are the Reynolds number, R_e; Mach number, M; Froude number, F_r; and Weber number, W_e. The first two (R_e and M) are of direct relevance to pumping machinery. The Reynolds number represents the ratio between inertia and viscous forces and it is important for achieving similarity in totally enclosed flows (such as flow in pipes and in air conditioning ducts). The Reynolds number is also important for achieving similarity for flow over fully submerged bodies (such as flow of air over a car or flow of water over a submarine). On the other hand, the Mach number represents the ratio between inertia and compressibility forces and is important for achieving similarity in high-speed flows (such as flow of steam in a steam turbine nozzle or flow of air over a supersonic aircraft).

1.5 Laminar and Turbulent Flows

In laminar flows, the fluid moves in layers, every layer sliding over the adjacent one. There is no interchange of momentum due to the movement of fluid particles between different layers. The only forces between fluid layers are the viscous shear forces in addition to pressure forces.

On the other hand, turbulent flows are characterized by a high degree of mixing due to the erratic movement of fluid particles between adjacent fluid layers, as shown in Figure 1.13. In addition to the viscous and pressure forces, there is a considerable turbulent shear force arising from the strong interchange of momentum between adjacent layers. In fact, the turbulent shear stress is much greater than the viscous shear stress in turbulent flows.

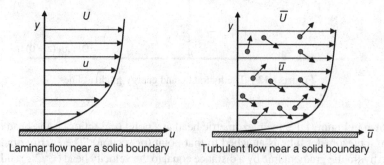

Laminar flow near a solid boundary   Turbulent flow near a solid boundary

Figure 1.13 The velocity profiles in laminar and turbulent flows

1.6 Flow Separation

The main stream may detach (separate) from the body surface as a result of the positive pressure gradient ($\partial p/\partial x$) due to the surface curvature (Figure 1.14). Separation may cause a transition to turbulence. The location of the point of separation depends on the flow Reynolds number, the body shape, and the surface roughness.

1.7 Cavitation

Cavitation is a phenomenon that is completely different from flow separation. It refers to the formation of vapor cavities in liquid flow as a result of the drop in liquid pressure below the vapor pressure. Cavitation always starts at the point of minimum pressure and it may cause severe damage in pump impellers and turbine runners due to the accompanying material erosion. Figure 1.15 shows a typical p–T diagram for a pure substance. In thermodynamics, the vapor pressure is normally referred to as the saturation pressure, and it increases with the increase in liquid temperature. At every temperature there is only one value for the vapor pressure.

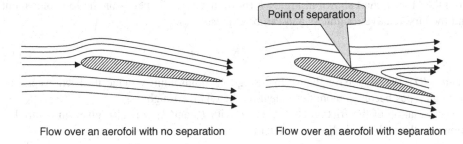

Flow over an aerofoil with no separation   Flow over an aerofoil with separation

Figure 1.14   Streamlines for flow over an aerofoil, showing the point of separation

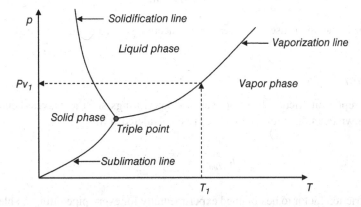

Figure 1.15   The p–T diagram for a pure substance showing the vaporization line

1.8   Friction Losses in Pipes and Pipe Fittings

Friction losses in piping systems are normally divided into two parts: major losses and minor losses. The major losses represent the friction losses in straight pipes while the minor losses represent the losses in various types of pipe fittings including bends, valves, filters, and flowmeters.

1.8.1   Major Losses

These losses represent the pipe friction losses and can be calculated using the Darcy–Weisbach formula which takes the form

$$h_L = \frac{fLV^2}{2gD} \tag{1.34}$$

where f is the coefficient of friction, D is the pipe internal diameter, L is the pipe length, and V is the average velocity in the pipe. Up to Reynolds number $R_e = 2000$, the flow can be considered laminar in normal engineering applications and f can be calculated from $f = 64/Re$. For values of $Re > 2000$, the flow can be considered turbulent and the friction coefficient can be obtained from the Moody chart shown in Figure 1.16. In this case, f depends on the Reynolds number and the pipe relative roughness and can be expressed as

$$f = f\left(R_e, k_s/D\right) \tag{1.35}$$

where k_s/D is the relative roughness. In the high Reynolds number range, the friction coefficient in rough pipes becomes more dependent on the relative roughness.

The variation of the friction coefficient, f, with R_e and k_s/D is also given in a correlation known as the Colebrook–White formula that can be written as:

$$f = \frac{0.25}{\left[\log_{10}\left(\dfrac{k_s}{3.7D} + \dfrac{5.74}{Re^{0.9}}\right)\right]^2} \tag{1.36}$$

The above formula can be used with reasonable accuracy in computational codes.

1.8.2   Minor Losses

These losses represent friction losses in various pipe fittings such as valves, bends, elbows, filters, and flowmeters. The minor losses can be calculated from

$$\left(h_L\right)_{minor} = \sum K \frac{V^2}{2g} \tag{1.37}$$

where K is a friction factor to be obtained experimentally for every pipe fitting. Tables for K are available for different pipe fittings.

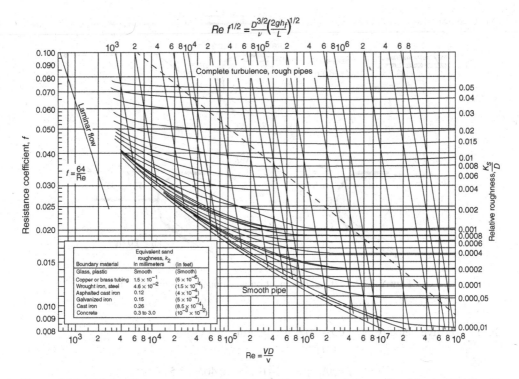

$$Re\ f^{1/2} = \frac{D^{3/2}}{\nu}\left(\frac{2gh_f}{L}\right)^{1/2}$$

Figure 1.16 Friction factor versus Re (Reprinted with minor variations after Moody (1944) with permission from ASME)

Example 1.1

The water jet emerging from a circular pipe along the x-axis has a velocity of 60 m/s. The water jet impacts a curved blade as shown in Figure 1.17.

a. Determine the x-component of the force exerted by the jet on the blade if the blade is stationary.
b. Determine the same force if the jet moves to the right at a speed of 20 m/s.

Solution

Assuming frictionless flow, the magnitude of the fluid velocity relative to the blade at exit will be the same as at inlet. Now, we can apply Eq. (1.6) as follows:

a. Equation (1.6) $\Rightarrow \sum \underline{F} = \sum \left(\dot{m}\underline{V} \right)_{out} - \sum \left(\dot{m}\underline{V} \right)_{in}$

$$\Rightarrow -F_x = \dot{m}\left(V_{2x} - V_{1x} \right) = \rho A_j V_j \left(-V_j \cos 45° - V_j \right)$$

$$\Rightarrow F_x = \rho A_j V_j^2 \left(\cos 45° + 1 \right)$$

$$\Rightarrow F_x = 10^3 \frac{\pi}{4}(0.05)^2 (60)^2 \left(\cos 45° + 1 \right) = \underline{12.1 \times 10^3 \text{N}}$$

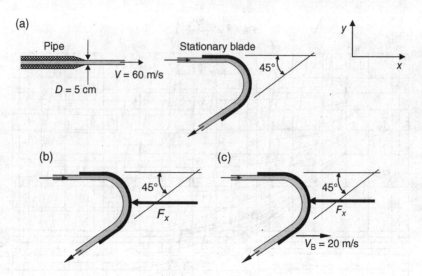

Figure 1.17 Jet impingement on a stationary or a moving blade

b. The relative velocity, $V_r = 60 - 20 = 40$ m/s Equation (1.6)

$$\Rightarrow \dot{m} = \rho A_j V_r = 10^3 \frac{\pi}{4}(0.05)^2 \times 40 = 78.5 \, \text{kg/s}$$

$$\Rightarrow -F_x = \dot{m}(-V_r \cos 45° - V_r)$$

$$\Rightarrow F_x = \dot{m} V_r(\cos 45° + 1) = 78.5 \times 40(\cos 45° + 1)$$

$$\Rightarrow F_x = 78.5 \times 40(\cos 45° + 1) = \underline{5.36 \times 10^3 \text{N}}$$

Example 1.2

Figure 1.18 shows a lawn sprinkler having three arms. Water enters the sprinkler from a vertical central pipe at a flow rate of 3×10^{-3} m³/s. The water jet issuing from each arm has a velocity of 12 m/s. Determine the torque required to hold the lawn sprinkler stationary.

Data $R = 20$ cm and $\theta = 30°$

Solution

The mass flow rate from each nozzle, $\dot{m}_n = \frac{1}{3}\dot{m}_T$. Now, apply the angular momentum conservation equation, Eq. (1.12),

Equation (1.12) $\Rightarrow \sum \underline{M} = \sum\left(\dot{m}\,\underline{V} \times \underline{r}\right)_{out} - \sum\left(\dot{m}\,\underline{V} \times \underline{r}\right)_{in}$

$$\Rightarrow T = \sum\left(\dot{m}\,\underline{V} \times \underline{r}\right)_{out} - \overbrace{\sum\left(\dot{m}\,\underline{V} \times \underline{r}\right)_{in}}^{=0} = 3\dot{m}_n\left(V_{exit} \times r_{exit}\right)$$

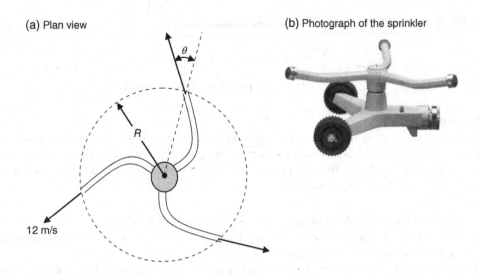

Figure 1.18   A plan view and a photograph of a water sprinkler

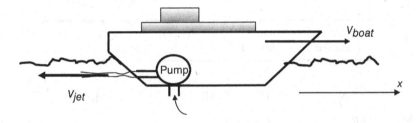

Figure 1.19   A schematic of a boat powered by a water jet

$$\Rightarrow T = \dot{m}_T \left(\underline{V_{exit}} \times \underline{r_{exit}} \right) = \rho Q \left(\underline{V_{exit}} \times \underline{r_{exit}} \right) = 10^3 \times 3 \times 10^{-3} \left(12 \times 0.2 \sin 30° \right)$$

$$\Rightarrow T = \underline{3.6\,\text{N.m}}$$

Example 1.3

A boat is powered by a water jet as shown in Figure 1.19. The pump sucks water through a 10 cm diameter pipe and discharges it through a 5 cm diameter pipe at a rate of 0.047 m³/s. Knowing that the boat is moving at a constant speed of 10 m/s, determine the total resistance to the motion of the boat.

Solution

$$Q = V_j A_j = V_j \times \frac{\pi}{4}(0.05)^2 = 0.047\,\text{m}^3/\text{s}$$

$$\Rightarrow V_j = 23.9\,\text{m/s} \Rightarrow \dot{m} = \rho V_j A_j = 47\,\text{kg/s}$$

Apply the momentum equation in the x-direction,

$$\sum F_x = \sum \dot{m}_{ex} V_{ex} \Rightarrow -F_{resist} = \dot{m}_{ex}\left(V_{boat} - V_j\right)$$

$$\Rightarrow \quad -F_{resist} = 47(10-23.9) \Rightarrow F_{resist} = 653.3\text{N}$$

Example 1.4

An explosion occurs in the atmosphere when an anti-aircraft missile meets its target as shown in Figure 1.20. A shock wave spreads out radially from the explosion. The pressure difference across the wave (ΔP) is a function of time (t), speed of sound (C), and the total amount of energy released by the explosion (E).

1. Using the above variables, obtain a dimensionless parameter for ΔP.
2. Knowing that the radial distance of the blast wave (r) from the center of explosion depends on the same variables. Obtain another dimensionless parameter for r.
3. For a given explosion, if the time (t) elapsed after explosion doubles (while C and E are unchanged), by what factor will ΔP decrease?

Solution

1.

Variable	Δp	t	E	C
Dimensions	$ML^{-1}T^{-2}$	T	ML^2T^{-2}	LT^{-1}

$$\pi_1 = t^a E^b C^c \Delta p^1 \Rightarrow M^0 L^0 T^0 = (T)^a \left(ML^2T^{-2}\right)^b \left(LT^{-1}\right)^c \left(ML^{-1}T^{-2}\right)^1$$

$$\left.\begin{array}{l} M \Rightarrow 0 = b+1 \qquad\qquad \Rightarrow b = -1 \\ L \Rightarrow 0 = 2b+c-1 \qquad \Rightarrow c = +3 \\ T \Rightarrow 0 = a-2b-c-2 \Rightarrow a = +3 \end{array}\right\} \Rightarrow \pi_1 = t^3 E^{-1} C^3 \Delta p \Rightarrow \boxed{\pi_1 = \frac{\Delta p\, t^3 C^3}{E}}$$

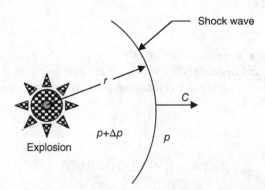

Figure 1.20   Schematic of a shock wave resulting from an explosion

2.

Variable	r	t	E	C
Dimensions	L	T	ML^2T^{-2}	LT^{-1}

$$\pi_2 = t^a \, E^b \, C^c \, r^1 \;\Rightarrow\; M^0 L^0 T^0 = (T)^a \, (ML^2 T^{-2})^b \, (LT^{-1})^c \, (L)^1$$

$$
\left.
\begin{array}{l}
M \;\Rightarrow\; 0 = b \qquad\qquad\qquad \Rightarrow\; b = 0 \\[4pt]
L \;\Rightarrow\; 0 = 2b + c + 1 \;\Rightarrow\; c = -1 \\[4pt]
T \;\Rightarrow\; 0 = a - 2b - c \;\Rightarrow\; a = -1
\end{array}
\right\}
\;\Rightarrow\; \pi_2 = t^{-1} E^0 C^{-1} r \;\Rightarrow\; \boxed{\pi_2 = \dfrac{r}{Ct}}
$$

3. $\pi_1 = \dfrac{\Delta p \, t^3 C^3}{E} = Const.$ and since C and E are constants, therefore $\Delta p = \dfrac{C}{t^3}$, accordingly, if t is doubled, Δp will decrease by a factor of 8.

Example 1.5

A model test is performed to study flow through a large valve having a 2 ft diameter at inlet and carrying water at a flow rate of 30 ft³/s. The model and prototype are geometrically similar, and the model inlet diameter is 0.25 ft. Knowing that the working fluid is the same for model and prototype, determine the flow rate in the model for dynamically similar conditions.

Solution

Data $\dfrac{D_m}{D_p} = \dfrac{0.25}{2.0} = \dfrac{1}{8}$, $Q_p = 30 \, \text{ft}^3/\text{s}$, same fluid $\Rightarrow \rho_m = \rho_p$, and $\mu_m = \mu_p$

For dynamic similarity, the Reynolds number must be the same for model and prototype.

Therefore, $\left(\dfrac{\rho VD}{\mu}\right)_m = \left(\dfrac{\rho VD}{\mu}\right)_p \;\Rightarrow\; \dfrac{V_m}{V_p} = \dfrac{D_p}{D_m} = 8$

Knowing that $Q = VA$, therefore,

$$\dfrac{Q_m}{Q_p} = \dfrac{V_m A_m}{V_p A_p} = \dfrac{V_m}{V_p}\left(\dfrac{D_m}{D_p}\right)^2 = 8\left(\dfrac{1}{8}\right)^2 = \dfrac{1}{8} \;\Rightarrow\; Q_m = 30 \times \dfrac{1}{8} = \underline{3.75 \, \text{ft}^3/\text{s}}$$

Example 1.6

Figure 1.21 shows a pumping system in which water ($\rho = 10^3 \, \text{kg/m}^3$, $\mu = 1.31 \times 10^{-3} \, \text{N} \cdot \text{s/m}^2$) is pumped from reservoir A to reservoir B at a rate of 0.07 m³/s. The water surface elevation in reservoir A is 20 m, and the surface elevation in reservoir B is 80 m while the pump elevation is 10 m. All pipes are made of commercial steel.

a. Show whether the flow in the pipe is laminar or turbulent and determine the friction coefficient.

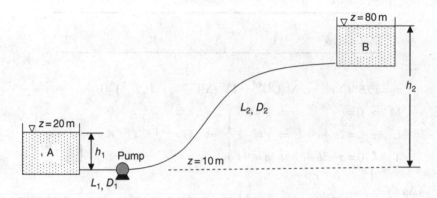

Figure 1.21   Pumping system

b. What is the maximum shear stress in the pipe?
c. What is the pump power consumption assuming a pump efficiency of 70%? Consider minor losses at the pipe entrance and exit sections, and assume $K_{ent.} = 0.5$.
d. Sketch the hydraulic and energy gradient lines.

Data $L_1 = 15$ m, $L_2 = 120$ m, $D_1 = D_2 = 15$ cm

Solution

a. Velocity of flow in the pipe, $V = Q/A = 4Q/\pi D^2 = 4 \times 0.07/\pi(0.15)^2 = 3.96$ m/s

$$Re = \frac{\rho VD}{\mu} = \frac{10^3 \times 3.96 \times 0.15}{1.31 \times 10^{-3}} = 4.53 \times 10^5 \Rightarrow \text{Flow is turbulent}$$

b. The friction coefficient f is a function of Re and k_s/D and can be determined using the Moody chart. For the given steel pipe, $k_s/D = 0.046/150 = 3.07 \times 10^{-4}$ and

$$Re = \frac{\rho VD}{\mu} = \frac{10^3 \times 3.96 \times 0.15}{1.31 \times 10^{-3}} = 4.53 \times 10^5$$

Now, using the Moody chart, we obtain $f = 0.017$.
c. The maximum shear stress in the pipe occurs at the pipe wall and

$$\tau_o = C_f \frac{1}{2}\rho V^2 \text{ and } C_f = f/4 \Rightarrow \tau_o = \frac{f}{8}\rho V^2$$

$$\tau_o = \frac{0.017}{8} \times 10^3 \times (3.96)^2 = 33.3\,\text{N/m}^2$$

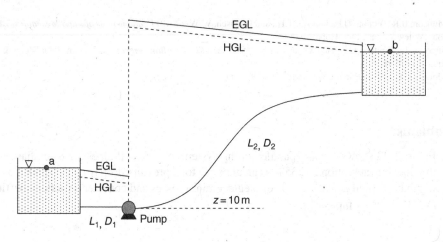

Figure 1.22 Hydraulic and energy gradient lines

d. Apply the energy equation between points **a** and **b** and neglect all minor losses, thus

$$\frac{p_a}{\gamma}+\frac{V_a^2}{2g}+z_a+h_p=\frac{p_b}{\gamma}+\frac{V_b^2}{2g}+z_b+\sum h_L \quad\Rightarrow\quad h_p=(z_b-z_a)+\sum h_L \tag{1}$$

In the above equation, $\sum h_L$ can be determined using Darcy's formula and considering minor losses at pipe inlet (from reservoir A) and at pipe exit (to reservoir B),

$$\Rightarrow \sum h_L=\sum f\frac{L}{D}\frac{V^2}{2g}+K_{ent.}\frac{V^2}{2g}+K_{exit}\frac{V^2}{2g} \tag{2}$$

Therefore, $\Rightarrow \sum h_L = f\frac{(L_1+L_2)}{D}\frac{V^2}{2g}+(K_{ent.}+K_{exit})\frac{V^2}{2g}$

$$\Rightarrow \sum h_L = \left(0.017\frac{(15+120)}{0.15}+0.5+1\right)\frac{3.96^2}{2\times9.81}=13.4\,\text{m}.$$

Substitute in Eq. (1) $\Rightarrow h_p=(80-20)+13.4=73.4$ m.

Power required to drive the pump $=\dfrac{\gamma Q h_p}{\eta_{pump}}=\dfrac{9.81\times0.07\times73.4}{0.7}=\underline{72\,\text{kW}}$

e. The hydraulic and energy gradient lines are shown in Figure 1.22.

References

1. Elger, D.F., Williams, B.C., Crowe, C.T., and Roberson, J.A. (2012) *Engineering Fluid Mechanics*, 10th edn. John Wiley & Sons, Inc., Hoboken, NJ.
2. White, F.M. (2011) *Fluid Mechanics*, 7th edn. McGraw Hill, New York.
3. Fox, R.W. and McDonald, A.T. (2010) *Introduction to Fluid Mechanics*, 8th edn. John Wiley & Sons, Inc., Hoboken, NJ.
4. Douglas, J.F., Gasiorek, J.M., Swaffield, J.A., and Jack, L.B. (2011) *Fluid Mechanics*, 6th edn. Prentice Hall Publishers, New York.

5. Munson, B.R., Young, D.F., Okiishi, T.H., and Huebsch, W.W. (2012) *Fundamentals of Fluid Mechanics*, 7th edn. John Wiley & Sons, Inc., Hoboken, NJ.
6. Street, R.L., Watters, G.Z., and Vennard, J.K. (1995) *Elementary Fluid Mechanics*, 7th edn. John Wiley & Sons, Inc., New York.
7. Moody, L.F. (1944) Friction factors for pipe flow. *Trans ASME*, 66, 671.

Problems

1.1 Figure 1.23 shows a series-parallel piping system in which all pipes are 8 cm diameter. If the flow rate at section 2 is 35 L/s calculate the total pressure drop $(p_1 - p_2)$ in kPa, assuming that the fluid is water at 20 °C. Neglect minor losses and consider a friction coefficient $f = 0.025$ for all pipes.

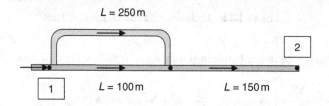

Figure 1.23 Series-parallel piping system

1.2 Figure 1.24 shows two water reservoirs connected by a pipe (A) branching to two pipes (B and C). Knowing that all pipes are 8 cm diameter, determine the flow rate from reservoir 1 to reservoir 2 if the valve in branch C is fully open. Consider $K_v = 0.5$ and assume a friction coefficient $f = 0.02$ for all pipes.

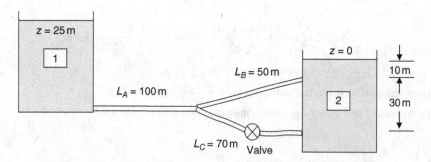

Figure 1.24 Two water reservoirs connected by a branching pipe

1.3 Water at 20 °C is to be pumped through 2000 ft of pipe from reservoir 1 to reservoir 2 at a rate of 3 ft³/s, as shown in Figure 1.25. If the pipe is cast iron of diameter is 6 in. and the pump is 75% efficient, what is the pump power consumption? Neglect minor losses.

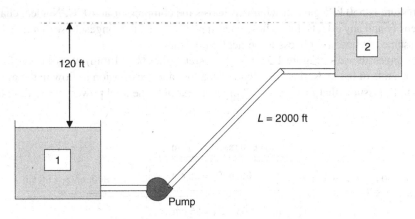

Figure 1.25 Simple pumping system

1.4 Figure 1.26 shows a U-tube which is filled with water at 20 °C. The tube is sealed at A and is open to the atmosphere at D. If the tube is rotated about the vertical axis AC, identify the point of minimum pressure and determine the maximum speed of rotation if cavitation is to be avoided.

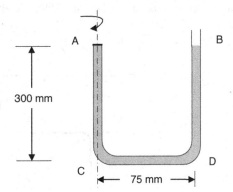

Figure 1.26 A U-tube rotating about axis AC

1.5 A fluid (ρ = 900 kg/m^3 and ν = 10^{-5} m^2 /s) flows at a rate of 0.2 m^3/s through 500 m cast iron pipe of diameter 20 cm. Knowing that the pipe has a 10° downward slope, determine (a) the friction head loss and (b) the pressure drop.

1.6 Water flows in a horizontal pipe with velocity 4 m/s and accelerating at a rate of 1.25 m/s^2. The pipe is 5 cm diameter and 15 m long. Assuming a friction coefficient f = 0.01, determine the pressure drop through the pipe.

1.7 A compressed air drill requires 0.2 kg/s of air at 500 kPa gage at the drill. The hose from the air compressor to the drill is 25 mm inside diameter. The maximum compressor discharge pressure is 550 kPa gage at which air leaves the compressor at 40 °C. Neglect changes in density and any effects due to hose curvature. Calculate the longest hose that may be used. Assume the hose roughness to be negligibly small.

1.8 The pipes shown in Figure 1.27 are all concrete pipes ($k_s = 1$ mm). Water flows from A to B at a rate of 0.8 m³/s, find the friction head loss and the division of flow in the pipes from A to B. Assume that $f = 0.03$ for all pipes. Determine the total power loss in fluid friction.

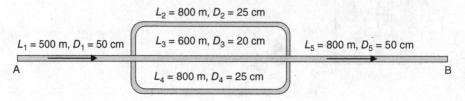

Figure 1.27 Series-parallel piping arrangement

1.9 Fuel oil (sp. gr. $= 0.9$ and $\nu = 5 \times 10^{-5}$ m²/s) is pumped from reservoir A to reservoir B at a rate of 0.05 m³/s as shown in Figure 1.28. All pipes used in the system are commercial steel pipes ($k_s = 0.046$ mm) of diameter 15 cm. The lengths of the suction and delivery pipes are 20 and 180 m, respectively. Determine the total head developed by the pump. Assuming a pump overall efficiency of 75%, determine the pump power consumption.

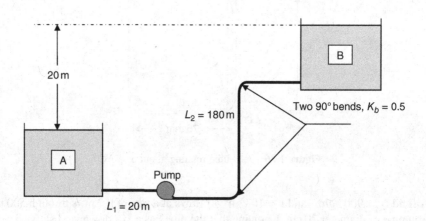

Figure 1.28 Schematic of the pumping system

1.10 Figure 1.29 shows a cylindrical tank full of water at 20 °C. The tank is completely closed, except for a vent at the center of its top plate. The tank is forced to rotate about its vertical axis at a speed of 300 rpm. Determine the pressure distribution on the top plate. Determine also the location and magnitude of the maximum pressure in the tank.

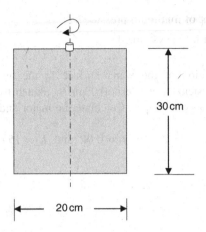

Figure 1.29 A cylindrical tank rotating about its own vertical axis

1.11 Figure 1.30 shows a closed tank 1 m diameter and 2 m long that is completely filled with water. If the tank rotates about its own horizontal axis at a speed of 120 rpm, determine the maximum pressure difference in the tank and identify the points of maximum and minimum pressures. Assume forced vortex motion.

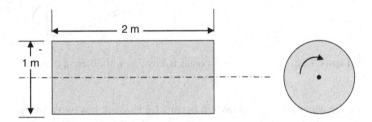

Figure 1.30 A cylindrical tank rotating about its own horizontal axis

1.12 Study the system shown in Figure 1.31 and answer the following questions:

    a. What is the flow direction?

    b. What kind of machine is at B?

    c. Do the pipes AB and BC have the same diameter?

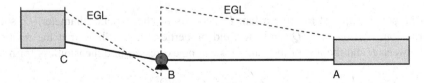

Figure 1.31 Schematic of a hydraulic system showing the energy gradient line

    d. Where is the point of minimum pressure?

    e. Sketch the HGL for the system.

1.13 Determine the elevation of the water surface in the upstream reservoir shown in Figure 1.32 if the system flow rate is 0.09 m³/s. Sketch the HGL and EGL, showing the relative magnitudes and slopes. Calculate the major and minor losses in the piping system.

    Data $L_1 = $ 20 m, $D_1 = 25$ cm, $f_1 = 0.02$ and $L_2 = 15$ m, $D_2 = 15$ cm, $f_2 = 0.015$, $K_{elbow} = 0.5$

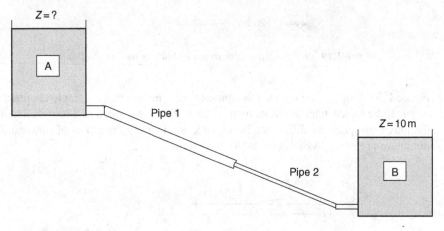

Figure 1.32 Two reservoirs connected by pipes of different diameters

1.14 Consider the water sprinkler shown in Figure 1.33, which has two arms rotating in a horizontal plane. Each arm has a radius $r = 20$ cm and ends with a nozzle of c.s. area 10 mm². Knowing that $\theta = 20°$ and the flow rate delivered by the sprinkler is 0.30 L/s, what torque (T) will be required to hold the sprinkler arms stationary?

1.15 The total shear force, F, acting on the walls of a pipe due to fluid flow depends on the tube diameter, D; tube length, L; surface roughness, k; fluid density, ρ; viscosity, μ; and the average flow velocity, V. Show that the relationship between these variables can be written in the following dimensionless form:

$$\frac{F}{\rho V^2 D^2} = f\left(\frac{\rho V D}{\mu}, \frac{L}{D}, \frac{k}{D}\right)$$

1.16 The power required to drive a fan, P, depends on the impeller diameter, D; speed of rotation, N; flow rate, Q; and the fluid properties (ρ, μ). Show that the relationship between P and the rest of the variables can be expressed in dimensionless form as

$$\frac{P}{\rho N^3 D^5} = f\left(\frac{Q}{ND^3}, \frac{\rho N D^2}{\mu}\right)$$

How does the term $\rho N D^2/\mu$ relate to the dimensionless parameters that you know?

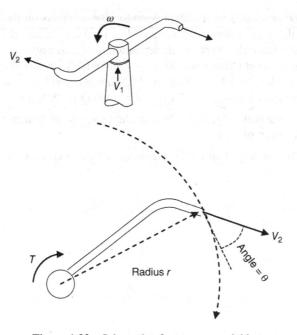

Figure 1.33 Schematic of a two-arm sprinkler

1.17 A three-nozzle water sprinkler discharges water while the sprinkler head rotates. The angular velocity, ω, depends on the water density, ρ; volume flow rate, Q; water velocity at the nozzle exit, V; nozzle area, A; and the resisting frictional torque, T_f. Use dimensional analysis to develop an expression for the angular velocity.

1.18 The valve at the end of the inclined pipe shown in Figure 1.34 is gradually closed. The resulting fluid deceleration during valve closure is 1.2 m/s^2. Neglecting viscous effects, determine the difference in pressure between points 1 and 2.

    Data $L = 20$ m, $D = 10$ cm, $V = 4$ m/s, $\theta = 30°$

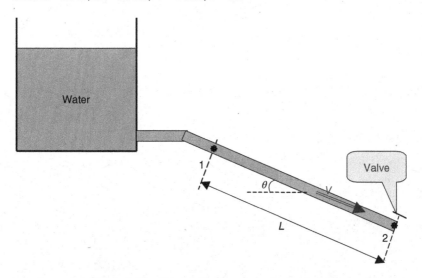

Figure 1.34 Decelerating flow in an inclined pipe

1.19 The power P generated by a certain wind turbine, depends on its diameter, D; the air density, ρ; the wind velocity, V; the speed of rotation, Ω; and the number of blades, N. Write the relationship between these variables in dimensionless form.

1.20 A model of the wind turbine mentioned in Problem 1.17 was built and tested in a wind tunnel. The model which has a diameter of 0.5 m was tested at a wind speed of 40 m/s and developed an output power of 2.7 kW when rotating at 4800 rpm.

   a. Determine the power developed by a similar prototype of diameter 5 m when operating at a wind speed of 12 m/s.

   b. What is the prototype speed for dynamically similar conditions?

2

Introduction and Basic Considerations

Fluid movers are very widely used in almost all industries, with the main task of moving fluids (liquids or gases) from one place to another. All fluid movers suck fluid at a low pressure and supply it at a higher pressure. The pressure increase through the machine may be small as in the case of axial flow fans used to circulate air for ventilation in air-conditioned buildings and may be high as in the compressors operating as part of the jet engine. Much higher pressures are developed by compressors used in petrochemical industries where chemical reactions may require very high pressures to take place. The fluid movers handling liquids are called pumps, and these are used in many applications, such as automobile engine cooling systems, city water networks, oil production from deep oil wells, steam power plants (e.g. boiler feed pumps), and oil transportation in pipelines. The design, manufacturing, operation, and maintenance of fluid movers require engineering experience and skills. The terminology being used in pumping machinery is very important for facilitating proper communication between engineers and scientists working in this field. In this chapter, the main definitions and terminology are introduced, together with the basic fundamentals relevant to pumping machinery.

2.1 Introduction

2.1.1 Definitions and Main Features of Fluid Movers

Fluid movers are classified into two main categories, liquid movers, normally referred to as pumps, and gas movers that are classified into three categories: fans, blowers and compressors, as shown in Figure 2.1. All fluid movers are used to increase the total energy content of a fluid. This energy increase may be in the form of pressure increase, velocity increase, or the combined effect of both. Liquid and gas movers have many common features, but the flow in liquid movers is always incompressible while the flow in gas movers may be incompressible (as in fans) or compressible (as in compressors). The performance characteristics of each of these

Pumping Machinery Theory and Practice, First Edition. Hassan M. Badr and Wael H. Ahmed.
© 2015 John Wiley & Sons, Ltd. Published 2015 by John Wiley & Sons, Ltd.

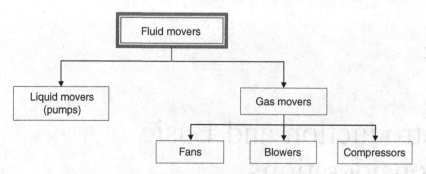

Figure 2.1 Classification of fluid movers

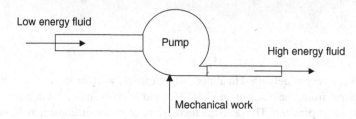

Figure 2.2 Pump input and output energies

machines depend mainly on the fluid properties, the shape and size of the machine, and its speed of rotation.

As shown in Figure 2.2, a pump is simply a liquid mover that utilizes mechanical work to increase the total energy content of a fluid. Pumps are used in almost all industries, including petroleum refineries, petrochemical industries, power stations, water desalination plants, and water and oil pipelines.

A fan is a gas mover that is used to develop a small increase in the fluid pressure. Accordingly, the density variation through the fan is very small (less than 5%) and the flow may be assumed incompressible. On the other hand, compressors are gas movers that are characterized by higher pressure increase through the machine. The term 'blower' is sometimes used *commercially* for gas movers when the pressure rise is between 2 and 10 psi considering atmospheric conditions at the inlet to the suction nozzle. The remainder of this chapter focuses on pumps, while detailed analyses of the performance of fans and compressors are presented in Chapter 8.

2.1.2   Classification of Pumps

The bases of pump classification are numerous, but the most common classifications are based on the way energy is added to the fluid, and the pump geometry. Accordingly, pumps are classified in the following two main categories:

a. **Dynamic pumps**: In these pumps, the fluid velocity is increased inside the pump to values higher than the discharge velocity. Velocity reductions within or after the pump create higher pressure.

b. **Displacement pumps**: In these pumps, energy is added to the fluid by the direct application of a force that moves the fluid from the low pressure side (suction) to the high pressure side (delivery).

Figure 2.3 shows a schematic of a dynamic pump, namely, a radial-type centrifugal pump. The fluid enters the impeller axially through the inlet eye in a direction perpendicular to the page, and is then forced to rotate by the impeller vanes. While rotating inside the impeller the fluid moves outward, thus gaining increase in pressure with a parallel increase in kinetic energy. The high velocity at the impeller exit is transformed to a pressure increase through the volute casing and discharge nozzle which has a diffuser shape.

Another example of dynamic pumps is the axial-flow pump shown in Figure 2.4. The fluid enters the pump axially through the pump suction nozzle and leaves also in the axial direction.

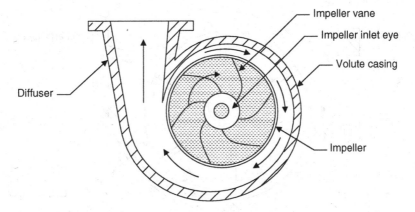

Figure 2.3   A sectional view of a radial-type centrifugal pump

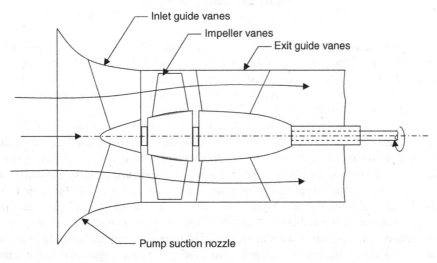

Figure 2.4   A sectional view of an axial-flow pump

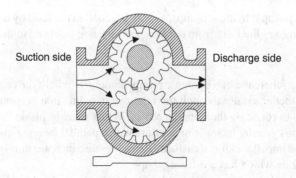

Figure 2.5 Spur gear pump

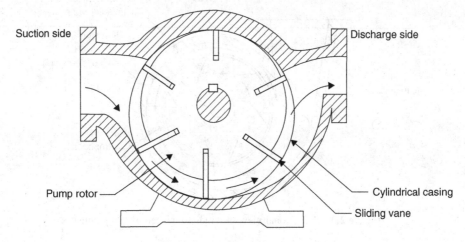

Figure 2.6 Sliding-vane pump

The inlet guide vanes direct the fluid to enter the pump impeller in the axial direction (removes any prerotation) and the exit guide vanes remove fluid rotation (swirl) after the impeller and straighten the flow in the axial direction. Sometimes, the inlet guide vanes are movable and are used for flow rate control. The axial flow pumps are used in applications where a large flow rate is required at low head.

Figure 2.5 shows a spur gear pump as an example of rotary displacement pumps. In general, gear pumps are widely used in fluid power systems for delivering hydraulic fluids at small flow rates and high pressures. The fluid enters the shown pump from the left side (suction side) and moves in segments occupying the space between adjacent teeth and the casing towards the discharge side. The flow rate supplied by this pump is governed by the speed of the prime mover.

The sliding vane pump shown in Figure 2.6 is another example of rotary displacement pumps with a performance similar to that of a gear pump. The rotor is a cylindrical block that has radial groves equally spaced around the circumference. The rotor is placed eccentrically in a cylindrical casing, as shown in the figure. Each grove has a sliding vane that is guided to move in and out by the cylindrical casing. Similar to the gear pump shown in Figure 2.5, the fluid enters the

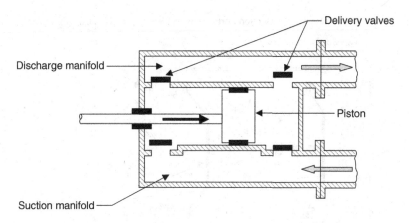

Figure 2.7   Schematic of a double-acting piston pump

sliding-vane pump from the left side (suction side) and moves in segments (pockets) occupying the space between pairs of vanes and the casing towards the discharge side. The pump flow rate is controlled by either the speed of rotation or the eccentricity between the rotor and the casing.

Figure 2.7 shows a schematic of a double-acting piston pump as an example of a displacement reciprocating pump. The cylinder functions as two cylinders; one in front of the piston (on the right side) and the other is in the back side (on the left side). This cylinder has two pairs of suction and delivery valves. During every stroke of the piston motion, one cylinder performs suction while the other performs delivery. Both cylinders suck the fluid from one suction manifold and both deliver to the same discharge manifold as shown in the figure. The volume supplied by the cylinder on the left side per stroke is slightly less than that supplied by the one on the right side due to the volume occupied by the piston rod. Figure 2.8 shows another example of a displacement reciprocating pump, in which a double-acting diaphragm is operated by compressed air. The two diaphragms are coupled by a connecting rod and move together in a reciprocating motion, thus operating the two cylinders. The compressed air is directed by a rotating control valve to push one of the diaphragms, thus creating the delivery stroke while the other diaphragm performs the suction stroke. It is clear from the figure that one complete rotation of the control valve creates four strokes for each diaphragm. Similar to the double-acting piston pump (Figure 2.7), the two cylinders use the same suction and discharge manifolds. The compressed air pressure should be high enough to overcome the force acting on the diaphragm due to the high pressure in the delivery side.

Special effect pumps are less common in the industry but are widely used in special applications. The mechanism of energy addition to the fluid varies from one pump to another. Figure 2.9 shows a vortex pump, as an example of special effect pumps. The pump impeller rotates in a cylindrical casing creating a vortical motion (close to a forced vortex). The discharge pipe is shaped like a stagnation tube that extracts part of the rotating fluid at a pressure higher than the suction pressure and at a higher velocity. The energy is added to the fluid in the form of increasing both pressure and kinetic energy.

Jet pumps represent another type of special effect pump. Figure 2.10 shows an annular-nozzle jet pump in which a small stream of a high pressure fluid (motive fluid) is used to move a large volume of the pumped fluid. The jet pump has no moving parts and the pumping process

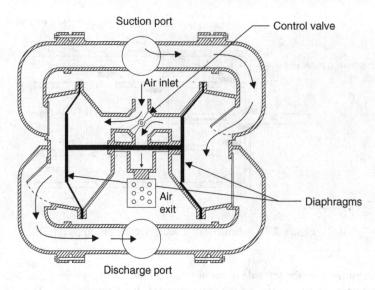

Figure 2.8   Air-operated double-acting diaphragm pump

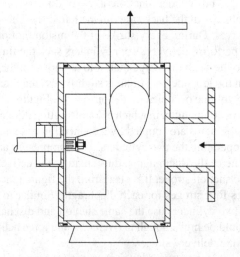

Figure 2.9   Vortex pump

is achieved by using a high-pressure fluid stream (motive fluid) to create a high velocity annular jet in a mixing chamber. Both low pressure and high velocity in the mixing chamber drive the pumped fluid (suction fluid) to enter the chamber and mix with the motive fluid and then discharge at a velocity between the motive and suction velocities. Following the mixing chamber, a diffuser is used to convert the resulting high kinetic energy of the mixture into pressure. The motive fluid can be air, steam, water, or the same pumped fluid. These pumps are useful in applications that require effective mixing of two fluids or transporting fluids that require

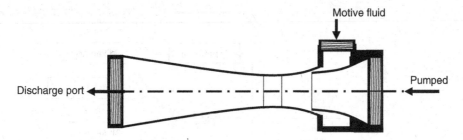

Figure 2.10 Annular-nozzle jet pump

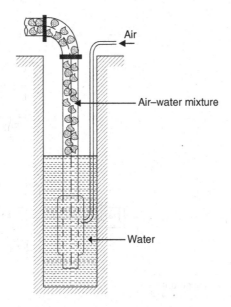

Figure 2.11 Schematic of a typical airlift pump

simultaneous dilution. It can also be used for increasing the suction lift of a centrifugal pump without causing cavitation.

The airlift pump is another type of special-effect pump that is widely used in many applications such as wastewater treatment plants and lifting solid particles from deep wells. Similar to jet pumps, airlift pumps do not have any moving parts and are powered by compressed air (or gas). Figure 2.11 shows a schematic of an airlift water pump that is made of a vertical pipe with a perforated section surrounded by an air jacket near the bottom. The injected air creates a column of two-phase mixture (air and water) of specific weight less than that of water. The buoyancy force moves the two-phase mixture upward toward the pipe exit, where water can be collected. The total head developed depends on the liquid submergence head and the injected air mass flow rate. Recently, Ahmed and Badr [1] investigated the effect of dual gas injection (in both radial and axial directions) on the performance characteristics of an airlift pump. The proposed injection method was found to improve the overall pump efficiency (see Ref. [2]).

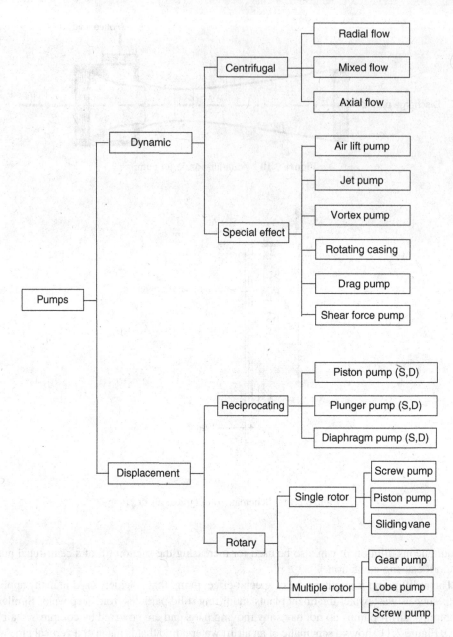

Figure 2.12 Classification of pumps

2.1.3 Additional Classifications

In addition to the classification shown in Figure 2.12, pumps may also be classified based on type of pumped fluid, number of stages, casing geometry, pump layout and so on. The following are the most common bases for pump additional classifications:

a. Shape of casing: volute shape, double volute, diffuser, annular, tubular, split casing, etc.
b. Inlet geometry: single suction, double suction, axial inlet, side inlet, top inlet, etc.
c. Layout: the pump shaft may be horizontal, vertical, or inclined.
d. Discharge pressure or the energy consumption: pumps are sometimes classified as low pressure, high pressure, or high energy.
e. Number of stages: in the cases of radial and mixed-flow centrifugal pumps, they may be classified as single-stage, double-stage, or multistage.
f. Liquid handled: the type of pumped fluid may necessitate some special design considerations. For example, gasoline pumps require special sealing system to avoid leakage in order to reduce fire hazard, and similarly for handling toxic liquids.
g. Material of pump parts: the material used for manufacturing the impeller and pump casing may differ based on the type of pumped fluid. Special materials or coatings are used when handling corrosive liquids (such as sulfuric acid). or liquids containing solid particles.
h. Type of prime mover: in most cases, pumps are driven by electric motors, but in some cases they can be driven by diesel engines or steam or gas turbines.
i. Operating condition: such as submersible pump, wet motor pump, standby pump, and auxiliary pump.

Examples
- radial-type vertical-axis water pump
- multistage mixed-flow submersible oil pump
- mixed-flow double-volute auxiliary gasoline pump
- single-stage double-suction centrifugal pump
- air-operated double-acting diaphragm pump

2.2 Basic Definitions and Terminology

Knowing the definitions and terminology used in pumping machinery is essential not only for ease of communication between engineers but also for avoiding misunderstanding of concepts and provided information. In the following, the basic definitions and terminology are introduced.

2.2.1 Pump Capacity 'Q'

This term is commonly used to express the actual volume flow rate delivered by the pump. Units of capacity include m^3/s, ft^3/s (cfs), gallons/minute (gpm), and barrels/day (*BPD*).

2.2.2 Pump Heads

There are several heads being used, such as pump suction head, delivery head, total head, and others. The definition and physical significance of each are given below:

2.2.2.1 Static suction and static delivery heads h_{ss} and h_{sd}

The static suction head (h_{ss}) and static delivery head (h_{sd}) represent the pressure heads at the suction and delivery nozzles, respectively, of the pump at zero flow rate. Considering

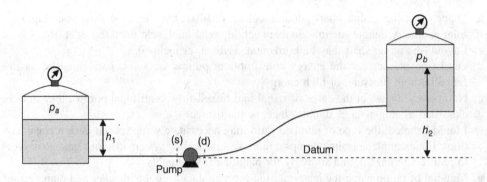

Figure 2.13 Schematic sketch of a simple pumping system

the simple pumping system shown in Figure 2.13, the defining equations of h_{ss} and h_{sd} can be written as

$$h_{ss} = h_1 + \frac{p_a}{\gamma} \quad \text{and} \quad h_{sd} = h_2 + \frac{p_b}{\gamma} \tag{2.1}$$

where p_a and p_b represent the gage pressures at the fluid-free surface in the suction and delivery reservoirs. The datum used is usually taken at the pump shaft centerline for horizontal pumps and the inlet eye of the first stage impeller for vertical pumps. The subscript s is normally used for the inlet to the suction nozzle and d for the exit of the delivery nozzle.

2.2.2.2 The total static head h_{st}

The total static head represents the difference between the static delivery and static suction heads. When both reservoirs are open to the atmosphere, h_{st} represents the difference in elevation between the fluid free surface in the two reservoirs.

$$h_{st} = h_{sd} - h_{ss} \tag{2.2}$$

2.2.2.3 The pump suction head h_s

The pump suction head represents the pressure head at the pump suction nozzle (s) when the pump is in operating condition. Figure 2.14 shows a schematic of the suction side of a pump together with the hydraulic and energy gradient lines indicating h_s. Applying the energy equation between the fluid free surface and the suction nozzle, we obtain

$$h_s = \frac{p_a}{\gamma} + h_1 - \frac{v_s^2}{2g} - h_{Ls} \tag{2.3}$$

where h_{Ls} represents the friction head loss (both major and minor losses) between the suction reservoir and the suction nozzle.

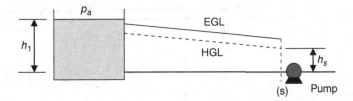

Figure 2.14   Schematic sketch of the suction side showing HGL and EGL

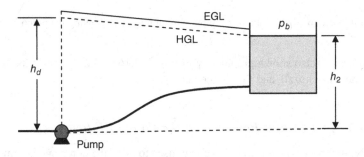

Figure 2.15   Schematic of the delivery side showing HGL and EGL

2.2.2.4   The pump delivery head h_d

The pump delivery head represents the pressure head at the pump delivery nozzle (d) when the pump is in operation. Figure 2.15 shows a schematic of the delivery side of a pump, together with the hydraulic and energy gradient lines indicating h_d. Applying the energy equation between the delivery nozzle and the fluid free surface in the delivery reservoir, we obtain

$$h_d = \frac{p_b}{\gamma} + h_2 - \frac{v_d^2}{2g} + h_{Ld} \tag{2.4}$$

where h_{Ld} represents the friction head loss (both major and minor losses) between the delivery nozzle and the delivery reservoir.

2.2.2.5   The pump total head H

The pump total head represents the energy added to the fluid by the pump (between the suction and delivery nozzles) per unit weight of fluid. This added energy must be in the form of either increase in kinetic energy or increase in pressure head or the combined effect of both. Applying the energy equation between the suction and discharge nozzles we obtain

$$H = (h_d - h_s) + \left(\frac{v_d^2 - v_s^2}{2g}\right) + (z_d - z_s) \tag{2.5}$$

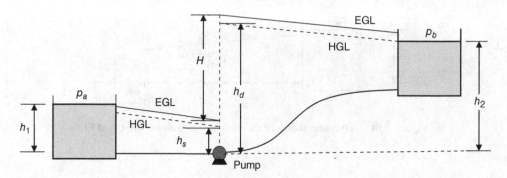

Figure 2.16   Schematic of a simple pumping system showing the pump heads

The total head is also shown graphically in Figure 2.16 as the difference between the total head at the delivery nozzle and that at the suction nozzle.

2.2.3   Input and Output Powers and the Overall Efficiency

The pump input power is the mechanical power used to drive the pump and is sometimes called the shaft power or brake power (*BP*). The term brake power is used since the pump operates as a brake for the prime mover. On the other hand, the pump output power represents the energy added to the fluid by the pump per unit time. It is sometimes called the pump hydraulic power or the pump fluid power. Accordingly, we can write

Pump input power = BP = Driving torque × angular velocity = $T\omega$

Pump output power = Pump fluid power = γQH

Pump overall efficiency = Output power/Input power $\Rightarrow \eta_o = \gamma QH/BP$ (2.6)

2.2.4   Pump Performance Characteristics

The pump performance characteristics is a term used by engineers referring to the relationship between each of the total head developed by the pump (*H*), the pump power consumption (*BP*), the pump overall efficiency (η_o), and the pump flow rate (*Q*). These are usually presented graphically in terms of the three curves of *H–Q*, *P–Q*, and η_o–*Q* when the pump operates at a constant speed *N*, considering clear water as the pumped fluid. Typical performance curves are shown in Figure 2.17 for a radial-type centrifugal pump.

The pump performance characteristics can also be presented in the form of a set of iso-efficiency curves. These curves can be constructed in one of the following forms:

i. Iso-efficiency curves for same pump operating at different speeds as shown in Figure 2.18
ii. Iso-efficiency curves for a set of geometrically similar pumps of different sizes operating at the same speed as shown in Figure 2.19

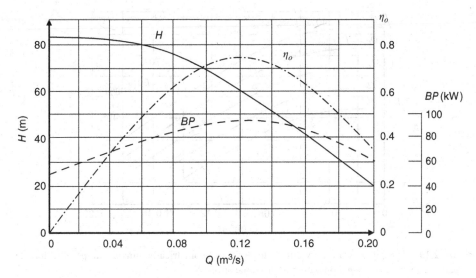

Figure 2.17 Typical $H - Q$, $BP - Q$, and $\eta_o - Q$ curves for a radial-type centrifugal pump operating at a constant speed

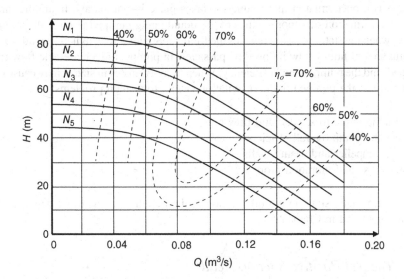

Figure 2.18 Typical iso-efficiency curves for a radial-type centrifugal pump operating at different speeds (N_1 to N_5)

It is worth mentioning that the three curves, $H - Q$, $BP - Q$, and $\eta_o - Q$, for a given pump can be deduced from Figure 2.18 at any selected speed within the given range (i.e. between N_1 and N_5). The same three curves can be extracted from Figure 2.19 for any impeller size between ϕ_1 and ϕ_5 at the same operating speed N.

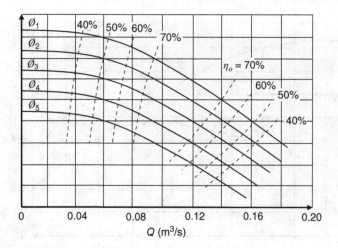

Figure 2.19   Typical iso-efficiency curves for a set of geometrically similar pumps of different sizes (impeller diameters ϕ_1 to ϕ_5) operating at the same speed N

2.2.5   Cavitation

Cavitation is a phenomenon that frequently takes place in pumps and hydraulic turbines. It refers to the formation of vapor cavities of the liquid being pumped at normal operating temperatures when the static pressure reaches the fluid vapor pressure. Cavitation tends to decrease the useful area of liquid flow in the flow passages. In pumps, it decreases the flow rate, head developed, and the pump overall efficiency. When a cavity grows in size, it may cause complete discontinuity of the pumped liquid. Cavitation causes four main problems:

a. reduces the pump total head and flow rate
b. reduces the pump overall efficiency
c. damages impeller and casing walls
d. creates noise and vibration problems

Cavitation usually starts in the inlet region of the impeller vanes. More details about cavitation are presented in Chapter 5.

2.2.6   The Net Positive Suction Head

In order to avoid cavitation, the minimum pressure inside the pump should be higher than the fluid vapor pressure. Since it is only possible to measure (or calculate) the static pressure up to the suction nozzle of the pump, an allowance should be made for the additional pressure drop occurring inside the pump. This allowance is called the net positive suction head (*NPSH*). The defining equation for *NPSH* is

$$NPSH = h_{sn} + \frac{v_{sn}^2}{2g} - \frac{p_v}{\gamma} \qquad (2.7)$$

where the subscript (sn) refers to the conditions at inlet of the pump suction nozzle. The minimum value of $NPSH$ needed to avoid cavitation is called the required $NPSH$ ($NPSHR$). This represents one of the important specifications of the pump and is always given by the manufacturer, together with the pump performance curves. The value of $NPSH$ calculated using Eq. (2.7) based on the actual working conditions is called the available $NPSH$ and labeled $NPSHA$. In order to avoid cavitation, the suction conditions must satisfy

$$NPSHA > NPSHR \qquad (2.8)$$

Figure 2.20 shows the suction side of a pump that sucks water from an open reservoir. The term suction lift refers to the depth of the fluid free surface in the suction reservoir below the pump level (e.g. below the datum). If we apply the energy equation between (a) and (sn), we can write

$$\frac{p_a}{\gamma} + \frac{v_a^2}{2g} + z_a = \frac{p_{sn}}{\gamma} + \frac{v_{sn}^2}{2g} + z_{sn} + h_{Ls} \qquad (2.9)$$

But $v_a = 0$ and $p_{sn}/\gamma = h_{sn}$, $z_{sn} - z_a = suction\ lift$, therefore we can simplify Eq. (2.9) to obtain

$$h_{sn} + \frac{v_{sn}^2}{2g} = \frac{p_a}{\gamma} - suction\ lift - h_{Ls} \qquad (2.10)$$

Based on the chosen datum, $z = 0$ at the pump suction nozzle, which can also be considered invariant inside the pump. Accordingly, the term $\left(h_{sn} + v_{sn}^2/2g\right)$ represents the *available total head* at the pump suction nozzle. Since the fluid pressure should not reach the vapor pressure to avoid cavitation, the allowable drop in the pressure head before the start of cavitation can be measured by the difference between the *available total head* and the head corresponding to the vapor pressure, therefore

$$\text{Allowable drop in pressure head before start of cavitation} = h_{sn} + \frac{v_{sn}^2}{2g} - \frac{p_v}{\gamma}$$

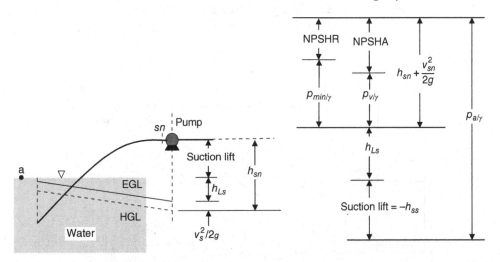

Figure 2.20   Graphical representation of *NPSHA*

The above term is referred to as the available *NPSH* (*NPSHA*) with the defining equation

$$NPSHA = h_{sn} + \frac{v_{sn}^2}{2g} - \frac{p_v}{\gamma} \qquad (2.11)$$

On the other hand, the required net positive suction head (*NPSHR*) represents the actual corresponding pressure drop occurring between the pump suction nozzle and the minimum pressure point inside the pump. The value of *NPSHR* is equal to the value of *NPSHA* at the start of cavitation. At this time the minimum pressure becomes equal to the vapor pressure. Accordingly, we can write

$$NPSHR + \frac{p_{min}}{\gamma} = NPSHA + \frac{p_v}{\gamma} \qquad (2.12)$$

From Eq. (2.12), it is clear that $p_{min} > p_v$ only when *NPSHA* > *NPSHR*, and this explains the basis of the condition given in Eq. (2.8) for avoiding cavitation. The same concept is shown graphically in Figure 2.20.

Example 2.1

Figure 2.21 shows a centrifugal pump used to pump gasoline ($\gamma = 8.5$ kN/m^3 and $p_v = 60$ kPa) from reservoir A to another reservoir B (not shown in the figure). It is required to determine the maximum flow rate that can be delivered by the pump if cavitation is to be avoided. You may neglect minor losses and assume that *NPSHR* = 0.75 m.

Data: $p_{atm} = 101$ kPa, $D_s = 0.15$ m, $L_s = 50$ m, $f = 0.018$

Solution

Apply the energy equation between points (a) and (sn) shown in the figure,

$$\frac{p_a}{\gamma} + \frac{v_a^2}{2g} + z_a = \frac{p_{sn}}{\gamma} + \frac{v_{sn}^2}{2g} + z_{sn} + h_{Ls} \qquad (1)$$

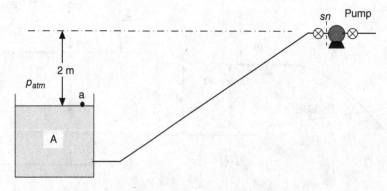

Figure 2.21   Schematic of the pumping system of Example 2.1

Knowing that $v_a = 0$, $p_a = p_{atm}$, and $p_{sn}/\gamma = h_{sn}$, the above equation can be simplified to

$$h_{sn} + \frac{v_{sn}^2}{2g} = \frac{p_{atm}}{\gamma} + (z_a - z_{sn}) - h_{Ls} \tag{2}$$

Using the defining equation of NPSHA (Eq. (2.9)), we can write

$$NPSHA = \frac{p_{atm}}{\gamma} + (z_a - z_{sn}) - h_{Ls} - \frac{p_v}{\gamma} \tag{3}$$

Using Darcy's formula,

$$h_{Ls} = \frac{fLV_s^2}{2gD_s} + \overbrace{Minor\ losses}^{=0} = \frac{fL(Q/A_s)^2}{2gD_s} = \frac{0.018 \times 50 \times Q^2}{2 \times 9.81 \times 0.15 A_s^2} = 979Q^2$$

Substitute in Eq. (3)

$$NPSHA = \frac{101}{8.5} + (-2) - 979Q^2 - \frac{60}{8.5} = 2.83 - 979Q^2 \tag{4}$$

In order to avoid cavitation, the condition $NPSHA > NPSHR$ must be satisfied, therefore

$$2.83 - 979Q^2 > 0.75$$

Finally, Q must be less than 0.046 m³/s in order to avoid cavitation. Such flow rate is called the critical flow rate.

2.3 Determination of Flow Rate in a Pumping System

A pumping system normally contains a pump (or a group of pumps arranged in series or in parallel) and a piping system. The piping system may include a single pipe or a group of pipes arranged in series or in parallel or a combination of both, in addition to various pipe fittings (bends, elbows, valves, flowmeters, etc.). We assume that the pump characteristics and speed of rotation are completely known and also assume that the piping system components are fully defined. Consider the simple pumping system shown in Figure 2.22 and also consider the pump to have the H–Q curve shown in Figure 2.23.

Apply the energy equation between points a and b shown in Figure 2.22,

$$\frac{p_a}{\gamma} + \frac{v_a^2}{2g} + z_a + H = \frac{p_b}{\gamma} + \frac{v_b^2}{2g} + z_b + \sum h_L \tag{1}$$

where $\sum h_L$ represents all losses in the piping system (major and minor losses between points a and b). Since $V_a = V_b = 0$, the above equation can be simplified to

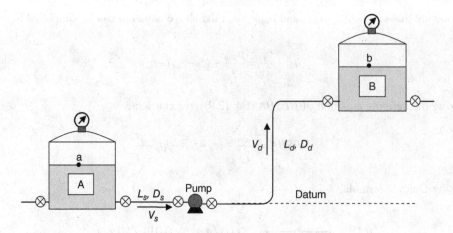

Figure 2.22   A typical pumping system containing a number of fittings

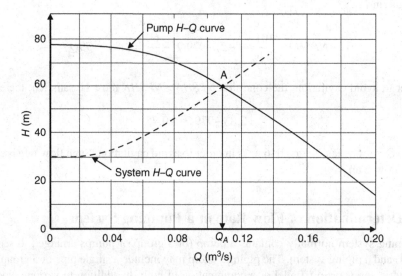

Figure 2.23   The pump H–Q characteristic curve

$$H = \frac{p_b - p_a}{\gamma} + (z_b - z_a) + \sum h_L$$

$$\text{or } H = h_{st} + \sum h_L \tag{2}$$

The term $\sum h_L$ can be expressed as

$$\sum h_L = \frac{f_s L_s V_s^2}{2gD_s} + \frac{f_d L_d V_d^2}{2gD_d} + \sum K \frac{V^2}{2g} \tag{3}$$

where the subscripts s and d are used for suction and delivery sides respectively and K is the loss coefficient for each pipe fitting. The first two terms on the RHS of Eq. (3) represent the major losses in the suction and delivery pipes and the third term represents all minor losses in the piping system.

In a wide range of engineering applications, the pipes used are rough pipes, and the Reynolds number is very high, leading to a friction coefficient that depends on the pipe roughness and is less dependent on the Reynolds number [4]. In such cases, the friction coefficient (f) may be assumed constant (as a rough approximation). Knowing that $V = Q/A$, Eq. (3) can be expressed in the form

$$\sum h_L = C_2 Q^2 \qquad (4)$$

Substituting in Eq. (2) and assuming that h_{st} is constant, we can write

$$H = C_1 + C_2 Q^2 \qquad (2.13)$$

Equation (2.13) describes what is called the piping system curve or simply 'system curve.' The equation depends mainly on the system specifications and has no relation to the pump characteristics. In order to determine the pump flow rate, we plot Eq. (2.13) on the same graph containing the pump $H–Q$ curve (Figure 2.23) and the point of intersection (point A) defines the point of pump operation.

Nowadays, plastic pipes are widely used, especially for water pipelines. The relative roughness for these pipes is very small and they can be practically considered as smooth pipes. In this case and also in the case of pipes with small roughness, the friction coefficient varies with Reynolds number and cannot be considered a function of the relative roughness only. To determine the friction coefficient for smooth pipes, we can use the following Blasius formula [3] for turbulent flow up to $Re = 10^5$:

$$f = 0.316/Re^{1/4} \quad \text{for} \quad 4 \times 10^3 < Re < 10^5 \qquad (2.14)$$

For higher Reynolds numbers, the following equation was derived by Prandtl in 1935 for smooth pipes, and this can be used for a wide range of Reynolds numbers:

$$\frac{1}{\sqrt{f}} = 2.0 \log \left(Re \sqrt{f} \right) - 0.8 \qquad (2.15)$$

Equation (2.15) can be solved iteratively for any given value of Re. For other pipes with different roughness, either the Moody chart or the Colebrook–White formula (Eq. (1.36)) can be used for the determination of the friction coefficient. Accordingly, a better prediction of the system curve requires detailed calculation of the total head, H, for different values of the flow rate, Q, using Eqs (2) and (3) and utilizing the appropriate equation – (2.14), (2.15), or (1.36) – for the friction coefficient. This is demonstrated in Example 2.3.

Example 2.2

A centrifugal pump having the shown $H–Q$ and $\eta_o–Q$ characteristics at $N = 1800$ rpm is used to pump water at $20\,°C$ ($\gamma = 9.79$ kN/m^3 and $\nu = 1.004 \times 10^{-6}$ m^2/s) from reservoir A to reservoir

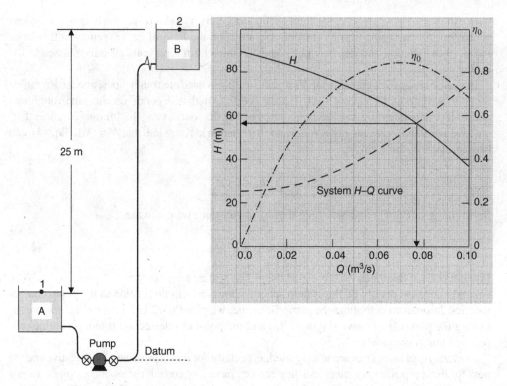

Figure 2.24 Schematic of the pumping system and pump characteristic curves

B as shown in Figure 2.24. The suction pipe is 15 cm diameter and 6.7 m long and the delivery pipe is 15 cm diameter and 205 m long. The piping system has four bends ($K_b = 0.2$) and two valves ($K_v = 2.0$). The loss coefficient at the inlet of the suction pipe, $K_i = 0.1$ and assume a pipe friction coefficient, $f = 0.018$. Plot the system H–Q curve and determine:

a. system flow rate
b. head developed by the pump
c. pump power consumption.

Solution
The equation of the system H–Q curve can be obtained by applying the energy equation between points (1) and (2) indicated in Figure 2.24.

$$\frac{p_1}{\gamma} + \frac{V_1^2}{2g} + Z_1 + H = \frac{p_2}{\gamma} + \frac{V_2^2}{2g} + Z_2 + \sum h_L \tag{1}$$

where $\sum h_L$ is the summation of all major and minor losses ($\sum h_L = \sum h_{Ls} + \sum h_{Ld}$).

Now since $p_1 = p_2 = p_{atm}$ and $V_1 = V_2 = 0$, therefore Eq. (1) can be reduced to

$$H = (z_2 - z_1) + \sum h_L \qquad (2)$$

and $\sum h_L$ can be expressed as

$$\sum h_L = \frac{fL_s V_s^2}{2gD_s} + \frac{fL_d V_d^2}{2gD_d} + 2K_b \frac{V_s^2}{2g} + 2K_b \frac{V_d^2}{2g} + K_v \frac{V_s^2}{2g} + K_v \frac{V_d^2}{2g} + K_i \frac{V_s^2}{2g} + K_e \frac{V_d^2}{2g}$$

where K_e is the loss coefficient at the delivery-pipe exit. However, since $D_s = D_d$, then $V_s = V_d = V$ and the above equation can be simplified to

$$\sum h_L = \frac{V^2}{2g} \left[\frac{f(L_s + L_d)}{D} + 4K_b + 2K_v + K_i + K_e \right]$$

$$= \frac{Q^2}{2gA^2} \left[\frac{f(L_s + L_d)}{D} + 4K_b + 2K_v + K_i + K_e \right]$$

$$= \frac{Q^2}{2 \times 9.81 (0.0177)^2} \left[\frac{0.018(205 + 6.7)}{0.15} + 4 \times 0.2 + 2 \times 2.0 + 0.1 + 1.0 \right] = 5093 Q^2$$

Substitute in Eq. (2) to obtain

$$H = 25 + 5093 Q^2 \qquad (3)$$

The above equation describes the system H–Q curve (or system characteristic curve). Equation (3) can be used to plot the system curve as shown in Figure 2.24. The point of intersection of the system and pump H–Q curves is the point of operation, and accordingly, $Q = 0.077$ m^3/s and the head developed by the pump, $H = 56$ m.

At the point operation, the pump overall efficiency, $\eta_o = 84\%$.

The pump power consumption, $BP = \dfrac{\gamma QH}{\eta_o} = \dfrac{9.79 \times 0.077 \times 56}{0.84} = \underline{50.3\,\text{kW}}$

Example 2.3
Solve the same problem given in Example 2.2, but without assuming a value for the friction coefficient and considering a commercial steel pipe.

Solution
The equation of the system H–Q curve $[H = (z_2 - z_1) + \sum h_L]$ will be exactly the same as in Example 2.2 with $\sum h_L$ representing all major and minor losses that can be expressed as

$$\sum h_L = \frac{Q^2}{2gA^2}\left[\frac{f(L_s + L_d)}{D} + 4K_b + 2K_v + K_i + K_e\right]$$

$$= \frac{Q^2}{2 \times 9.81(0.0177)^2}\left[\frac{f(205 + 6.7)}{0.15} + 4 \times 0.2 + 2 \times 2.0 + 0.1 + 1.0\right]$$

$$= 2.296 \times 10^5 f Q^2 + 960 Q^2$$

Therefore, we can write

$$H = 25 + (2.296 \times 10^5 f + 960) Q^2 \tag{1}$$

The above equation describes the system H–Q curve with f depending on Re and k_s/D. For commercial steel pipes, $k_s = 0.045$ mm and $k_s/D = 0.045/150 = 0.0003$. Now, we can use Eq. (1) to obtain numerical values for H for some selected values of Q utilizing the Colebrook–White formula for f (Eq. (1.36)) as shown in the table below.

Q (m³/s)	V (m/s)	Re	f (using Eq. (1.36))	H (m) (using Eq. (1))
0.00	0	0	—	25.0
0.02	1.13	1.69×10^5	0.0182	27.1
0.04	2.26	3.38×10^5	0.0169	32.7
0.06	3.40	5.06×10^5	0.0164	42.0
0.08	4.53	6.75×10^5	0.0161	54.8
0.10	5.66	8.44×10^5	0.0159	71.1

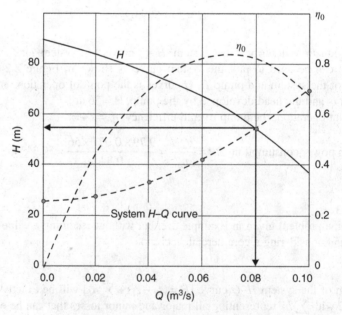

Figure 2.25 H–Q curve based on the tabulated data

The H–Q curve based on the above tabulated data is shown in Figure 2.25 and the point of intersection with the pump H–Q curves gives $Q = 0.08$ m^3/s and the head developed by the pump, $H = 53$ m.

At this point operation, the pump overall efficiency, $\eta_o = 82\%$.

The pump power consumption, $BP = \dfrac{\gamma QH}{\eta_o} = \dfrac{9.79 \times 0.08 \times 53}{0.82} = \underline{50.6\,\text{kW}}$

In comparison with the results obtained in Example 2.2, we find about 3.9% increase in the flow rate for the approximately the same power consumption.

2.4   Operation of Pumps in Parallel and in Series

Pumps are sometimes operated in parallel or in series, in order to increase the system flow rate or to develop the required delivery head.

2.4.1   Parallel Operation

Identical or different pumps may be connected in parallel in order to increase the volume flow rate through a piping system. Any two pumps can be considered identical if they have the same H–Q characteristic, and this happens only if they have the same design, size, and speed of rotation. Consider the general case of two different pumps P1 and P2 connected in parallel as shown in Figure 2.26. In the parallel mode of operation, the total flow rate is the summation of the flow rates delivered by the two pumps, while the head developed by the two pumps is exactly the same as the head developed by each pump. Accordingly, the combined H–Q curve (H_c–Q_c) for the two pumps can be plotted using the simple equations,

$$Q_c = Q_1 + Q_2 \quad \text{and} \quad H_c = H_1 = H_2 \tag{2.16}$$

The system is analogous to the operation of two batteries connected in parallel. The flow rate is similar to the current and the total head developed is similar to the potential difference. The arrangement shown in Figure 2.26 allows the use of pump P1 alone, pump P2 alone, or the two pumps operating in parallel.

The H_c–Q_c curve is constructed by choosing any value of H_c ($H_c = H_1 = H_2$) and determine Q_1 and Q_2 at that value as shown in Figure 2.27. Q_c is then obtained by using Eq. (2.16). These values of H_c and Q_c enable plotting one point on the H_c–Q_c curve and similarly other points can

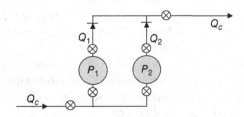

Figure 2.26   Schematic of two centrifugal pumps operating in parallel

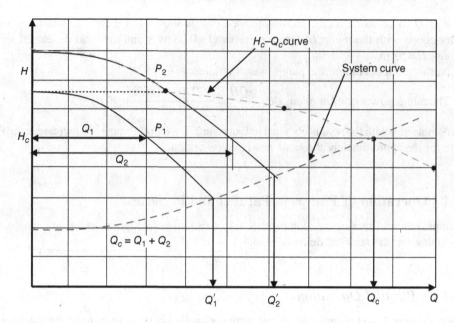

Figure 2.27 The $H\text{–}Q$ curves of the two different pumps P1 and P2 operating in parallel and the combined $H_c\text{–}Q_c$ curve

be obtained. If the system curve is as shown in the figure, the flow rate obtained when P1 is only operating in the system is Q_1' and that obtained when P2 is only operating is Q_2'. When the two pumps are operating in parallel, the system flow rate is Q_c. It is now clear that $Q_c > Q_1'$ and $Q_c > Q_2'$ but $Q_c \neq Q_1' + Q_2'$.

2.4.2 Series Operation

Consider the two different pumps P1 and P2 connected in series, as shown in the figure. The combined $H\text{–}Q$ curve ($H_c\text{–}Q_c$) for the two pumps can be plotted using the equations

$$Q_c = Q_1 = Q_2 \quad \text{and} \quad H_c = H_1 + H_2 \qquad (2.17)$$

The system is analogous to the operation of two batteries connected in series. The arrangement shown in Figure 2.28 allows the use of pump P1 alone, Pump P2 alone, or the two pumps operating in series or in parallel using the three valves at A, B, and C.

The $H_c\text{–}Q_c$ curve is constructed by choosing any value of Q_c ($Q_c = Q_1 = Q_2$) and determine H_1 and H_2 at that value as shown in Figure 2.29. H_c is then obtained by using Eq. (2.17). The obtained values of H_c and Q_c enable plotting one point on the $H_c\text{–}Q_c$ curve and similarly other points can be obtained. Considering the system curve shown in Figure 2.29, the flow rate obtained when P1 is only operating in the system is Q_1' and that obtained when P2 is only operating is Q_2'. When the two pumps are operating in series, the system flow rate is Q_c. It is again clear that $Q_c > Q_1'$ and $Q_c > Q_2'$ but $Q_c \neq Q_1' + Q_2'$.

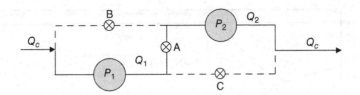

Figure 2.28 Schematic of two centrifugal pumps operating in series

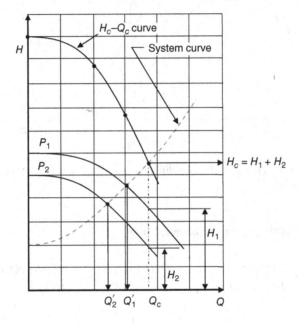

Figure 2.29 The H–Q curves of the two different pumps P1 and P2 connected in series and the combined H_c–Q_c curve

Example 2.4

Two pumps, A and B, are arranged to operate in parallel. The H–Q and η_o–Q characteristics of each pump are as shown in Figure 2.30. Obtain the combined H–Q and η_o–Q curves.

Solution

At any head H, the flow rate of each pump (Q_A and Q_B) can be easily determined from the Figure 2.30. Also, the corresponding efficiency of each pump can be obtained from the same figure. The table below shows Q_A, Q_B, Q_C, η_A, and η_B at selected values of H. The efficiency of the combined system is obtained as follows:

$$(BP)_A = \frac{\gamma Q_A H}{\eta_A} \quad \text{and} \quad (BP)_B = \frac{\gamma Q_B H}{\eta_B}$$

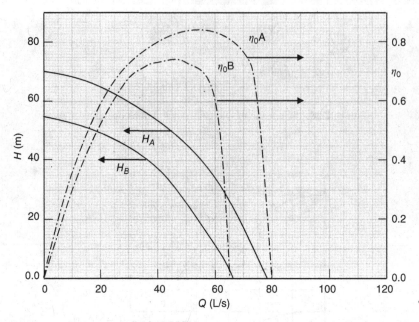

Figure 2.30   The H–Q curves of the two different pumps P1 and P2 connected in series and the combined H_c–Q_c curve

The combined efficiency η_C can be obtained from

$$\eta_C = \frac{Total\ output\ power}{Total\ input\ power} = \frac{\gamma Q_A H + \gamma Q_B H}{(BP)_A + (BP)_B} = \frac{\gamma Q_A H + \gamma Q_B H}{\gamma Q_A H / \eta_A + \gamma Q_B H / \eta_B}$$

Simplify to obtain $\eta_C = \dfrac{\eta_A \eta_B (Q_A + Q_B)}{\eta_B Q_A + \eta_A Q_B}$.

Based on the above equation, the value of η_C is calculated at all points. The combined H_c–Q_c and η_c–Q_c curves are shown in Figure 2.31.

Note: It may be more efficient to operate the two pumps only when the flow rate exceeds Q_D (see the efficiency curves).

H (m)	54.0	53.0	50.0	40.0	30.0	20.0	10.0
Q_A (L/s)	38.0	40.0	43.0	54.0	62.0	69.5	75.5
Q_B (L/s)	5.0	10.0	18.0	35.0	45.5	54.0	60.5
$Q_C = Q_A + Q_B$	43.0	50.0	61.0	89.0	107.5	123.5	136.0
η_A	0.79	0.8	0.82	0.85	0.82	0.72	0.55
η_B	0.16	0.29	0.48	0.70	0.74	0.70	0.55
η_C	0.54	0.59	0.68	0.79	0.78	0.71	0.55

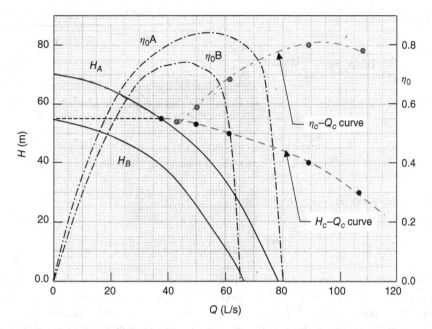

Figure 2.31 The H–Q curves of the two different pumps P1 and P2 connected in series and the combined H_c–Q_c curve

2.5 Similitude Applied to Centrifugal and Axial Flow Pumps

The performance of centrifugal and axial-flow pumps (Figure 2.32) depends on the pump shape (geometry) and size, the type of fluid pumped (fluid density and viscosity), and the operating condition.

The pump operating conditions can be adjusted by controlling either the speed of rotation N (by adjusting the speed of the prime mover), or the flow rate Q (by adjusting the delivery valve opening), or the combined effect of both.

The rest of the variables such as the total head H, input power P, and overall efficiency η_o are dependent on the operating conditions (N, Q), pump size and geometry, and type of fluid pumped (ρ and μ). Accordingly, we can write

$$gH = \phi_1 \left(\overbrace{Q, N}^{\text{Operating condition}}, \quad \overbrace{\rho, \mu}^{\text{Fluid properties}}, \quad \overbrace{D, L_1, L_2, \ldots}^{\text{Size and geometry}} \right) \qquad (2.18)$$

where the term gH represents the work done/unit mass of fluid and the dimensions L_1, L_2, and so on describe the pump geometry.

Using the Buckingham π-theorem [5], we can obtain the following dimensionless terms

$$\frac{gH}{N^2D^2}, \frac{Q}{ND^3}, \frac{\rho ND^2}{\mu}, \frac{L_1}{D}, \frac{L_2}{D}, \ldots$$

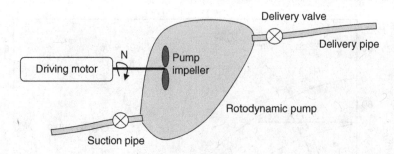

Figure 2.32 Schematic of main pump components

Now, we can write Eq. (2.18) in the form

$$\frac{gH}{N^2D^2} = f_1\left(\frac{Q}{ND^3}, \frac{\rho ND^2}{\mu}, \frac{L_1}{D}, \frac{L_2}{D}, \ldots\right) \qquad (2.19)$$

The term gH/N^2D^2 is called the head coefficient and labeled C_H; the term Q/ND^3 is called the flow coefficient, C_Q; the term $\rho ND^2/\mu$ represents the ratio between inertia and viscous forces (Reynolds number); and L_1/D, L_2/D, ... represent pump geometrical characteristics.

If we repeat the above analysis twice, once with the input power P replacing the term gH and once again using η_o, we obtain

$$\frac{P}{\rho N^3 D^5} = f_2\left(\frac{Q}{ND^3}, \frac{\rho ND^2}{\mu}, \frac{L_1}{D}, \frac{L_2}{D}, \ldots\right) \qquad (2.20)$$

$$\eta_o = f_3\left(\frac{Q}{ND^3}, \frac{\rho ND^2}{\mu}, \frac{L_1}{D}, \frac{L_2}{D}, \ldots\right) \qquad (2.21)$$

where $P/\rho N^3 D^5$ is called the power coefficient and labeled C_P.

If we are studying the performance of the same pump or a series of geometrically similar pumps, the terms L_1/D, L_2/D, ... can be treated as constants and can be omitted from the right-hand side of Eqs (2.19), (2.20), and (2.21).

When handling liquids of low viscosity (such as water, benzene, kerosene, and diesel fuel), the effect of viscosity on the pump performance at normal operating conditions is very small and can be neglected. This is because the role of viscosity inside the pump is closely related to friction losses in various flow passages, including the suction nozzle, impeller flow passages, and the volute casing, followed by the delivery nozzle. On the other hand, the head developed by the pump comes mainly from the work done by the impeller vanes on the fluid. Now, when the viscosity is very small, the Reynolds number ($Re = \rho ND^2/\mu$) becomes very high, and the friction coefficient becomes independent of Re. The same behavior is experienced in pipe flow where the friction head loss becomes independent of viscosity at very high values of Re, especially when the pipe roughness is not very small (see the Moody chart in Figure 1.16). Therefore, the Reynolds number can be dropped in Eqs (2.19)–(2.21) when handling liquids of low viscosity. Now, if we apply Eqs (2.19)–(2.21) to cases in which the same pump is operating at different speeds or a group of geometrically similar pumps, then L_1/D, L_2/D, ... can be treated as constants and the above equations can be simplified to

$$\frac{gH}{N^2D^2} = f_1\left(\frac{Q}{ND^3}\right) \quad \text{or} \quad C_H = f_1(C_Q) \tag{2.22}$$

$$\frac{P}{\rho N^3 D^5} = f_1\left(\frac{Q}{ND^3}\right) \quad \text{or} \quad C_P = f_2(C_Q) \tag{2.23}$$

$$\eta_o = f_3\left(\frac{Q}{ND^3}\right) \quad \text{or} \quad \eta_o = f_3(C_Q) \tag{2.24}$$

The importance of Eqs (2.22)–(2.24) is because the H–Q characteristics of a group of geometrically similar pumps driven at different speeds can be all presented as *one curve* if we plot them using C_H–C_Q. The same applies to C_P–C_Q and η_o–C_Q. To examine this concept experimentally, Figure 2.33 shows the γH–Q curves for a given pump when operating at two different speeds, and Figure 2.34 shows the same experimental data plotted using the coordinates $\gamma H/N^2$ and Q/N. The difference between these coordinates and the desired C_H–C_Q are just constant multipliers. Figure 2.34 shows only one curve for the two speeds and this proves the accuracy of the assumptions made in the derivation of Eq. (2.22).

The mathematical relationship between C_P, C_Q, C_H, and η_o can be obtained as follows:

$$\text{We know that } P = \frac{\gamma QH}{\eta_o} = \frac{\rho g QH}{\eta_o} = \left(\frac{1}{\eta_o}\right)\left(\frac{gH}{N^2D^2}\right)\left(\frac{Q}{ND^3}\right)(\rho N^3 D^5)$$

$$\text{Rearrange } \frac{P}{\rho N^3 D^5} = \left(\frac{1}{\eta_o}\right)\left(\frac{gH}{N^2D^2}\right)\left(\frac{Q}{ND^3}\right)$$

$$\text{Or simply, } C_P = \frac{C_Q C_H}{\eta_o} \tag{2.25}$$

If two geometrically similar pumps are operating at dynamically similar conditions, then

$$C_{Q1} = C_{Q2}, \ C_{H1} = C_{H2}, \ C_{P1} = C_{P2}, \text{ and } \eta_{o1} = \eta_{o2} \tag{2.26}$$

2.5.1 The Locus of Similarity

The locus of similarity represents a set of points of operation of the same pump operating at different speeds or a set of points of operation of a number of geometrically similar pumps of different sizes operating at the same speed. Based on Eqs (2.19)–(2.21), the same pump or geometrically similar pumps can operate at dynamically similar conditions if C_Q and Re are the same. However, when handling fluids of low viscosities, the Reynolds number can be dropped (as in Eqs (2.22)–(2.24)) and C_Q will be sufficient.

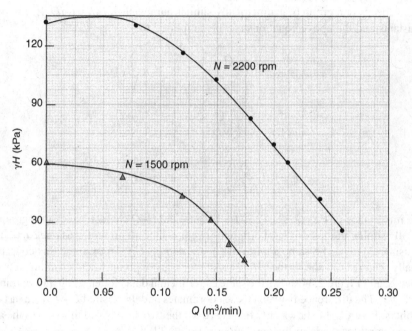

Figure 2.33 γH–Q curves for a radial-type centrifugal pump operating at two different speeds

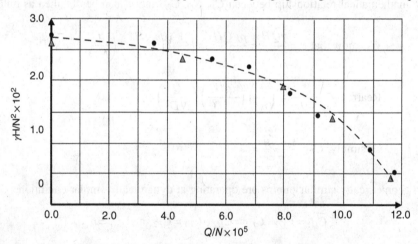

Figure 2.34 Graphical presentation of experimental data in Figure 2.33 using a form of C_H–C_Q coordinates

2.5.1.1 The Locus of Similarity for the Same Pump Operating at Different Speeds

Consider the case of a given pump that can operate at different speeds. Let the H–Q curves shown in Figure 2.35 to represent the pump performance at speeds N_1, N_2, and N_3. If we select point 1 at speed N_1, and wish to find point 2 at speed N_2 (or point 3 at speed N_3) that is dynamically similar to point 1, we can proceed as follows:

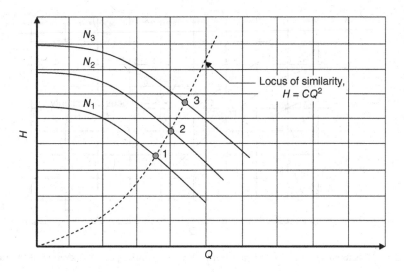

Figure 2.35 Locus of similarity for the same pump operating at different speeds

Since the two points are dynamically similar, therefore $C_{Q1} = C_{Q2}$, and it follows that

$$\frac{Q_1}{N_1 D_1^3} = \frac{Q_2}{N_2 D_2^3} \text{ but } D_1 = D_2, \text{therefore } \frac{Q_1}{N_1} = \frac{Q_2}{N_2} \text{ or } Q \propto N \qquad (2.27)$$

Based on Eq. (2.26), we can write $C_{H1} = C_{H2}$, and it follows that

$$\frac{gH_1}{N_1^2 D_1^2} = \frac{gH_2}{N_2^2 D_2^2} \text{ but } D_1 = D_2, \text{therefore } \frac{H_1}{N_1^2} = \frac{H_2}{N_2^2} \text{ or } H \propto N^2 \qquad (2.28)$$

Combining Eqs (2.27) and (2.28), we can write

$$H \propto Q^2 \quad \Rightarrow \quad H = CQ^2 \qquad (2.29)$$

Equation (2.29) describes the locus of similar points and the constant C can be determined for any chosen point (such as point 1 selected in this analysis).

2.5.1.2 The Locus of Similarity for a Group of Geometrically Similar Pumps Driven at the Same Speed

Consider the case of a set of geometrically similar pumps operating at the same speed (N). Let the H–Q curves shown in Figure 2.36 represent the performance of three geometrically similar pumps having impeller diameters D_1, D_2, and D_3. If we select point 1 for pump D_1 and wish to find point 2 for pump D_2 that is dynamically similar to point 1, we can proceed as follows:

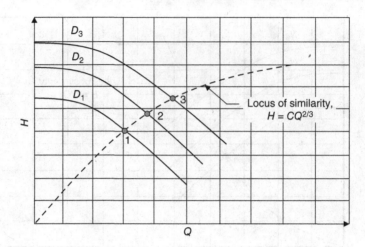

Figure 2.36 Locus of similarity for a set of geometrically similar pumps operating at the same speed

Since the two points are dynamically similar, therefore $C_{Q1} = C_{Q2}$, and it follows that

$$\frac{Q_1}{N_1 D_1^3} = \frac{Q_2}{N_2 D_2^3} \text{ but } N_1 = N_2, \text{therefore } \frac{Q_1}{D_1^3} = \frac{Q_2}{D_2^3} \text{ or } Q \propto D^3 \qquad (2.30)$$

Based on Eq. (2.26), we can write $C_{H1} = C_{H2}$, and it follows that

$$\frac{gH_1}{N_1^2 D_1^2} = \frac{gH_2}{N_2^2 D_2^2} \text{ but } N_1 = N_2, \text{therefore } \frac{H_1}{D_1^2} = \frac{H_2}{D_2^2} \text{ or } H \propto D^2 \qquad (2.31)$$

Combining Eqs (2.30) and (2.31), we can write

$$H \propto Q^{2/3} \quad \Rightarrow \quad H = CQ^{2/3} \qquad (2.32)$$

Equation (2.32) describes the locus of similar points for a set of geometrically similar pumps operating at the same speed.

Example 2.5
The performance curves (H–Q and η_o–Q) for a centrifugal pump are given in Figure 2.37 when operating at a speed of 1500 rpm. It is required to predict similar pump performance curves when operating the pump at a speed of 1800 rpm.

Solution
Let us select one point on the given performance curves (at speed $N_1 = 1500$ rpm); say $Q_1 = 0.01$ at which $H_1 = 48.5$ m and $\eta_{o1} = 0.25$ and try to find the *similar* point (Q_2, H_2, and η_{o2}) on the required performance curves at speed $N_2 = 1800$ rpm.

Similarity of the two points requires that $C_{Q1} = C_{Q2} \Rightarrow \frac{Q_1}{N_1 D_1^3} = \frac{Q_2}{N_2 D_2^3}$ and since D is constant,

therefore, $Q_2 = Q_1(N_2/N_1) = 0.01(1800/1500) = 0.012$ m³/s

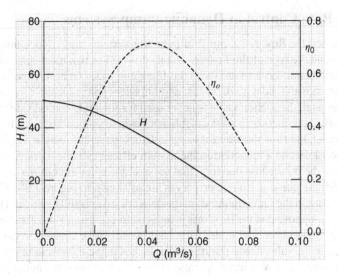

Figure 2.37 The performance curves for a centrifugal pump operating at a speed of 1800 rpm

Table 2.1 Values of Q, H, and η_o at 1500 rpm extracted from Figure 2.37

Q_1 (m³/s)	0.00	0.02	0.04	0.06	0.08
H_1 (m)	50.0	46.0	36.0	23.5	10.0
η_{o1}	0.00	0.48	0.71	0.59	0.30

Table 2.2 Values of Q, H, and η_o at 1800 rpm

Q_1 (m³/s)	0.00	0.024	0.048	0.072	0.096
H_1 (m)	72.0	66.2	51.8	33.8	14.4
η_{o1}	0.00	0.48	0.71	0.59	0.30

Also, $C_{H1} = C_{H2} \Rightarrow \dfrac{gH_1}{N_1^2 D_1^2} = \dfrac{gH_2}{N_2^2 D_2^2}$ and $D = $ Const., therefore,

$$H_2 = H_1(N_2/N_1)^2 = 48.5(1800/1500)^2 = 69.8\,\text{m}$$

The overall efficiency at point 2 will be exactly the same as that at point 1 since C_Q is the same (see Eq. (2.24)). Accordingly, $\eta_{o2} = 0.25$.

The above procedure can be repeated for all points, and the final results are given in Table 2.2, based on the given data shown in Table 2.1.

The obtained data (Table 2.2) can be presented graphically similar to Figure 2.37.

2.6 Flow Rate Control in Dynamic Pump Systems

The methods used for flow rate control in pumping systems vary, depending on the system characteristics and the type of fluid mover and its driver. The two most commonly used methods in centrifugal pump systems are valve throttling and speed control. Other methods such as using a bypass, movable inlet guide vanes, and impellers with adjustable vane angles are less common. In general, the suction valve is never used for flow rate control since valve throttling creates a significant reduction in the available net positive suction head, leading to cavitation. On the other hand, the flow rate control using delivery valve throttling is very common, but it results in power loss and a reduction in the system efficiency.

To determine the power loss due to partial valve closure, let us consider a simple pumping system in which the pump performance curves are as shown in Figure 2.38. The system H–Q curve is also shown in the same figure when the delivery valve is fully open and the system flow rate is Q_A. Now, suppose that the flow rate is reduced to Q_B, using valve throttling. Because of the increase of friction head loss in the valve, the system curve changes to curve 2. The additional head loss in the valve due to throttling is indicated in the figure and labeled h_{Lv}. It is clear from the figure that the new point of operation (B) is characterized by higher pump total head, lower efficiency and the power consumption may be the same or slightly less. In this case, the ratio between the power loss in the valve and the pump power consumption is equal to h_{Lv}/H_B. In conclusion, the use of the delivery valve for flow rate control should be limited to small variations in the pump capacity in order to avoid considerable power loss and significant reduction in system efficiency.

Using a variable speed prime mover for flow rate control represents a better alternative. In this case, the flow rate can be increased or decreased by increasing or decreasing the pump speed. Using the affinity laws, we can determine the new pump speed required to achieve the chosen flow rate. We need to keep in mind that the pump speed range is limited by the

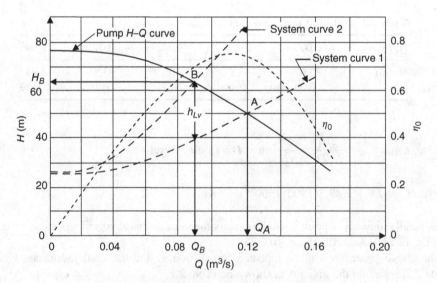

Figure 2.38 Flow rate control using valve throttling

pump design features and its critical speeds as specified by the manufacturer. The following example demonstrates the use of pump speed for flow rate control.

Example 2.6
The performance curves (H–Q and η_o–Q) for a centrifugal water pump are given in Figure 2.39 when operating at a speed of 1750 rpm. The system H–Q curve is also shown in the same figure. Determine the pump speed required to reduce the flow rate by 25%.

Solution
The initial point of operation is point A and the flow rate, $Q_A = 0.12$ m³/s. The required flow rate will be $Q_B = 0.75 \times 0.12 = 0.09$ m³/s and the new point of operation will be point B shown in Figure 2.40.

The equation of the locus of similarity passing by point B can be obtained as follows:

$$H = CQ^2 \quad \Rightarrow \quad C = H_B/Q_B^2 = 38.5/(0.09)^2 = 4753 \quad \Rightarrow \quad H = 4753\,Q^2 \tag{1}$$

By plotting the locus of similarity (Eq. (1)), the point of operation C (at speed 1750 rpm) that is dynamically similar to point B can be obtained as shown in the figure. Now, apply the affinity lows between points B and C to obtain

$$(C_H)_B = (C_H)_C \quad \Rightarrow \quad \frac{gH_B}{N_B^2 D_B^2} = \frac{gH_C}{N_C^2 D_C^2} \quad \Rightarrow \quad \frac{H_B}{N_B^2} = \frac{H_C}{N_C^2} \quad \Rightarrow \quad N_B = N_C\sqrt{H_B/H_C}$$

Therefore, $N_B = 1750\sqrt{38.5/57} = \underline{1438\ \text{rpm}}$

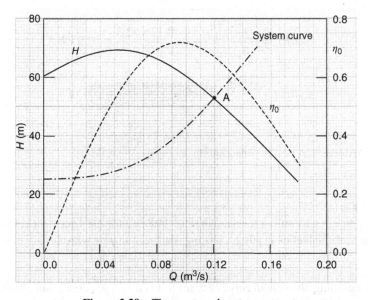

Figure 2.39 The pump and system curves

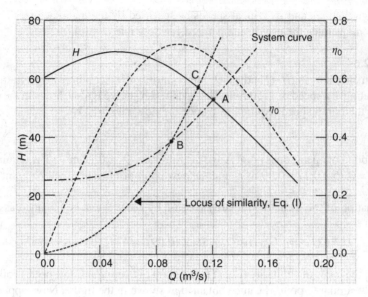

Figure 2.40   *H–Q* graph for Example 2.6

Example 2.7

Considering the data given in Example 2.6 determine the percentage reduction in the pump power consumption, assuming that the pump overall efficiency at the new point of operation (point B) is 0.65.

Solution

$$(BP)_A = \frac{\gamma Q_A H_A}{\eta_{oA}} = \frac{9.81 \times 0.12 \times 52.5}{0.67} = 92.2\,\text{kW}$$

$$(BP)_B = \frac{\gamma Q_B H_B}{\eta_{oB}} = \frac{9.81 \times 0.09 \times 38.5}{0.65} = 52.3\,\text{kW}$$

Percentage reduction in power consumption $= \dfrac{92.2 - 52.3}{92.2} = \underline{43.3\%}$

Example 2.8

Consider using valve throttling for reducing the flow rate for the same data given in Example 2.6. Determine the reduction in power consumption and compare with the case of speed control.

Solution

After valve throttling, the system curve changes from curve 1 to curve 2, and the point of pump operation changes from A to B, as shown in Figure 2.41. The additional friction head loss in the

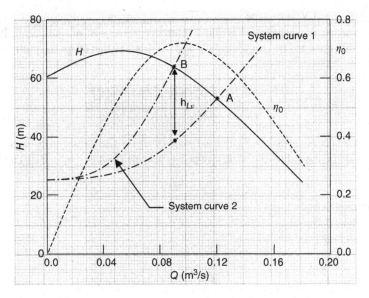

Figure 2.41 New system curve after valve throttling

delivery valve due to valve throttling is approximately $h_{Lv} = 26$ m. Also, the pump overall efficiency changes from 67% at point A to 71% at point B.

$$(BP)_B = \frac{\gamma Q_B H_B}{\eta_{oB}} = \frac{9.81 \times 0.09 \times 64}{0.71} = 79.6 \, \text{kW}$$

Percentage reduction in power consumption $= \dfrac{92.2 - 79.6}{92.2} = 13.7\%$

It is clear that the power saving when using speed control (43.3%) is much higher than that using valve throttling (13.7%).

2.7 Pump Specific Speed

One of the problems facing engineers, is the selection of the type of pump (or shape of pump) suitable for use in a specific application. In general, the pump shape has a strong effect on the pump flow rate and the total head developed. The rule of thumb is that the area of impeller flow passages has a direct impact on the flow rate, while the diameter of the impeller affects the total head developed. Different pump shapes makes them suitable for use in different applications, as, for example,

- radial-type pumps are most suitable for *high head* and *low flow rate* requirements
- mixed-flow pumps are most suitable for *moderate head* and *moderate flow rate* requirements
- axial-flow pumps are most suitable for *low head* and *high flow rate* requirements.

The specific speed N_s is a shape factor that helps the engineer to determine the type of pump to be used in a specific application without reference to the size of that pump. It is a dimensionless quantity obtained by combining C_H and C_Q in a way that drops the pump size (impeller diameter D) as follows:

$$n_s = \frac{C_Q^{1/2}}{C_H^{3/4}} = \frac{(Q/nD^3)^{1/2}}{(gH/n^2D^2)^{3/4}} = \frac{n\sqrt{Q}}{(gH)^{3/4}} \tag{2.33}$$

where n_s is the specific speed in SI and is dimensionless. The values to be used in the above equation should be the values at the design point (best efficiency point – BEP) and the units to be used are

n in revolutions/second
Q in cubic meters/second
g in meters/second^2
H in meters.

In the American system of units, the term $g^{3/4}$ has been traditionally dropped and the specific speed (N_s) becomes

$$N_s = \frac{N\sqrt{Q}}{H^{3/4}} \tag{2.34}$$

where N is the pump rated speed in revs/minute (rpm)
Q is the pump rated capacity in gallons/minute
H is the rated total head/stage in feet.

Note that N_s in Eq. (2.34) is a dimensional quantity but is normally written without units. It is worth mentioning that every pump has only one BEP (or design point) and accordingly, one specific speed. The data for the BEP is sometimes preceded with the word 'rated.' So, the rated capacity indicates the flow rate at the BEP and the same applies for the rated head and rated power.

The specific speed is sometimes called the shape number, since it gives the designer a guide to the type of machine to be used for a specific application. Table 2.3 gives typical values of N_s for various types of pumps.

Table 2.3 The specific speed range for different types of pumps

Type of pump	N_s range
Displacement pumps	<500
Radial-type centrifugal pumps	500–5000
Mixed-flow pumps	4000–10 000
Axial-flow pumps	9000–15 000

Notes
[1] If the specific speed calculated based on the actual working conditions falls in the overlapping areas (4000–5000 or 9000–10 000), the type of pump having higher N_s range is better to be selected. This will lead to smaller size pump which is also of lower cost.
[2] The conversion factor between N_s and n_s is 17 200 (i.e. $N_s = 17\,200\,n_s$).

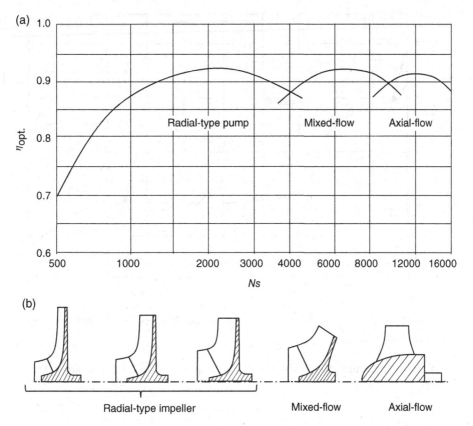

Figure 2.42 Variation of optimum efficiency (a) and impeller shape (b) of rotodynamic-pumps with specific speed (N_s)

The variation of the impeller shape with N_s is shown in Figure 2.42. The figure also shows the expected optimum efficiency (topmost efficiency at BEP) and its variation with N_s. Figure 2.43 shows typical C_H–C_Q curves for four different impeller shapes.

Example 2.9
What type of pump should be used to pump kerosene at a rate of 0.35 m³/s under a head of 60 m assuming that $N = 1450$ rpm.

Solution
We will assume that the given operating condition corresponds to the pump design point (BEP), therefore

$$n_s = \frac{n\sqrt{Q}}{(gH)^{3/4}} = \frac{(1450/60)\sqrt{0.35}}{(9.81 \times 60)^{3/4}} = 0.12 \quad \Rightarrow \quad N_s = 17\,200\,n_s = 2060$$

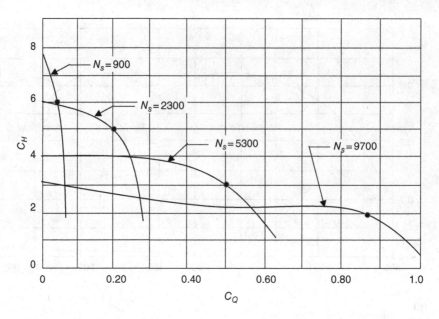

Figure 2.43 Combined performance curves for four different impellers with the circles showing the best efficiency point for each

Since N_s is in the range 500–5000, therefore the pump should be a *radial-type centrifugal pump.*

Example 2.10
A dynamic pump with a 24 in. impeller is designed to deliver 12 000 gpm of water at a total head of 100 ft when operating at a speed of 850 rpm. A 1/4 scale model of the pump is tested at 1750 rpm.

a. What are the corresponding capacity and head for the model when operating at homologous conditions?

b. If the overall efficiency is 84% for both model and prototype, what horsepower will be required to drive each?

c. If the operating condition given above corresponds to the BEP, what type of pump is this?

  Note: 1 gallon = 0.1336 cu. ft

Solution

Prototype data: $D_p = 24''$, $Q_p = 12\,000$ gpm = 26.7 ft³/s, $H_p = 100$ ft, $N_p = 850$ rpm
Model data: $D_m = 6''$, $Q_m = ?$, $H_m = ?$, $N_m = 1750$ rpm

a. Since the operating conditions are dynamically similar, then

$$\frac{Q_p}{N_p D_p^3} = \frac{Q_m}{N_m D_m^3} \Rightarrow Q_m = 12000(1750/850)(1/4)^3$$

Therefore, $Q_m = 386 \text{ gpm} = 0.86 \text{ ft}^3/\text{s}$

Also $\left(\dfrac{gH}{N^2 D^2}\right)_p = \left(\dfrac{gH}{N^2 D^2}\right)_m \Rightarrow H_m = 100 (1750/850)^2 (1/4)^2$

Therefore, $H_m = 26.5 \text{ ft}$

b.
$$P_m = \frac{\gamma Q_m H_m}{\eta_o} = \frac{62.4 \times 0.86 \times 26.5}{0.84 \times 550} = 3.08 \text{ hp}$$

$$P_p = \frac{\gamma Q_p H_p}{\eta_o} = \frac{62.4 \times 26.7 \times 100}{0.84 \times 550} = 361 \text{ hp}$$

c. In order to determine the type of pump, we first calculate the specific speed as follows:

$$N_s = \frac{N \sqrt{Q}}{H^{3/4}} = \frac{850 \sqrt{12{,}000}}{(100)^{3/4}} = 2954$$

Therefore the pump is a radial-type centrifugal pump since $500 < N_s < 5000$.

Example 2.11

Figure 2.44 shows the H–Q and η_o–Q curves for a centrifugal pump having an impeller of diameter 32 cm when operating at its rated speed of 1800 rpm. A schematic of the pumping system

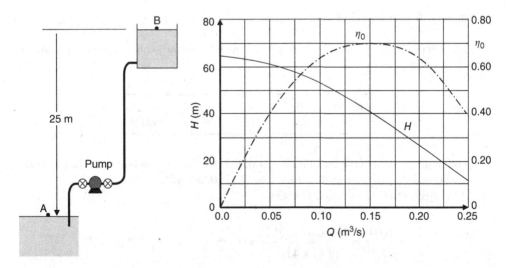

Figure 2.44 A schematic of the piping system and the performance curves for the pump when operating at 1800 rpm

is also shown in the figure. The minor losses in the suction and delivery sides of the pump amount to $6.8\,V^2/2g$, where V is the average pipe velocity.

a. Determine the system flow rate Q.
b. Calculate the pump specific speed N_s, and determine the impeller shape.
c. What power is necessary to drive this pump?
d. If a geometrically similar pump is required to develop a total head of 50 m at a flow rate of $0.2\,\text{m}^3/\text{s}$ when operating at the same speed, estimate the impeller diameter of the pump.

Data: $L_s = 25$ m, $L_d = 120$ m, $D_s = D_d = 0.15$ m, friction coefficient $f = 0.02$.

Solution

a. Apply the energy equation between points (A) and (B) shown in the figure,

$$\frac{p_A}{\gamma} + \frac{v_A^2}{2g} + z_A + H = \frac{p_B}{\gamma} + \frac{v_B^2}{2g} + z_B + \sum h_L \tag{1}$$

Knowing that $p_A = p_B = p_{atm}$ and $V_A = V_B = 0$, the above equation can be simplified to

$$H = (z_B - z_A) + \sum h_L \tag{2}$$

Substituting from the given data, we can write

$$H = 25 + \frac{Q^2}{2gA^2}\left(\frac{f(L_s + L_d)}{D} + 6.8\right) = 25 + \frac{Q^2}{2 \times 9.81\left(\frac{\pi}{4}0.15^2\right)^2}\left(\frac{0.02(25 + 120)}{0.15} + 6.8\right)$$

Simplify, $H = 25 + 4265\,Q^2$ \hfill (3)

Equation (3) describes the system curve, and the table below is obtained from this equation for selected values of Q.

Q (m³/s)	0.0	0.05	0.10	0.15
H (m)	25.0	35.7	67.7	121.0

By plotting the system curve (see Figure 2.45), the point of intersection with the pump H–Q curve (point A) gives the system flow rate $Q_{sys} = 0.085\,\text{m}^3/\text{s}$.

b. At the BEP, $Q = 0.15\,\text{m}^3/\text{s}$ and $H = 43$ m.

$$n_s = \frac{n\sqrt{Q}}{(gH)^{3/4}} = \frac{(1800/60)\sqrt{0.15}}{(9.81 \times 43)^{3/4}} = 0.125 \quad \Rightarrow \quad N_s = 17{,}200\,n_s = 2147$$

Therefore the pump is a *radial-type centrifugal pump*.

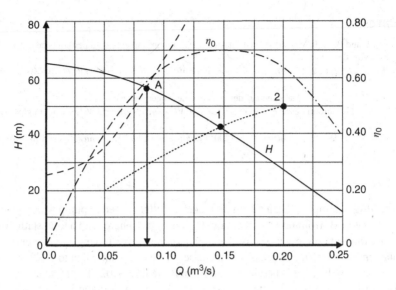

Figure 2.45 *H–Q* graph for Example 2.11

c. The pump brake power is obtained from

$$B.P. = \frac{\gamma Q H}{\eta_o} = \frac{9.81 \times 0.085 \times 54}{0.57} = 78\,\text{kW}$$

d. The locus of similarity for a group of geometrically similar pumps of different sizes running at the same speed is given by $H = C\,Q^{2/3}$.

For $H_2 = 50$ m when $Q_2 = 0.2$ m³/s, the value of C will be

$$C = 50/(0.2)^{2/3} = 146.3$$

The equation describing the locus of similarity will be $H = 146.3\,Q^{2/3}$, and this is tabulated below

Q (m³/s)	0.05	0.10	0.15	0.20
H (m)	19.8	31.5	41.3	50.0

By drawing the locus of similarity in Figure 2.45, the point of intersection with the pump *H–Q* curve (point 1) is similar to point 2 and the two points must have the same C_Q and the same C_H .

Therefore $C_{Q1} = C_{Q2} \Rightarrow \dfrac{Q_1}{N_1 D_1^3} = \dfrac{Q_2}{N_2 D_2^3}$ and $N = \text{Const.}$, therefore, $\dfrac{Q_1}{Q_2} = \dfrac{D_1^3}{D_2^3}$

Or $\dfrac{0.153}{0.2} = \left(\dfrac{0.32}{D_2}\right)^3 \Rightarrow \underline{D_2 = 0.35\,\text{m}}$

References

1. Ahmed, W.H. and Badr, H.M. (2013) Dual-injection airlift pumps: an enhanced performance. *J Part Sci Technol*, **30** (6), 497–516.
2. Badr, H.M. and Ahmed, W.H. (2013) Dual injection airlift pump. US Patent 8,596,989 B2, Issued on December 3, 2013.
3. Blasius, H. (1913) Forschungsarbeiten auf dem Gebiete des Ingenieusersens, 131.
4. Nikuradse, J. (1933) Stromungsgesetze in rauhen rohren, *VDI-Forschungsheft*, 361, see the English Translation in NACA TM 1292.
5. Crowe, C.T., Elger, D.F. and Roberson, J.A. (2005) *Engineering Fluid Mechanics*. 8th edn. John Wiley & Sons, Inc., New York.

Problems

2.1 A centrifugal pump having a 42 cm diameter impeller is used to pump water from reservoir A to reservoir B. Both reservoirs are open to the atmosphere with a total static lift of 20 m. The suction and delivery pipes used are old concrete pipes having a diameter of 15 cm and a total length of 390 m. The losses in the pipe fittings are equivalent to $12\ V^2/2g$, where V is the velocity in the pipe. The pipe friction coefficient $f = 0.02$. The H–Q and η_o–Q characteristics of the pump at its rated speed of 1500 rpm are as given in the table below.

Q (m³/s)	0.0	0.02	0.04	0.06	0.08	0.10	0.12
H (m)	95.0	93.0	87.5	77.5	62.5	44.0	19.0
η_o	0.0	0.55	0.78	0.85	0.79	0.61	0.33

    a. Determine the system flow rate, Q_1, and the pump power consumption, BP_1.
    b. In order to increase the system flow rate, a young engineer suggested replacing the old concrete pipe by a commercial steel pipe having a friction coefficient, $f = 0.01$, and minor losses equivalent to $8\ V^2/2g$. Determine the new flow rate, Q_2, and the new power consumption, BP_2.

2.2 In the system described in Problem 2.1, another engineer suggested keeping the old concrete pipe and increasing the flow rate by installing a second identical pump that can operate either in parallel or in series with the first one. Determine the percentage increase in the system flow rate for parallel operation.

2.3 The suction reservoir shown in the Figure 2.46 contains kerosene ($\gamma = 7.96$ kN/m³ and $p_v = 18$ kPa abs.) to a height $h = 6$ m. The pump is required to supply the fluid to the delivery line at a rate of 0.06 m³/s. Neglecting minor losses, determine the available net positive suction head, *NPSHA*. What is the maximum flow rate that can be delivered without cavitation.

Data: $L_s = 20$ m, $D_s = 10$ cm, friction coeff. $f = 0.01$, *NPSHR* $= 1.4$ m

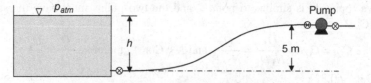

Figure 2.46 System details for Problem 2.3

2.4 The data given below corresponds to a pumping system that contains one single-stage centrifugal pump used for moving water at 50°C ($p_v = 12.4$ kPa abs.) from a suction reservoir that is open to the atmosphere to a chemical reactor.

Flow rate = 0.12 m³/s	Total static head = 15 m
Static suction head = 1.0 m	Diameter of all pipes = 15 cm
Length of suction pipe = 10 m	Length of delivery pipe = 120 m
Minor losses in the suction side = 1.6 $V^2/2\,g$	Minor losses in the delivery side = 4.8 $V^2/2\,g$
NPSHR = 1.2 m	Coefficient of friction for all pipes = 0.015

Investigate the system to show whether cavitation will occur or not.

2.5 A small pumping station is equipped with two identical centrifugal pumps (the characteristics of each pump are tabulated below) and used to deliver water from a river at elevation $z = -5$ m to a reservoir having a water level $z = 30$ m, through a 300 mm diameter pipe with a total length of 1250 m.

Q (L/s)	0.0	50.0	100.0	150.0	200.0
H (m)	60.0	58.0	52.0	41.0	25.0
η_o (%)	0	44	65	64	48

The minor losses in valves, bends, and other fittings amount to 10 $V^2/2\,g$, where V is the pipe flow velocity. Assuming a friction coefficient of 0.01, calculate the system flow rate and the total power consumption for the following two cases:

a. two pumps are connected in series
b. two pumps are connected in parallel

2.6 A dynamic pump is tested at its rated speed (1200 rpm) and the following data is obtained:

Q (L/s)	8.0	15.0	24.0	35.0	40.0	45.0	60.0
H (m)	37.0	32.7	27.0	22.2	20.4	18.0	9.0
BP (kW)	11.2	10.6	9.8	9.8	10.0	10.5	19.6

Knowing that the fluid used is water ($\gamma = 9.81$ kN/m³, $\mu = 10^{-3}$ N·s/m²), calculate the pump specific speed (n_s). State whether the pump is of radial-type, mixed-flow, or axial-flow. Justify your answer.

If the pump is required to deliver a flow rate of 35 L/s against a total head of 32 m, at what speed should it be driven?

2.7 A centrifugal pump is designed to operate at 1450 rpm and to develop a total head of 60 m at a flow rate of 250 L/s. The following H–Q characteristic was obtained for a 1/4 scale model of the pump when running at 1800 rpm.

Q (L/s)	0.0	2.0	4.0	6.0	8.0
H (m)	8.0	7.6	6.4	4.2	1.0

Obtain the corresponding characteristic of the prototype and state whether or not it meets the design requirements.

2.8 A centrifugal pump, with the characteristics tabulated below, delivers water from a river at an elevation of 4 m below the standard datum, to a reservoir with a water level at elevation $z = 32$ m. The pipes used are 2000 m long and 350 mm diameter. The minor losses in valves, bends, elbows, and other fittings amount to $10\ V^2/2g$, where V is the average pipe velocity. Assuming a friction coefficient of 0.015, calculate the system flow rate and the pump power consumption.

Q (L/s)	0.0	40.0	80.0	120.0	160.0	200.0
H (m)	80.0	75.0	67.0	56.5	42.5	20.0
η_o (%)	0	50	74	80	72	48

2.9 In the system described in Problem 2.8, it is required to increase the system flow rate by installing a second pump having the characteristics tabulated below.

Q (L/s)	0.0	40.0	80.0	120.0	160.0	200.0
H (m)	70.0	65.0	55.5	42.0	24.0	0.0
η_o (%)	0	43	74	85	72	0

a. Determine the new system flow rate if the two pumps operate in parallel.
b. Determine the new system flow rate if the two pumps operate in series.

2.10 In the arrangement described in Problem 2.9a (parallel operation), the system flow rate is reduced to 80% of its original value by partially closing the delivery valve. Calculate the total power consumption and the combined efficiency of the two pumps.

2.11 The pump shown in the Figure 2.47 sucks kerosene (sp. wt. $= 8$ kN/m^3) from reservoir A through a 15 cm diameter pipe at a flow rate of 0.12 m^3/s. The kerosene level in the reservoir $h = 2.4$ m and the length of the suction pipe is 18 m. Assuming that the minor losses between reservoir A and the pump to be $1.2\ V_s^2/2g$, determine the minimum value of h if cavitation is to be avoided.

Data: $p_v = 6.4$ kPa abs., $f = 0.012$, $NPSHR = 1.5$ m

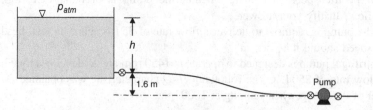

Figure 2.47   System details for Problem 2.11

2.12 A small community with a population of 12 000 is to be supplied with sweet water from a desalination plant 4 km away. The average water consumption is 200 L/person/day. The total daily consumption must be supplied in 8 hours. Knowing that the total static head is 15 m and the pump will be driven by a diesel engine operating at a speed of 1500 rpm, select a suitable pump for this system.

Hint: Assume that the pipes used are commercial steel pipes and assume a reasonable pipe flow velocity (5–6 m/s). The standard dimensions of pipes can be obtained from the internet or from *Mark's Standard Handbook for Mechanical Engineers*.

2.13 A centrifugal pump running at 1000 rpm gave the following data for the flow rate Q versus the total head developed H:

Q (m³/min)	0.0	4.5	9.0	13.5	18.5	22.5
H (m)	22.5	22.2	21.6	19.5	14.1	0.0

The pump is connected to 300 mm suction and delivery pipes with a total length of 69 m and the discharge to the atmosphere is 15 m above sump level. The entrance losses are equivalent to an additional 6 m of pipe friction and the friction coefficient is assumed to be 0.006.

a. Calculate the flow rate in m³/min.
b. If it is required to adjust the flow rate by regulating the pump speed, estimate the speed required to reduce the flow rate to half its original value.

2.14 A centrifugal pump having the characteristics tabulated below is used to move a fluid (sp. gr. = 0.85, $p_v = 20$ kPa abs.) from reservoir A to reservoir B. The reservoirs are open to the atmosphere and the fluid free surface level in each reservoir is given by $Z_A = 5$ m and $Z_B = 25$ m. The minor losses in the suction and delivery sides of the pump are equivalent to 1.8 $V^2/2g$ and 4.2 $V^2/2g$ respectively, where V is the average pipe velocity.

Q (m³/s)	0.0	0.02	0.04	0.06	0.08	0.10	0.12	0.16
H (m)	80.0	78.6	77.0	74.0	70.0	65.2	59.0	40.0

a. Determine the system flow rate.
b. Determine the available net positive suction head.
c. If the flow rate is reduced by 25% by using the delivery valve for flow rate control, determine the power loss in the valve.

Data: $L_s = 20$ m, $L_d = 1.2$ km, $D_s = D_d = 0.2$ m, $f = 0.012$ for all pipes, $p_{atm} = 101$ kPa

2.15 Figure 2.48 shows the H–Q and η_o–Q curves for a radial-type centrifugal pump. The pump has an impeller of diameter 25 cm and runs at a speed of 1800 rpm. Geometrically similar pumps are available with different impeller sizes.

a. Assuming the same operating speed, determine the pump size that will be suitable for delivering water at a rate of 0.2 m³/s against a total head of 40 m.
b. What will be the overall efficiency at the point of operation for the pump selected in (a)?
c. Determine the input power at the point of operation for the pump selected in (a).

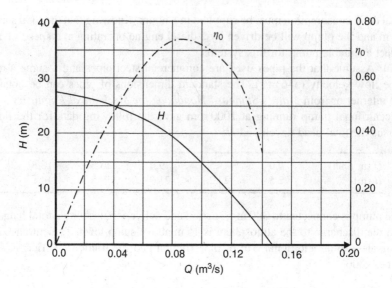

Figure 2.48   *H–Q* graph for Problem 2.15

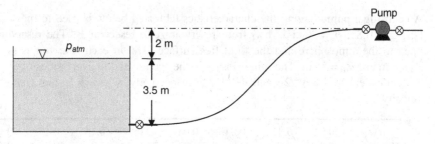

Figure 2.49   System details for Problem 2.17

2.16   In the system described in Problem 2.15, the flow rate is to be increased by the installation of a second identical pump that can operate either in parallel or in series with the first one. Determine the percentage increase in flow rate in each of the two cases.

2.17   The suction side of a pumping system is as shown in Figure 2.49. The fluid pumped is gasoline ($\gamma = 8.5$ kN/m³ and $p_v = 60$ kPa abs.). It is required to calculate *NPSHA* when the flow rate is 0.06 m³/s. What will be the maximum flow rate if cavitation is to be avoided? You may neglect minor losses and assume *NPSHR* = 0.75 m.

Data: $L_s = 50$ m, $D_s = 0.15$ m, $f = 0.018$

2.18   A centrifugal water pump running at 1000 rpm has the following *H–Q* characteristic:

Q (m³/min)	0.0	4.5	9.0	13.5	18.5	22.5
H (m)	26.5	26.2	25.6	23.5	18.1	4.0

The pump is connected to 300 mm smooth suction and delivery pipes the total length of which is 85 m and discharges to the atmosphere at a level 15 m above fluid level in the suction reservoir. The minor losses are equivalent to an additional 8 m of pipe friction, and the friction coefficient is related to Reynolds number by $f = 0.316/(Re)^{1/4}$. Assume a water temperature of 20°C.

a. Based on the above information, calculate the flow rate in m^3/minute.
b. If it is required to adjust the flow rate to 50% of the value obtained in (a) by partially closing the delivery valve, determine the amount of friction head loss in the valve as a percentage of the head developed by the pump.

2.19 A centrifugal pump with 20 cm impeller diameter is driven at a speed of 5000 rpm and used to pump kerosene. The pump performance curves are as shown in Figure 2.50. Estimate the flow rate, pressure rise across the pump, and the pump input power when operating at its BEP. What type of pump is it?

2.20 Water is to be pumped at a rate of 1.2 m^3/s from a lower reservoir to an upper one with a total static head of 3 m. The suction and delivery pipes have a total length of 20 m and a diameter of 100 cm. What type of pump would you suggest if the normal operating speed of the prime mover is 750 rpm? Note clearly any necessary assumptions.

2.21 The table below gives the H–Q characteristic for a centrifugal pump having 24 cm diameter impeller when driven at 1450 rpm. Geometrically similar pumps are available for impeller diameters of 16, 20, 28, 32, 40, and 50 cm.

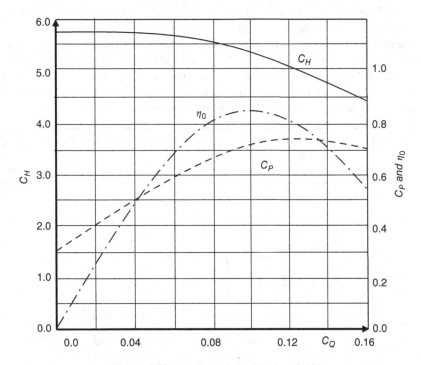

Figure 2.50 H–Q graph for Problem 2.19

Q (L/s)	0.0	10.0	20.0	30.0	40.0	50.0
H (m)	25.0	29.5	29.5	25.5	17.0	5.0

   a. What pump size is suitable for supplying a flow rate of 30 L/s against a required
system head of 35 m when driven at the same speed (1450 rpm)?
Hint: If the calculated impeller size is not available, select the next larger size.
   b. Predict the *H–Q* curve for the selected pump.
   c. The selected pump is to be used in a pumping system that utilizes the delivery valve for
flow rate control. Determine the loss of power occurring in the valve as a percentage of
the output fluid power when the operating condition is as specified in (a).

2.22 Plot the *H–Q* curve for the pump given in Problem 2.21 (impeller diameter = 24 cm) when
driven at a speed of 1750 rpm.

2.23 Water at 20°C is to be pumped at a rate of 0.8 m³/s from a lower reservoir to an upper one
through a concrete pipe of diameter 0.5 m and total length of 200 m. The elevations of the
fluid free surfaces in the two reservoirs are 16 and 24 m. What type of pump would you
suggest for this system if the normal operating speed of the prime mover is 900 rpm?
Neglect minor losses.

2.24 A centrifugal pump draws a fluid ($\gamma = 8.5$ kN/m³, $p_v = 15$ kPa abs.) from the sealed reservoir shown in Figure 2.51 at a fixed rate of 0.04 m³/s. The minor losses in the suction
pipe amount to $1.2 V_s^2/2g$, where V_s is the velocity in the suction pipe. Determine the
lowest height of the liquid in the reservoir (h) if cavitation is to be avoided. Consider
$NPSHR = 1.4$ m and the friction coefficient, $f = 0.018$.

2.25 A centrifugal pump having the characteristics tabulated below delivers water from a river
at elevation $z = 52$ m to a reservoir with a water level of 85 m, through a 350 mm diameter
smooth pipe of a total length of 2000 m. The pipeline has six bends ($K_b = 0.2$), four valves
($K_v = 0.12$), and one filter ($K_f = 0.8$). Assuming a water temperature of 20°C, determine
the system flow rate and the pump power consumption.

Q (L/s)	0.0	40.0	80.0	120.0	160.0	200.0
H (m)	80.0	75.0	67.0	56.5	42.5	20.0
η_o (%)	0	50	74	80	72	48

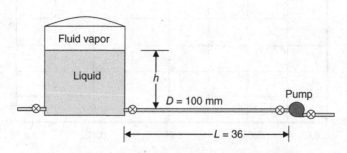

Figure 2.51   System details for Problem 2.24

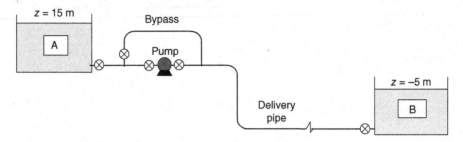

Figure 2.52   *H–Q* graph for Problem 2.26

2.26 Oil (sp. gr. = 0.85) is being transported from reservoir A to reservoir B using the pumping system shown in Figure 2.52. The pump is of centrifugal type and has the following *H–Q* characteristic when driven at 1200 rpm:

Q (m³/s)	0.0	0.02	0.04	0.06	0.08	0.10	0.12	0.16
H (m)	80.0	78.6	77.0	74.0	70.0	65.2	59.0	40.0

The minor losses in the system are equivalent to $9\,V^2/2\,g$, where V is the average pipe velocity.

a. Determine the system flow rate.
b. The shown bypass is used in case of power failure. If the additional minor losses in the bypass amounts to $3V^2/2g$, what will be the system flow rate in case of power failure?

Data: $L_s = 20$ m, $L_d = 2.4$ km, $D_s = D_d = 0.2$ m, $f = 0.012$ for all pipes.

3

Fundamentals of Energy Transfer in Centrifugal Pumps

Understanding the process of energy transfer from mechanical input work to output fluid power occurring inside the pump is essential, not only for designing a new pump but also for optimizing the operation of an existing one. As we know, the main function of a pump is to increase the energy content of the pumped fluid in the form of increasing pressure, kinetic energy, or both. The only place where energy is added to the fluid is inside the impeller through the work done by the impeller vanes to rotate the fluid. Accordingly, the number and shape of these impeller vanes have direct effect on the pump performance characteristics and its overall efficiency. A detailed analysis of the mechanism of energy transfer and the associated flow processes is important for reducing various types of losses. In principle, the amount of energy loss occurring inside the pump depends on the fluid properties, pump size and geometry, pump speed, and flow rate. These losses are divided to three main types: hydraulic, leakage, and mechanical. Each one of these will be studied in detail because of their direct effect on the pump overall efficiency. In this chapter, the main pump components are first introduced and the details of energy transfer from the pump impeller to the fluid are presented. In addition, various means for improving the overall pump efficiency are discussed.

3.1 Main Components of the Centrifugal Pump

The design features and main components of each pump depend on the required performance and the actual operating conditions. Some of the pump components are common for all types of pumps and some others are added either to solve an operational-type problem or to add

Pumping Machinery Theory and Practice, First Edition. Hassan M. Badr and Wael H. Ahmed.
© 2015 John Wiley & Sons, Ltd. Published 2015 by John Wiley & Sons, Ltd.

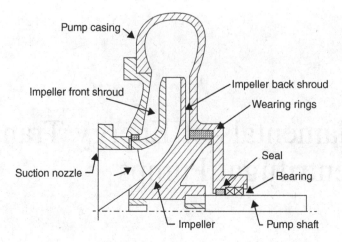

Figure 3.1   Main components of a single-stage centrifugal pump

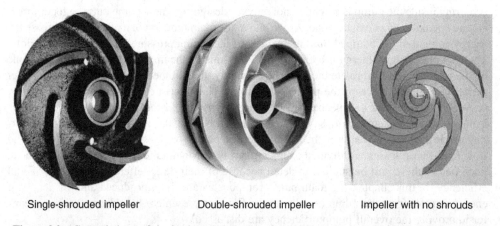

Single-shrouded impeller   Double-shrouded impeller   Impeller with no shrouds

Figure 3.2   General views of single-shrouded, double-shrouded impellers, and impeller with no shrouds

a required feature. The main components of a centrifugal pump and the function of each component are given below:

1. The pump shaft is an essential component and is used for transferring mechanical power from the prime mover to the pump and also for supporting the impeller. Figure 3.1 shows a sectional view of a typical single-stage centrifugal pump.
2. The pump impeller is the component that converts the input mechanical power to fluid power through the work done on the fluid. The fluid gains higher pressure and higher kinetic energy during its course of motion through the impeller. Impellers may be single-shrouded, double-shrouded, or with no shrouds, as shown in Figure 3.2. In the case of no shrouds, the

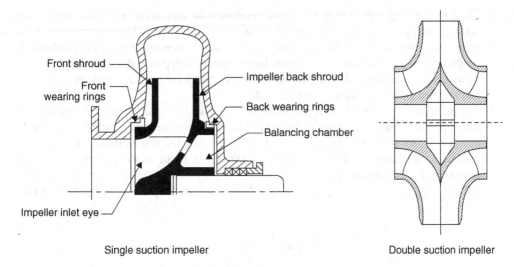

Single suction impeller   Double suction impeller

Figure 3.3   Sectional views of single- and double-suction impellers

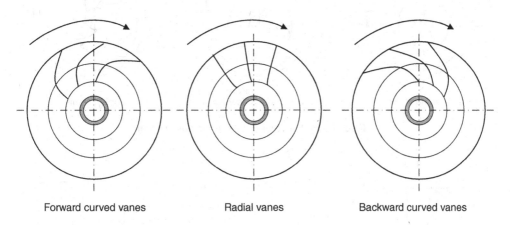

Forward curved vanes   Radial vanes   Backward curved vanes

Figure 3.4   Impeller vane shapes

vanes are structurally weak since they are mounted on the hub without enough support from the front and back sides. The local pressure drop at vane inlet is relatively high, resulting in higher values of NPSHR (net positive suction head, required). Such impellers cannot easily get clogged and accordingly are suitable for handling liquids with suspended materials/ solids. By contrast, the double-shrouded impellers provide maximum support to the vanes and are widely used for pumping liquids with less suspended solids. The use of double-shrouded impellers results in higher pump efficiency and lower NPSHR. The pump impeller may be single suction (suction from one side) or double suction (suction from opposite sides) as shown in Figure 3.3.

3. The impeller vanes are the most important elements in the pump. The work done on the fluid and the energy transfer from mechanical power to fluid power only occur because of the vanes. Also, the pump performance characteristics and the overall efficiency depend mainly on the vane shape and number of vanes. Impellers are equipped with vanes that may have various different shapes. The most common vane shapes are the backward curved, radial, and forward curved as shown in Figure 3.4.

4. The pump casing is an essential part of the pump which is important not only for housing the impeller(s) and sealing the system, but also for supporting the suction and delivery nozzles. In addition, the casing is used for collecting the fluid discharging from the impeller and conveying it to the delivery nozzle. The shape of the casing differs from one pump to another depending on the design requirements. These shapes include concentric volute, semi-concentric volute, and spiral volute (Figure 3.5). The casing may

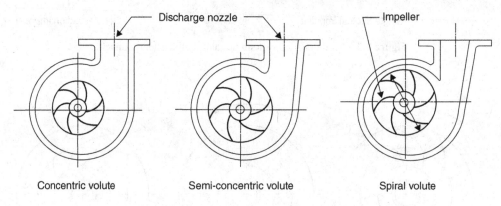

Concentric volute   Semi-concentric volute   Spiral volute

Figure 3.5   Different shapes of the volute casing

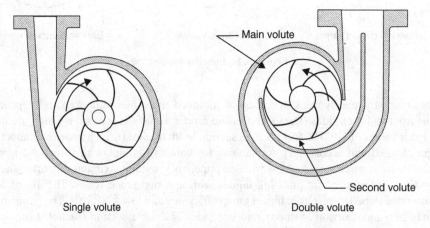

Single volute   Double volute

Figure 3.6   Casings with single and double volutes

(a)   (b)

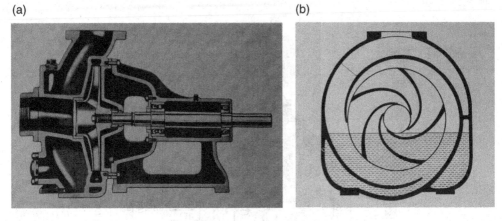

Figure 3.7 Double volute casing with self-priming capabilities (a) elevation and (b) side view (Gorman-Rupp pump; from Stepanoff [1])

have a single or double volute (Figure 3.6), and may be designed to give the pump a self-priming capabilities (Figure 3.7). The casing used for a multistage pump (sometimes called a barrel casing) has a special design to enable the pumped fluid to move from one stage to another with the minimum amount of loss while maintaining compact design (Figure 3.8).

5. The pump suction nozzle is used to direct the fluid from the suction pipe until it enters the impeller. These nozzles may have single entry or double entry (see Figure 3.9 for double entry) and may have an axial inlet or a side inlet (Figure 3.10). The suction nozzle may also be equipped with inlet guide vanes that are used for flow rate control.

6. The discharge nozzle directs the fluid from the casing to the discharge pipe. It also acts as a diffuser that converts the fluid's high velocity into pressure.

7. Bearings are needed for supporting the pump shaft. Various types are used such as journal bearings, ball and roller bearings, and thrust bearings.

8. The seals are very important for every pump since they prevent the pumped fluid from leaking out of the pump. The type of seal depends on the pumped fluid. For example, a stuffing box with compression packing is commonly used in water pumps, while mechanical seals are widely used in pumps handling toxic or flammable liquids, in order to avoid fire and environmental hazards. Figure 3.11 shows the configuration of a compression packing seal.

9. Wearing rings are commonly used for reducing internal fluid leakage from the high-pressure side (volute casing) to the low-pressure side (suction nozzle). A typical set of wearing rings can be seen in Figure 3.1.

10. Some pumps contain diffusing vanes in addition to the volute casing (Figure 3.12). The main function of these vanes is to streamline the flow at the impeller exit and convert the high velocity into pressure. This will lead to a reduction of friction losses in the volute casing.

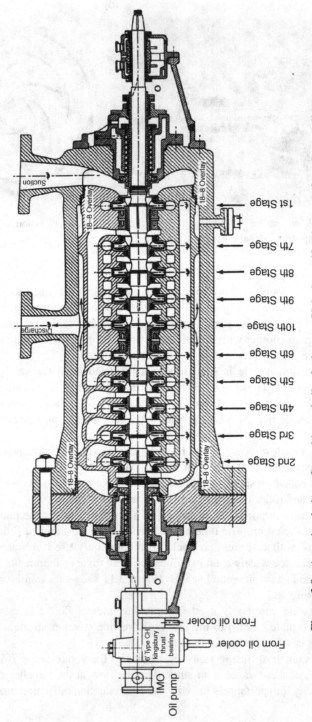

Figure 3.8 A sectional view of a multistage boiler feed pump (United centrifugal pumps; from Stepanoff [1])

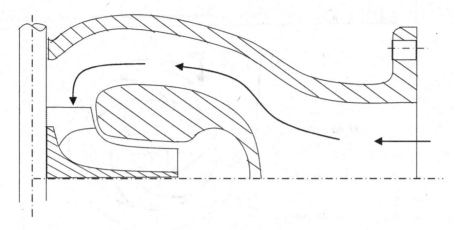

Figure 3.9　Schematic of a suction nozzle with double entry

(a)　　　　　　　　　　　　　　　　　(b)

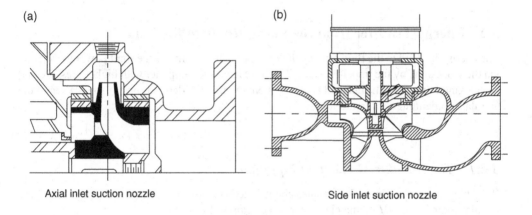

Axial inlet suction nozzle　　　　　　　Side inlet suction nozzle

Figure 3.10　Suction nozzle configurations (United centrifugal pumps, from Stepanoff [1])

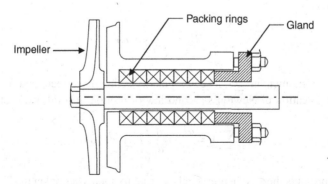

Figure 3.11　Typical stuffing box with compression packing

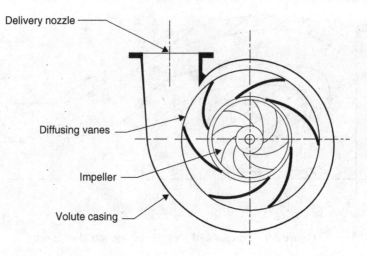

Figure 3.12   A pump volute casing equipped with diffusing vanes

3.2   Energy Transfer from the Pump Rotor to the Fluid

The energy is added to the fluid only by the impeller, which conveys the shaft power to the fluid via the work done by the impeller vanes. The fluid leaves the impeller with higher pressure and higher kinetic energy. The detailed analysis of energy transfer depends on the application of the angular momentum conservation equation.

3.2.1   The Angular Momentum Equation

If the body shown in Figure 3.13 rotates about an axis through O (normal to the page), then the angular momentum $d\underline{H}$ of the elementary mass dm will be

$$d\underline{H} = dm\,\underline{V} \times \underline{r} \quad and \quad dm = \rho\,d\Psi$$

where $d\Psi$ is the elementary volume of mass dm. Integrate the above expression over the entire volume to obtain

$$\underline{H} = \int_V \rho\left(\underline{V} \times \underline{r}\right) d\Psi \tag{3.1}$$

where the angular momentum $\underline{H}$ is a vector quantity acting in a direction normal to the plane of $\underline{r}$ and $\underline{V}$. The application of Newton's second law of motion to solid body rotation yields

$$\sum \underline{M} = \frac{d\underline{H}}{dt} = \frac{d}{dt}\int_V \rho\left(\underline{V} \times \underline{r}\right) d\Psi \tag{3.2}$$

where $\sum \underline{M}$ represents the summation of all external moments acting on the body and $d\underline{H}/dt$ is the rate of change of angular momentum.

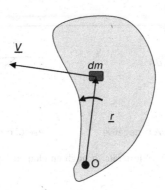

Figure 3.13 A solid body rotating about an axis at O

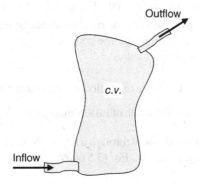

Figure 3.14 A control volume with inflow and outflow ports

For the case of a fluid body contained in the control volume (c.v.) shown in Figure 3.14, the change of angular momentum includes the change of angular momentum of fluid within the c.v. as well as the effects of the inflow and outflow of the angular momentum. The basic c.v. equation (Reynolds transport equation) for any physical quantity can be written as

$$\frac{dB}{dt} = \frac{\partial}{\partial t} \int_{c.v} \rho\beta \, d\Psi + \sum_{c.s.} \rho\beta \left(\underline{V} \cdot \underline{n} \right) A \tag{3.3}$$

where B is any extensive property, β is the corresponding intensive property, and n is a unit vector normal to the area A. If B represents the angular momentum, then $\beta = \underline{V} \times \underline{r}$ and applying Eq. (3.2) gives

$$\sum \underline{M} = \frac{\partial}{\partial t} \int_{c.v} \rho \left(\underline{V} \times \underline{r} \right) d\Psi + \sum_{c.s.} \rho \left(\underline{V} \times \underline{r} \right) \left(\underline{V} \cdot \underline{n} \right) A \tag{3.4}$$

For steady flow, the integral term in Eq. (3.4) vanishes and the equation can be reduced to

$$\sum \underline{M} = \sum_{c.s.} \rho \left(\underline{V} \times \underline{r} \right) \left(\underline{V} \cdot \underline{n} \right) A \tag{3.5}$$

Case of inflow: $(\underline{V} \cdot \underline{n})\, dA$ is negative   Case of outflow: $(\underline{V} \cdot \underline{n})\, dA$ is postive

Figure 3.15   Flow rate through an elementary surface area

The term $\left[(\underline{V}.\underline{n}) A \right]$ in the above equation represents the flow rate Q, which is positive for outflow and negative for inflow, as clarified in the Figure 3.15. The term $\sum \underline{M}$ represents the summation of all moments acting on the c.v. The right hand side of Eq. (3.5) represents the net rate of outflow of angular momentum through the control surface.

In words, Eq. (3.5) may be expressed as

$$\sum \underline{M} = \text{Total rate of outflow of angular momentum}$$

$$- \text{Total rate of inflow of angular momentum}$$

Assuming one-dimensional steady flow through a c.v. with one inlet and one exit, and making use of the mass conservation principle, Eq. (3.5) takes the form

$$\sum \underline{M} = \dot{m} \left[\left(\underline{V} \times \underline{r} \right)_{exit} - \left(\underline{V} \times \underline{r} \right)_{inlet} \right] \tag{3.6}$$

where $\dot{m}$ is the mass flow rate.

3.2.2   Application of Angular Momentum Equation to the Centrifugal Pump Impeller

To deduce the relationship between the torque applied to the impeller (driving torque) and the flow velocities at the impeller inlet and exit sections, we will apply the angular momentum equation considering the c.v. to occupy the space inside the impeller. We also assume one-dimensional flow through the impeller. Figure 3.16 shows the main dimensions of a typical radial-type impeller, as well as the vane inlet and exit vane angles (β_1 and β_2). The flow velocity diagrams are shown in Figure 3.16 at inlet and exit of a typical vane. Note that the velocity of the fluid relative to the vane (V_r) is tangential to the vane (shockless flow) at inlet when operating at the design point. Figure 3.17 shows the same velocity diagrams indicating the whirl component (V) and flow component (Y) of the flow velocity. The whirl component (or swirl-component) is a measure of the flow rotational velocity, while the flow component (Y) is the component of the velocity normal to the area (i.e. in the radial direction) and gives a measure of the flow rate.

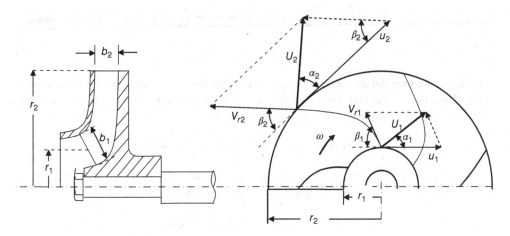

Figure 3.16 Main dimensions and vane angles for a typical impeller

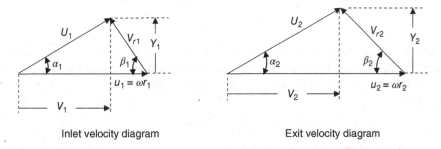

Inlet velocity diagram   Exit velocity diagram

Figure 3.17 Velocity diagrams at inlet and exit of impeller vanes showing the whirl component (V) and flow component (Y) of the flow velocity

The variables used and their definitions are as follows; the subscript 1 is used to label variables at the inlet of the impeller vane and 2 at the vane exit.

b	width of the impeller vane
r	average radius of the vane
u	velocity of the impeller vane $(u = \omega r)$
U	absolute flow velocity
V	whirl (swirl) component of the absolute velocity
V_r	fluid velocity relative to the vane
Y	radial component (flow component) of the absolute velocity
α	flow angle
β	vane angle

Now applying Eq. (3.6) between the impeller inlet and exit sections to obtain an expression for the driving torque T gives

$$\sum \underline{M} = T = \dot{m} \left[\underline{U_2} \times \underline{r_2} - \underline{U_1} \times \underline{r_1} \right]$$

But since $\dot{m} = \rho Q$, $\underline{U_1} \times \underline{r_1} = V_1 r_1$, and $\underline{U_2} \times \underline{r_2} = V_2 r_2$, the above equation can be simplified to

$$T = \rho Q \left(V_2 r_2 - V_1 r_1 \right) \tag{3.7}$$

where T is the impeller driving torque. It is important here to mention that we have assumed a shockless flow at the vane inlet and exit (i.e. V_r is tangential to the vane at inlet and exit). The power input to the impeller is given by

$$P = T\omega = \rho Q \omega \left(V_2 r_2 - V_1 r_1 \right) = \rho Q \left(V_2 u_2 - V_1 u_1 \right) \tag{3.8}$$

In the ideal case in which there are no energy losses anywhere in the pump we can equate the input mechanical power to the output fluid power. Accordingly, we can write

$$P = T\omega = \rho Q \left(V_2 u_2 - V_1 u_1 \right) = \gamma Q H_e$$

where H_e is the theoretical head developed by the impeller, and therefore

$$H_e = \frac{V_2 u_2 - V_1 u_1}{g} \tag{3.9}$$

The term H_e obtained from Eq. (3.9) represents the ideal value of the impeller input head and it is called the *Euler head*. In most cases, the fluid enters the impeller with no whirl component (i.e. $V_1 = 0$) when the pump operates at its design point (bep). In which case, Eq. (3.9) can be reduced to

$$H_e = V_2 u_2 / g \tag{3.10}$$

The relationship between the radial velocity components Y_1 and Y_2 and the flow rate, Q, can be obtained from the continuity equation as

$$Q = A_1 Y_1 = A_2 Y_2 \quad \text{or} \quad Q = 2\pi r_1 b_1 Y_1 = 2\pi r_2 b_2 Y_2 \tag{3.11}$$

The effect of the vane thickness on the impeller inlet and exit areas is neglected in this equation.

Example 3.1
A centrifugal pump has an impeller with an external diameter of 250 mm and an internal diameter of 150 mm. The impeller width at inlet is 15 mm. The vanes are backward curved and have an angle of 45° at exit. The fluid enters the impeller with no prerotation (no whirl) and the radial velocity component is constant throughout the impeller. The flow rate is 2.7 m³/min, when the speed of rotation is 1100 rpm.

a. Calculate the vane angle at inlet.
b. Determine the theoretical head developed by the pump (neglect all losses).

Solution

Data : $r_1 = 0.075\,\text{m}$, $b_1 = 0.015\,\text{m}$, $Y_1 = Y_2$, $N = 1100\,\text{rpm}$,

$r_2 = 0.125\,\text{m}$, $\beta_2 = 45°$, $Q = 2.7\,\text{m}^3/\text{min} = 0.045\,\text{m}^3/\text{s}$

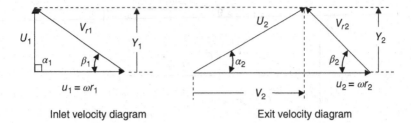

Inlet velocity diagram   Exit velocity diagram

Figure 3.18   Velocity diagrams at the vane inlet and exit sections

$$\omega = 2\pi N/60 = 2\pi \times 1100/60 = 115 \, \text{rad/s}$$

$$u_1 = \omega r_1 = 115 \times 0.125 = 8.64 \, \text{m/s} \quad \text{and} \quad u_2 = \omega r_2 = \cdots = 14.4 \, \text{m/s}$$

$$Y_1 = Y_2 = \frac{Q}{2\pi r_1 b_1} = \frac{0.045}{2\pi \times 0.075 \times 0.015} = 6.37 \, \text{m/s}$$

From the velocity diagram shown in Figure 3.18, we can write

$$\tan \beta_1 = \frac{Y_1}{u_1} = \frac{6.37}{8.64} = 0.737 \quad \Rightarrow \quad \underline{\beta_1 = 36.4°}$$

Also, $V_2 = u_2 - Y_2 \cot \beta_2 = 14.4 - 6.37 \cot 45° = 8.03 \, \text{m/s}$

Using Eq. (3.9) $\Rightarrow H_e = \dfrac{V_2 u_2 - V_1 u_1}{g} = \dfrac{14.4 \times 8.03 - 0}{9.81} = \underline{11.8 \, \text{m}}$

3.3   Theoretical Characteristic Curves

The pump theoretical characteristic curves (H–Q and BP–Q) can be deduced for any impeller vane shape using the laws of conservation of mass, momentum, and energy, and utilizing the inlet and exit velocity diagrams. In this analysis, we will assume one-dimensional steady flow and neglect friction and other losses. In general, the pump performance is always presented for a constant speed.

3.3.1   Theoretical H-Q Characteristics

The theoretical H–Q can be determined using Eq. (3.9) together with the velocity diagrams given in Figure 3.17 as follows:

From Figure 3.17, we can write

$$V_1 = u_1 - Y_1 \cot \beta_1 \quad \text{and} \quad V_2 = u_2 - Y_2 \cot \beta_2 \tag{3.12}$$

Also, using continuity (Eq. (3.11)),

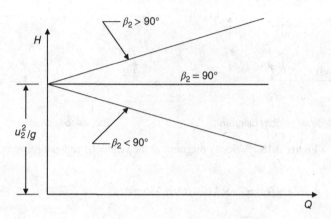

Figure 3.19   Theoretical $H_e - Q$ characteristic curves for various vane shapes—no prerotation at inlet

$$Y_1 = \frac{Q}{2\pi r_1 b_1} \quad \text{and} \quad Y_2 = \frac{Q}{2\pi r_2 b_2}$$

Therefore, Eq. (3.12) can be expressed as

$$V_1 = u_1 - \frac{Q}{2\pi r_1 b_1}\cot\beta_1 \quad \text{and} \quad V_2 = u_2 - \frac{Q}{2\pi r_2 b_2}\cot\beta_2$$

Substitute in Eq. (3.9) to obtain

$$H_e = \frac{1}{g}\left(u_2^2 - \frac{Qu_2}{2\pi r_2 b_2}\cot\beta_2\right) - \frac{1}{g}\left(u_1^2 - \frac{Qu_1}{2\pi r_1 b_1}\cot\beta_1\right) \tag{3.13}$$

Now, consider a constant speed of rotation and assume that the fluid enters the impeller with no prerotation ($V_1 = 0$). In this case Eq. (3.13) can be reduced to

$$H_e = \frac{u_2^2}{g} - \frac{Qu_2}{2\pi g r_2 b_2}\cot\beta_2 \quad \text{or} \quad H_e = C_1 - C_2 Q \cot\beta_2 \tag{3.14}$$

where C_1 and C_2 are easily identifiable constants. Figure 3.19 shows the H_e–Q characteristics for the case of no prerotation at inlet when using (a) forward curved vanes ($\beta_2 > 90°$), (b) radial vanes ($\beta_2 = 90°$), and (c) backward curved vanes ($\beta_2 < 90°$) based on Eq. (3.14).

If there is prerotation at the vane inlet, Eq. (3.13) may be expressed as,

$$H_e = \left(\frac{u_2^2 - u_1^2}{g}\right) - \left(\frac{u_2}{2g\pi r_2 b_2}\cot\beta_2 - \frac{u_1}{2\pi g r_1 b_1}\cot\beta_1\right)Q$$

$$\text{Simplify} \Rightarrow H_e = C_3 - (C_2\cot\beta_2 - C_4\cot\beta_1)Q \tag{3.15}$$

The H_e–Q relationship given in the above equation is plotted in Figure 3.20 (dotted line) for the case when both β_1 and β_2 are less than 90°. The point of intersection of the solid and

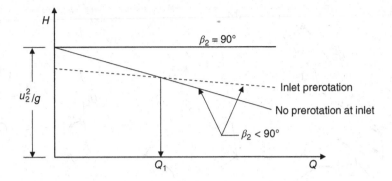

Figure 3.20   Theoretical $H_e - Q$ characteristic for $\beta_2 < 90°$ showing the effect of inlet prerotation

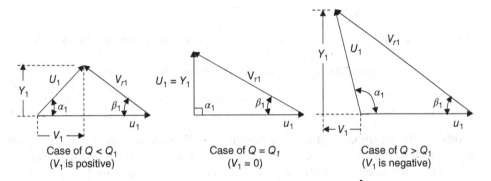

Figure 3.21   Inlet velocity diagrams for different flow rates

dotted lines exists at $Q_1 = 2\pi\, r_1\, b_1\, u_1\, \tan\beta_1$. This can be easily verified using Eqs. (3.14) and (3.15).

Also, by using Eq. (3.12), we can write

$$V_1 = u_1 - \frac{Q}{2\pi r_1 b_1}\cot\beta_1 = u_1 - \frac{Q u_1 \tan\beta_1 \cot\beta_1}{Q_1}$$

Simplifying:

$$V_1 = u_1\left(1 - \frac{Q}{Q_1}\right) \tag{3.16}$$

Accordingly, when $Q < Q_1$, V_1 becomes positive, and when $Q > Q_1$, V_1 becomes negative.

Finally, by using Eq. (3.9) we can see that negative values of V_1 result in higher theoretical head H_e. The velocity diagrams for the three cases of $Q < Q_1$, $Q = Q_1$, and $Q > Q_1$ are shown in Figure 3.21 for the same values of u_1 and β_1. The exit velocity diagrams for backward curved, radial, and forward curved vanes are shown in Figure 3.22.

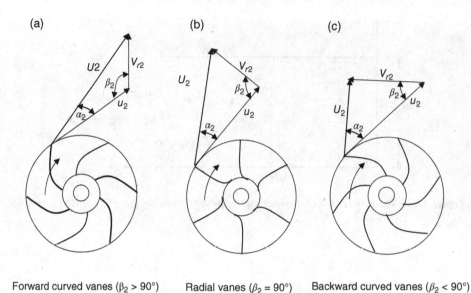

Forward curved vanes ($\beta_2 > 90°$)   Radial vanes ($\beta_2 = 90°$)   Backward curved vanes ($\beta_2 < 90°$)

Figure 3.22   Exit velocity diagrams for different vane shapes

Although the performance of impeller type (a) provides maximum head for all flow rates, such impellers are rarely used because of their low efficiency and unstable performance especially at low flow rates. Practically, top efficiencies are obtained by using type (c) impellers. The relationship between the impeller vane shape and the pump efficiency is discussed in the following section.

3.3.2   Relationship between Impeller Vane Shape and Pump Efficiency

The energy added to the fluid by the impeller is a combined effect of velocity head increase and pressure head increase. This energy addition occurs between the vane inlet section (1) and the vane exit section (2), as shown in Figure 3.23. Therefore, the head developed by the impeller can be divided into two main parts which are

a. pressure head, $H_p = (p_2 - p_1)/\gamma$
b. velocity head, $H_v = \left(U_2^2 - U_1^2\right)/2g$

Accordingly, we can express H_e as

$$H_e = (p_2 - p_1)/\gamma + \left(U_2^2 - U_1^2\right)/2g$$

From Figure 3.17, we can write

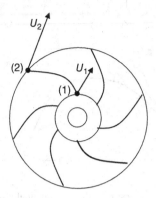

Figure 3.23 Absolute velocities at inlet and exit sections of a typical vane

$$U_1^2 = V_1^2 + Y_1^2 \text{ and } U_2^2 = V_2^2 + Y_2^2$$

Therefore,

$$U_2^2 - U_1^2 = \left(V_2^2 - V_1^2\right) + \left(Y_2^2 - Y_1^2\right)$$

In most impeller designs, the radial velocity component (Y) is almost constant and, assuming no prerotation at inlet ($V_1 = 0$), then

$U_2^2 - U_1^2 \approx V_2^2$, which leads to $H_e \approx H_p + V_2^2/2g$

In general, the ratio H_p/H is called the pump degree of reaction, where H is the pump total head. If we neglect all losses and accordingly consider $H = H_e$ then the pump degree of reaction (λ) can be expressed as

$$\lambda = H_p/H \approx \left(H_e - V_2^2/2g\right)/H_e \text{ or } \lambda \approx 1 - \frac{V_2^2}{2gH_e}$$

Using Eq. (3.10), the above expression can be simplified to

$$\lambda \approx 1 - \frac{V_2}{2u_2}$$

Using Eq. (3.12), together with the above equation, we can write

$$\lambda \approx 1 - \frac{u_2 - Y_2 \cot\beta_2}{2u_2}$$

$$\text{Finally,} \quad \lambda = \frac{H_p}{H_v + H_p} \approx 0.5 \left(1 + \frac{Y_2}{u_2} \cot\beta_2\right) \tag{3.17}$$

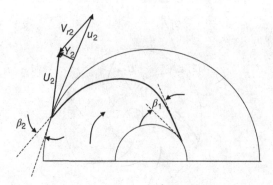

Figure 3.24   The exit velocity diagram for the hypothetical case of a very small vane exit angle

Equation (3.17) shows the relationship between the impeller vane shape (represented by the vane angle β_2) and the impeller degree of reaction, λ.

The increase of H_v means an increase of the fluid velocity at the impeller exit, which leads to a higher level of turbulence and accordingly higher friction losses. On the other hand, increasing pressure is the main objective of the pump, and this does not cause any losses. Therefore, the better (more efficient) impeller is the one that develops higher H_p and lower H_v, which means a higher degree of reaction, based on Eq. (3.17).

In conclusion, it is now clear from Eq. (3.17) that using backward curved vanes ($\beta_2 < 90°$) tends to produce a higher degree of reaction and accordingly higher efficiency (due to lower hydraulic losses) in comparison with radial or forward curved ones. In addition, a smaller value of β_2 results in higher degree of reaction. The value of β_2 normally ranges from 15 to 35°; however, higher values of β_2 are sometimes used for developing higher heads and flow rates. It is worth mentioning that decreasing β_2 to values much less than 15° leads to a higher degree of reaction but very small flow rates. Figure 3.24 shows the exit velocity diagram for a radial-type impeller equipped with backward-curved vanes having a very small exit vane angle, β_2. It is clear from the figure that the flow rate supplied by this impeller will be very small since the flow component of the velocity (Y_2) becomes very small.

3.3.3   Theoretical Relationship between Impeller Vane Shape and Pump Power Consumption

The effect of vane shape on the pump P–Q curve can be deduced based on the assumption of no losses. In this case, the pump output power (fluid power) will be exactly the same as the pump input power (brake power). The relationship between the fluid power, P_f, and the flow rate, Q, at a constant speed of rotation, N, can be obtained as follows.

We know that $(P_f)_{th.} = \gamma Q H_e$, and by using Eq. (3.14a), we can write

$$(P_f)_{th.} = \gamma Q (C_1 - C_2 Q \cot \beta_2) \;\Rightarrow\; (BP)_{th.} = K_1 Q - K_2 Q^2 \cot \beta_2 \qquad (3.18)$$

The variation of $(BP)_{th.}$ with Q is presented graphically in Figure 3.25 for the three cases of forward-curved, radial, and backward-curved vanes. It is clear from the figure that there is

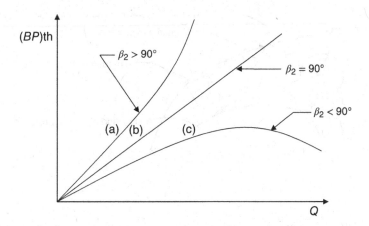

Figure 3.25   Variation of theoretical brake power with flow rate for various vane shapes at constant speed

an upper limit to $(BP)_{th.}$ in the case of using backward-curved vanes—impeller type (c)—while there is no upper limit in the two cases of radial and forward-curved vanes. This is a second advantage of using backward-curved vanes since the engineer can select a prime mover and guarantee that it will never be overloaded.

3.4   Deviation from Theoretical Characteristics

The actual performance of a centrifugal pump differs from the theoretical performance presented in Section 3.3. This is mainly due to various types of losses in addition to the relative circulation occurring inside the impeller due to the inertia effect. The effect of each type of loss, as well as the effect of relative circulation will be discussed in detail in the following sections.

3.4.1   Effect of Circulatory Flow on the Impeller Input Head

When the fluid enters the rotating impeller, it experiences a circulatory motion relative to the impeller vanes due to the inertia effect. As a result, a relative fluid circulation occurs in addition to the main through flow, as shown in Figure 3.26. This circulation tends to increase the velocity at the back of the vane, while decreasing it in front of the vane. The relative circulation is affected by the number of vanes and the impeller width. Increasing the number of vanes tends to increase surface friction and suppresses the relative circulation. The same effect occurs when decreasing the impeller width.

For a better understanding of this relative circulation, let us consider a simple experiment in which a small cylindrical container, full of an ideal fluid, is allowed to rotate (in a horizontal plane) in a clockwise direction about a center at O, as shown in Figure 3.27. The container orientation will be identified by points A and B on its walls. We also consider a fluid body (CD) shown inside the container. After rotating the container through an angle of 90°, the fluid body CD keeps the same orientation due to the absence of any moment that may cause rotation, since the fluid is frictionless. In the final position, despite the fact that the fluid body has not

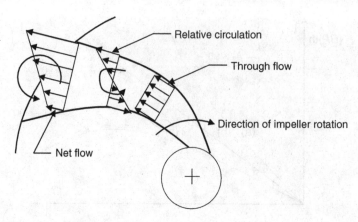

Figure 3.26   Effect of circulatory flow on the velocity distribution between adjacent vanes

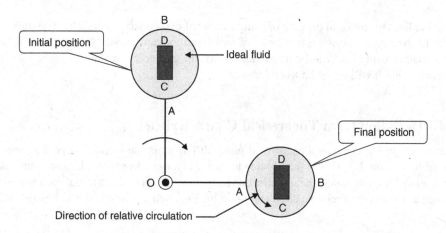

Figure 3.27   A simple experiment to demonstrate the relative circulation

performed any absolute rotation, there is a relative rotation with respect to the container. This relative rotation is in the counterclockwise direction (opposite to the direction of container rotation). We can also observe that the angle of container rotation (90°) is exactly the same as the angle of the fluid relative rotation. In case of a real fluid, the frictional moment will have some effect, and the two angles will not be the same.

The circulatory flow results in lower whirl component of velocity at the impeller exit and higher whirl component at inlet. Both effects lead to a reduction in the head developed by the pump. The actual velocity diagrams (dotted lines) are shown in Figure 3.28.

Using Eq. (3.9) and considering the circulatory flow effect, the impeller input head H_i (based on the actual whirl components V_1' and V_2') is given by

$$H_i = \frac{u_2 V_2' - u_1 V_1'}{g} \tag{3.19}$$

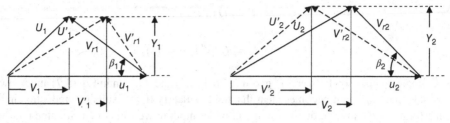

Ideal (——) and actual (---) inlet velocity diagrams   Ideal (——) and actual (---) exit velocity diagrams

Figure 3.28   Ideal and actual velocity diagrams

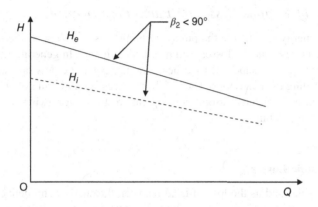

Figure 3.29   Effect of relative circulation on the head developed by an impeller equipped with backward-curved vanes

If we neglect prerotation at the inlet, the above equation becomes

$$H_i = u_2 V_2' / g \tag{3.20}$$

The vane efficiency, η_{vane}, is a measure of the loss of head due to the circulatory flow effect and is defined as

$$\eta_{vane} = \frac{H_i}{H_e} = \frac{u_2 V_2' - u_1 V_1'}{u_2 V_2 - u_1 V_1} \tag{3.21}$$

where H_e is the Euler head. The effect of circulatory flow on the H–Q characteristic is shown in Figure 3.29.

Although there is a loss of head due to the circulatory flow effect, it should not be regarded as loss of energy. This is mainly because of the parallel decrease in the impeller input power. In order to prove this, we can derive expressions for the actual torque, T', and the actual impeller input power, P', similar to the derivations of Eqs. (3.7) and (3.8). Accordingly, we can write

$$T' = \rho Q \left(V_2' r_2 - V_1' r_1 \right) \tag{3.22}$$

$$P' = \rho Q \left(V_2' u_2 - V_1' u_1 \right) \tag{3.23}$$

A simple comparison between T' given in Eq. (3.22) and T given in Eq. (3.7) and knowing that $V_2' < V_2$ and $V_1' > V_1$, it becomes clear that the circulatory flow results in a reduction in the driving torque and similar reduction in the impeller input power. In conclusion, the relative circulation results in a reduction in the impeller input head; however, there is a parallel reduction in the impeller input power.

3.4.2   Effect of Various Losses on Pump Performance

The sources of energy loss inside the pump are numerous, but they are normally categorized under hydraulic losses, leakage losses, and mechanical losses. In general, the factors affecting these losses are the pump shape and size, the operating speed, the flow rate, and the fluid properties. Understanding the behavior of each of these losses is important not only for design considerations but also for achieving optimum operation. In the following, each type of loss is investigated in some detail.

3.4.2.1   Hydraulic Losses

Hydraulic loss is defined as the loss of head between the suction and delivery nozzles of the pump. The loss of energy due to these hydraulic losses has a direct effect on the pump overall efficiency. Hydraulic losses can be divided into friction losses and eddy and separation losses.

3.4.2.2   Friction Losses

Similar to pipe flow, friction losses occur in various flow passages of the pump including the suction nozzle, the impeller flow passages, the volute casing, and the discharge nozzle. In general, friction losses depend on

1. flow passage geometry (shape and size)
2. surface roughness
3. flow velocity
4. fluid properties.

Since centrifugal pumps are normally used to pump liquids of low viscosity, the Reynolds number (Re) is very high in all pump flow passages. The behavior of friction losses in these passages follows the same features as losses in straight pipes or pipe fittings. In the high Re range, the friction coefficient in straight pipes depends mainly on the surface roughness and is less dependent on Re (see Figure 1.12). The same occurs in the pump flow passages since eddy friction greatly outweighs viscous friction. Accordingly, we can use Darcy's formula to write

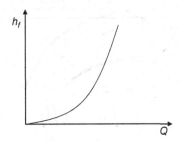

Figure 3.30 Effect of flow rate on the friction losses in the flow passages of a typical centrifugal pump

$$h_f = f LV^2/2gD, \text{ and } V = Q/A \Rightarrow h_f = fLQ^2/(2gDA^2)$$

In this case, if f can be assumed constant, the above equation can be reduced to

$$h_f = K_1 Q^2 \qquad (3.24)$$

where K_1 is a constant for a given pump. The above equation is plotted in Figure 3.30.

Based on the above, it is essential to reduce the roughness of all pump inner surfaces (including impeller vanes and shrouds) in order to reduce hydraulic losses.

3.4.2.3 Eddy and Separation Losses (Shock Losses)

The impeller vanes are always designed for a certain operating condition (speed and flow rate). This condition is referred to as the design point. Now, if the pump operates away from that point (say at partial capacity), the vane angles (β_1 and β_2) will not be correct, and this results in shock losses (i.e. the direction of flow relative velocity at vane inlet is not tangential to the vane) as shown Figure 3.31.

Now, let Q_s be the flow rate at the design point (no shocks). At this flow rate the relative velocity at inlet, V_{rs1}, will be tangential to the vane at inlet and there will be no shock losses. Let the velocity diagram in this case be as shown by solid lines in Figure 3.32, assuming that the flow enters the impeller with no prerotation ($\alpha_1 = 90°$). In case of operating at partial capacity ($Q < Q_s$) at the same speed of rotation and with no change in α_1, the inlet velocity diagram will be as shown in the same figure by dotted lines. This will result in a relative velocity, V_{r1}, in a direction, β_1', that differs from the inlet vane angle, β_1, as shown in the figure.

The shock losses at inlet and exit can be expressed as

$$h_s = C_1 (\Delta Y_1)^2/2g + C_2 (\Delta Y_2)^2/2g$$

where ΔY_1 and ΔY_2 represent the difference in the normal velocity component between the design and actual operating conditions.

But, since $\Delta Y \propto \Delta Q$, by using Eq. (3.11), the above equation can finally be written as

$$h_s = K_2 (\Delta Q)^2 = K_2 (Q - Q_s)^2 \qquad (3.25)$$

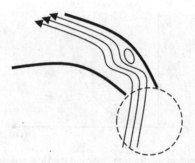

Figure 3.31   Direction of relative velocity at vane inlet when operating away from rated capacity

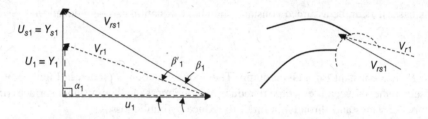

Figure 3.32   Effect of shock losses on velocity diagrams

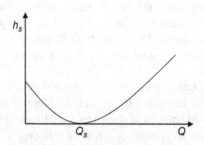

Figure 3.33   Variation of h_s with Q

The variation of h_s with Q according to Eq. (3.25) is presented in Figure 3.33.

Using Eqs. (3.24) and (3.25), the net head developed by the pump, H, can be expressed as

$$H = \text{Impeller input head} - \text{Hydraulic losses}$$

$$= H_i - (h_f + h_s) = H_i - K_1 Q^2 - K_2 (Q - Q_s)^2$$

The hydraulic efficiency η_h is defined as the ratio between the actual head developed by the pump (H) and the impeller input head (H_i), and can be written in the form

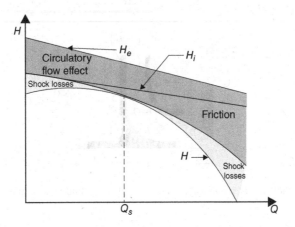

Figure 3.34 Effect of various losses on head-capacity curve

$$\eta_h = \frac{H}{H_i} = \frac{H_i - K_1 Q^2 - K_2(Q-Q_s)^2}{H_i} \tag{3.26}$$

Figure 3.34 shows the effect of circulatory flow, friction losses, and shock losses on the H–Q characteristic of a centrifugal pump. It is important here to mention that the point of zero shock losses is not necessarily the same point of maximum efficiency (since $(h_f + h_s)$ may not be minimum at that point).

The manometric efficiency η_{man} is defined as the ratio between the actual head developed by the pump and the Euler head, thus

$$\eta_{man.} = \frac{H}{H_e} = \frac{H}{H_i} \frac{H_i}{H_e} = \eta_h \eta_{vane} \tag{3.27}$$

3.5 Leakage Losses

The leakage losses represent the loss of energy due to the fluid leaking from the high pressure side of the impeller to the low pressure side through the clearance space between the impeller and the casing, as shown in Figure 3.35. Leakage may also occur through balancing chambers or the sealing system. Let the flow rate of the leaking fluid be Q_L, then the actual flow rate through the impeller will be $Q + Q_L$, where Q is the flow rate supplied by the pump. The volumetric efficiency, $\eta_{vol.}$, is defined as the ratio between the flow rate delivered by the pump and the actual flow rate through the impeller:

$$\eta_{vol.} = \frac{Q}{Q + Q_L} \tag{3.28}$$

For a new pump, the volumetric efficiency is usually greater than 95%. Erosion (or erosion/corrosion) causes an increase in the clearance space between the wearing rings, resulting in

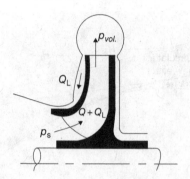

Figure 3.35   A schematic of a radial-type impeller showing fluid leakage through the wearing rings

higher leakage. Replacing or maintenance of the wearing rings is essential for reducing leakage losses. The power loss due to leakage can be expressed as

$$P_L = \gamma Q_L H_i = \frac{\gamma Q_L H}{\eta_h} \qquad (3.29)$$

The rate of leakage depends on the shape and size of the clearance space between the impeller and casing, including the clearance between the wearing rings. It also depends on the pressure difference between the volute casing and the suction nozzle ($\Delta p = p_{vol.} - p_s$) as well as the properties of the pumped fluid (ρ, μ).

Figure 3.35 shows the clearance space between the impeller and the casing with no wearing rings, while three types of wearing rings are shown in Figure 3.36. While the L-shaped wearing ring shown in Figure 3.36a has a simple design, the T-shaped ring shown in Figure 3.36c provides a clearance space with longer flow passage between the high and low pressure sides of the rings. This creates higher resistance to the motion of the leaking fluid. The presence of relief chambers adds more resistance and both effects tend to reduce Q_L. On the other hand, the initial cost and the cost of maintenance of the wearing rings increase as the design gets more complicated. Due to the presence of small solid particles in almost all liquids, the surface of the wearing rings erodes with time, causing larger clearance. This effect tends to produce a higher rate of leakage and lower volumetric efficiency. In addition, the rate of leakage gets higher as Δp increases.

3.6   Mechanical Losses

Mechanical losses represent the loss of power due to friction in the bearings and the sealing systems, and the fluid friction between the impeller and the casing. The last is called the disk friction loss. Normally, the power loss from disk friction forms the major part of the mechanical power losses. In order to find the relationship between the disk friction power loss and other parameters, we first consider a double-shrouded impeller that can be simulated by a circular disk of diameter D rotating in a concentric cylindrical casing as shown in Figure 3.37.

The flow regime in the clearance space between the disc and the casing may be laminar or turbulent, depending on the fluid properties (ρ, μ), size of the disc (D), the clearance space (B), and the speed of rotation (ω). These variables can be combined into the well-known Reynolds number ($Re = \rho ND^2/\mu$) and the ratio (B/D). If we assume laminar flow in the presence of a very

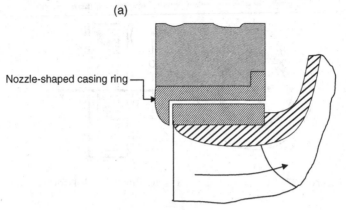

L-type wearing ring with a nozzle-shaped entrance

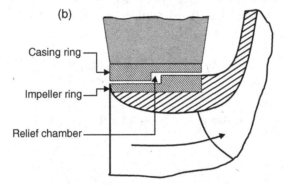

Step-type wearing ring with pressure relief chamber

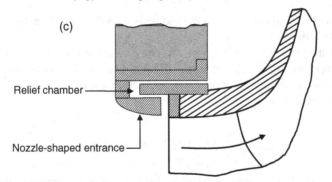

T-type wearing ring with nozzle-shaped entrance

Figure 3.36 Different shapes of wearing rings

small clearance (B), the power loss due to disc friction can be obtained by first calculating the frictional torque acting on the shown elementary area as

$$dT = dF \times r = \mu dA \frac{du}{dy} r \approx \mu (2\pi r dr) \frac{\omega r}{B} r \tag{I}$$

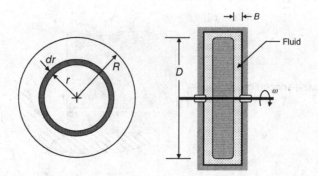

Figure 3.37 A disk rotating in a casing, simulating disk friction losses

The velocity distribution in the clearance space is assumed to be linear, based on the assumption that B is very small. Integrate Eq. (I) and consider friction on both sides of the disc to obtain

$$T \approx \frac{4\pi\mu\omega}{B} \int_0^R r^3 \, dr = \frac{\pi\mu\omega}{B} R^4 \tag{II}$$

Accordingly, the power loss due to disc friction for the special case of laminar flow can be written in the form

$$P_{DF} \approx \frac{\pi\mu\omega^2}{B} R^4 \quad \Rightarrow \quad P_{DF} \approx K_1 N^2 D^4 \tag{3.30}$$

where K_1 is a constant that depends on the geometry as well as the fluid viscosity, and N is the speed of rotation in rpm.

Because of the high speed of rotation, combined with the low viscosity of fluids pumped using centrifugal pumps, the flow regime in the clearance space between the impeller and the casing becomes turbulent in most cases. In such cases, the shear stress at the surface of the disk can be expressed as

$$\tau_o = C_f \frac{1}{2}\rho V^2 = \frac{1}{2} C_f \rho (\omega r)^2 \tag{III}$$

where C_f is the skin friction coefficient. In the high Re range, C_f becomes more dependent on the surface roughness than on the Reynolds number. Accordingly, C_f in Eq. (3.30) can be assumed constant for a given geometry.

The frictional torque on the shown elementary area will be

$$dT = dF \times r = \tau_o dA \times r = \frac{1}{2} C_f \rho (\omega r)^2 (2\pi r \, dr) r \tag{IV}$$

Integrate both sides of Eq. (IV) and consider friction on both sides of the disc to obtain

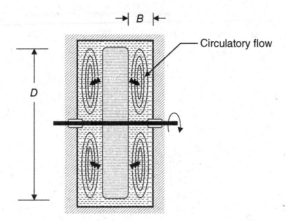

Figure 3.38 Circulatory flow between impeller and casing

$$T \approx 2\pi C_f \rho \omega^2 \int_0^R r^4 \, dr \approx \frac{2\pi}{5} C_f \rho \omega^2 R^5 \tag{V}$$

The loss of power due to disk friction can now be written as

$$P_{DF} = T\omega \approx \frac{2\pi}{5} C_f \rho \omega^3 R^5 \quad \Rightarrow \quad P_{DF} \approx K_2 N^3 D^5 \tag{3.31}$$

where K_2 is a constant that depends on the geometry of the pump as well as the roughness of its internal surfaces.

Another phenomenon that may take place in the clearance space between the impeller and the casing is the secondary vortex (donut-shaped vortical motion) shown in Figure 3.38. This secondary vortex occurs as a result of the centrifugal force acting on the fluid adjacent to the rotating impeller and starts to appear when the clearance, B, is large enough. This motion consumes larger amounts of energy and tends to increase the frictional torque. As the clearance B increases, the power loss due to this secondary vortex increases. Equation (3.31) can still be used for calculating the power loss due to disc friction, but the constant K_2 should also depend on B/D. In large pumps, the heat generation due to disk friction may create a considerable fluid temperature rise, especially at low flow rates.

Stepanoff [1] suggested the use of the following equation for calculating the disk friction power loss P_d based on the work by Pfleiderer [2]:

$$P_{DF} = K \gamma D^2 u^3 \tag{3.32}$$

where K is a numerical coefficient that depends on Re, surface roughness, and the ratio B/D, as shown in Figure 3.39, and the variables involved are defined as follows:

R_e is the Reynolds number ($=uR/\nu$)
u is outside peripheral velocity in ft/s

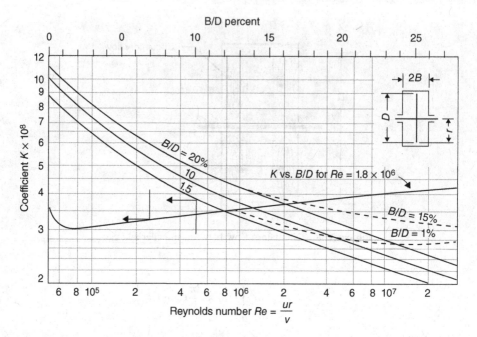

Figure 3.39 Variation of *K* with R_e and *B/D* for a smooth disk (—) and a rough disk (----) (from Stepanoff [1])

ν is the kinematic viscosity in ft²/s
R is the impeller radius in ft
D is the impeller diameter in ft
γ is the fluid sp. weight in lb/ft³
P_{DF} is the power loss in friction in horsepower

It is clear from Eqs. (3.31) to (3.32) that the disc friction power loss for a given pump does not depend on the pump flow rate, *Q*. The same applies to other sources of mechanical loss, such as the power loss in the bearings, seals, and other mechanical components.

Experimental measurements of disk friction were carried out by Mikhail *et al.* [3] in an attempt to obtain an expression for the power loss in terms of the pump specific speed, Reynolds number, relative roughness, the clearance space between the impeller and the casing, and the flow rate. They reported the following correlation for the power loss in disk friction:

$$P_{DF} = K\rho N^3 D^5 \tag{3.33}$$

and

$$K = C N_s^a (Q/Q_r)^b (k_s/D)^c (B/D)^d Re^e \tag{3.34}$$

where Q_r is the rated capacity, $c = 0.25$, $d = 0.1$, and $e = -0.2$ and the remaining constants depend on the flow rate range as follows:

For $Q/Q_r < 0.035$, $C = 10.6$, $a = 1.3$, and $b = -0.47$

For $Q/Q_r > 0.035$, $C = 15.9$, $a = 1.9$, and $b = -0.32$

Based on a thorough experimental study, Nemdili and Hellmann [4] developed an empirical equation for the disk friction coefficient K defined as

$$K = \frac{P_{DF}}{\rho \omega^3 R^5} \tag{3.35}$$

where ω is the angular velocity in rad/s and R is the disk radius. The coefficient K was found to depend on the relative roughness (k_s/R), the dimensionless gap width (B/R), and the dimensionless volute width (w/B) and can be obtained from the following correlation:

$$K = (k_s/R)^{0.25}(B/R)^{0.1}(w/B)^{0.2}Re^{-0.2} \tag{3.36}$$

In another study, Nemdili and Hellmann [5] investigated the relationship between the disk friction losses and the geometry of the disk with and without modified outlet sections simulating the impeller geometry.

For a single-shrouded impeller, the removal of the front shroud tends to decrease the power loss by disk friction while keeping the hydraulic losses almost the same as in double-shrouded impeller [1]. Despite the removal of the front wearing rings in the single-shrouded impeller, the leakage losses are approximately the same.

Finally, the mechanical efficiency, $\eta_{mech.}$, is defined as the ratio between the impeller input power and the pump mechanical input power (brake power or shaft power), thus

$$\eta_{mech.} = \frac{\text{Brake power} - \text{Mechanical losses}}{\text{Brake power}} = \frac{\text{Impeller input power}}{\text{Brake power}}$$

Or

$$\eta_{mech.} = \frac{\gamma(Q+Q_L)H_i}{\text{Brake power}} \tag{3.37}$$

3.7   Relationship between the Overall Efficiency and Other Efficiencies

The overall efficiency represents the ratio between the output fluid power and the input mechanical power (shaft power). The difference between the two represents the amount of heat generation due to all types of losses. These losses include hydraulic losses, leakage losses, and mechanical losses. The loss of head due to the circulatory flow effect is not included as energy loss. This is mainly because of the accompanying decrease in the input power due to that circulatory flow (see Section 3.4.1 for details). Therefore we can write

Brake power = Fluid power + Hydraulic losses + Leakage losses + Mechanical losses

or $BP = P_{fluid} + P_{hyd.} + P_L + P_{mech.}$

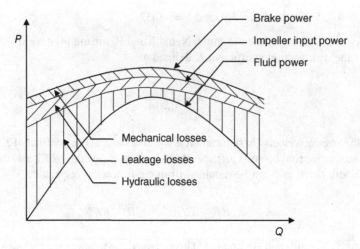

Figure 3.40 P–Q characteristic curve for a centrifugal pump, showing various losses ($N = $ const.)

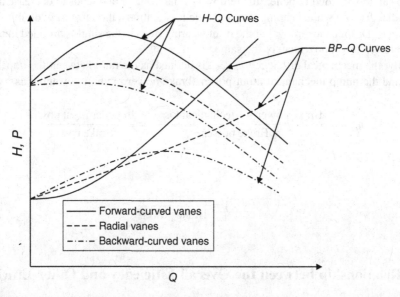

Figure 3.41 Effect of vane shape on the actual H–Q and BP–Q curves for a centrifugal pump operating at constant speed

The relationship between the overall efficiency and other efficiencies can be deduced as follows:

$$\eta_o = \frac{\gamma Q H}{B.P.} = \frac{H}{H_i} \frac{Q}{Q+Q_L} \frac{\gamma(Q+Q_L)H_i}{B.P.} = \eta_{hyd.} \eta_{vol.} \eta_{mech.} \tag{3.38}$$

Figure 3.40 shows the relationship between various types of power loss and the flow rate, Q, for a centrifugal pump operating at constant speed and has an impeller equipped with backward-curved vanes.

The actual H–Q and BP–Q curves depend on the impeller vane shape. Figure 3.41 shows typical performance curves when using impellers with backward-curved, radial, and forward-curved vanes.

Example 3.2

A radial type centrifugal pump is driven at 1200 rpm and consumes 20.3 kW of mechanical power. The pump impeller has the following dimensions:

$D_1 = 15$ cm, $b_1 = 4$ cm, $\beta_1 = 40°$
$D_2 = 30$ cm, $b_2 = 2.5$ cm, $\beta_2 = 30°$

The fluid enters the impeller vanes with no whirl. Knowing that the vane efficiency = 80%, volumetric efficiency = 95%, and the hydraulic efficiency = 88%, determine:

a. actual head developed by the pump
b. loss of head due to the combined effect of friction and shock losses
c. mechanical power losses
d. power loss due to fluid leakage

Solution

Some preliminary calculations

$u_1 = \omega R_1 = 9.42$ m/s	$u_2 = \omega R_2 = 18.9$ m/s
$Y_1 = u_1 \tan \beta_1 = 9.42 \tan 40° = 7.91$ m/s	$Y_2 = Y_1 A_1/A_2 = Y_1 D_1 b_1/D_2 b_2 = 6.33$ m/s
$Y_2 = u_2 - Y_2 \cot \beta_2 = 7.89$ m/s	$Q_{th} = Y_1 A_1 = Y_1 \times \pi D_1 b_1 = 0.149$ m³/s

a. $H_e = \frac{V_2 u_2 - V_1 u_1}{g} = \frac{18.9 \times 7.89 - 0}{9.81} = 15.2$ m $\quad H_i = \eta_{vane} H_e = 0.8 \times 15.2 = 12.1$ m $\quad H = \eta_{hyd.} H_i = 0.88 \times 12.1 = 10.7$ m

b. Shock and friction losses = $h_{hyd.} = H_i - H = 12.1 - 10.7 = 1.4$ m

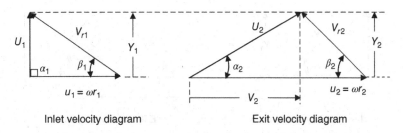

Inlet velocity diagram   Exit velocity diagram

Figure 3.42 Velocity diagrams for Example 3.2

c. $\eta_{mech.} = \gamma(Q + Q_L)H_i/BP. = 9.81 \times 0.149 \times 12.1/20.3 = 0.87$
   Mechanical power loss, $P_{mech.} = BP(1 - \eta_{mech.}) = 20.3(1 - 0.87) = \underline{2.55\,kW}$

d. $P_L = \gamma Q_L H_i = \gamma(1 - \eta_{vol.})Q_{th}H_i = 9.81(1 - 0.95) \times 0.149 \times 12.1 = \underline{0.83\,kW}$

Example 3.3

Consider the same data given in Example 3.1 and assume $\eta_{hyd.} = 85\%$, $\eta_{vane} = 90\%$, $\eta_{vol.} = 97\%$. Knowing that the power loss due to mechanical losses amounts to 0.48 kW, determine the vane angle at inlet and the pump power consumption.

Solution

Data: $r_1 = 0.075\,m$, $b_1 = 0.015\,m$, $Y_1 = Y_2$, $N = 1100\,rpm$,

$$r_2 = 0.125\,m, \beta_2 = 45°, Q = 2.7\,m^3/min. = 0.045\,m^3/s$$

$$\omega = 2\pi N/60 = 115.2\,rad/s, \; u_1 = \omega r_1 = 8.64\,m/s, \; u_2 = \omega r_2 = 14.4\,m/s$$

$$Y_2 = Y_1 = \frac{Q/\eta_{vol.}}{2\pi r_2 b_2} = \frac{0.045/0.97}{2\pi \times 0.125 \times 0.015} = 3.94\,m/s$$

$$\tan \beta_1 = Y_1/u_1 = 3.94/8.64 \;\Rightarrow\; \beta_1 = 24.5°$$

$$V_2 = u_2 - Y_2 \cot \beta_2 = 14.4 - 3.94 \cot 45° = 10.46\,m/s$$

$$H_e = \frac{u_2 V_2 - u_1 V_1}{g} = \frac{14.4 \times 10.46 - 0}{9.81} = 15.35\,m, \; H_i = \eta_{vane}H_e = 0.9 \times 15.35 = 13.8\,m$$

$$B.P. = \gamma(Q + Q_L)H_i + P_{mech.} = \gamma\frac{Q}{\eta_{vol.}}H_i + P_{mech.} = 9.81\frac{0.045}{0.97}13.8 + 0.48 = 6.77\,kW$$

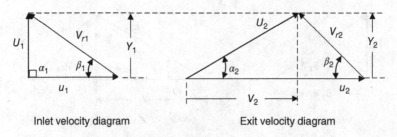

Inlet velocity diagram   Exit velocity diagram

Figure 3.43   Velocity diagrams for Example 3.3

Example 3.4

A town with a population of 30 000 is to be supplied with water at a rate of 200 L/person/day. Half of the total consumption is to be supplied in 8 h. The total static head is 40 m and the friction head loss in suction and delivery pipes amounts to 15 m. Select a suitable pump for this purpose and find the impeller power consumption. Assuming a motor speed of 1450 rpm, determine the impeller outer diameter. What will be the turning moment of the driving shaft?

Note: If needed consider $\eta_{hyd.} = 88\%$, $\eta_{vane} = 91\%$, $\eta_{vol.} = 95\%$, $\eta_{mech.} = 90\%$ and velocity coefficient $C_v = u_2 / \sqrt{2gH} = 0.95$.

Solution

$$n_s = \frac{n\sqrt{Q}}{(gH)^{3/4}} \quad \text{and} \quad n = N/60 = 24.2 \text{ rev/s}$$

$$Q = \frac{30{,}000 \times 0.2 \times 0.5}{8 \times 3600} = 0.104 \text{ m}^3/\text{s and } H = 40 + 15 = 55\text{m}$$

$$n_s = \frac{24.2\sqrt{0.104}}{(9.81 \times 55)^{3/4}} = 0.07 \quad \Rightarrow N_s = 17{,}200 \times 0.07 = 1200$$

Therefore, the pump should be *a radial-type centrifugal pump.*

Impeller power consumption, $P_{imp.} = \gamma \, (Q + Q_L) \, H_i$ $\qquad$ (I)

$$H_i = H/\eta_{hyd} = 55/0.88 = 62.5 \text{ m} \quad \text{and}$$

$$Q + Q_L = Q/\eta_{vol} = 0.104/0.95 = 0.109 \text{ m}^3/\text{s}$$

Substitute in Eq. (I) $\Rightarrow P_{imp} = 9.81 \times 0.109 \times 62.5 = 67$ kW

$$BP = P_{imp}/\eta_{mech.} = 67/0.9 = 74.4 \text{kW}$$

But $BP = T\omega \Rightarrow 74.4 = T \, (2\pi \times 1450/60) \Rightarrow T = 0.5 \text{ kN.m}$

To determine the impeller diameter, we have

$$C_v = u_2 / \sqrt{2gH} = 0.95 \Rightarrow u_2 = 0.95 \sqrt{2 \times 9.81 \times 55} = 31.2 \text{ m/s}$$

But $u_2 = \pi D_2 N/60 \Rightarrow D_2 = 0.41 \text{ m}$

Example 3.5

Determine the mechanical and leakage power loss in Example 3.4.

Solution

$$\text{Mechanical power loss} = \text{Brake power} - \text{Impeller input power}$$

$$= 74.4 - 67 = 7.4\,\textbf{kW}$$

$$\text{Power loss in leakage} = \gamma\,Q_L\,H_i = \frac{\gamma\,Q_L\,H}{\eta_{hyd}}$$

$$P_L = \frac{9.81\,(0.109-0.104)\,55}{0.88} = 3.07\,\text{kW}$$

Example 3.6

If the impeller in the previous problem has a uniform width of 20 mm from inlet to exit, find the absolute velocity of water leaving the impeller. Calculate also the degree of impeller reaction, and determine the vane angles at inlet and at exit. Assume $D_2 = 2D_1$ and the flow enter the impeller with no prerotation and consider a contraction factor of 0.9 due to vane thickness.

Solution

The absolute velocity at impeller exit $= U_2' = \sqrt{Y_2^2 + V_2'^2}$ (I)

But $Q + Q_L = \pi\,D_2\,b_2\,Y_2\,C_c \Rightarrow 0.109 = \pi\,0.41 \times 0.02 \times Y_2 \times 0.9$

Therefore, $Y_2 = 4.7$ m/s

$$H_e = \frac{u_2 V_2 - u_1 V_1}{g} = \frac{H_i}{\eta_{vane}} \quad \text{and since } V_1 = 0 \text{ we can write}$$

$$V_2 = \frac{gH_i}{u_2\eta_{vane}} = \frac{9.81 \times 62.5}{31.2 \times 0.91} = 21.6\,\text{m/s}$$

We also know that $H_i = \dfrac{u_2 V_2' - u_1 V_1'}{g}, \; V_1' = 0 \;\Rightarrow\; V_2' = 19.7$ m/s

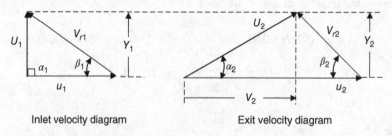

Inlet velocity diagram   Exit velocity diagram

Figure 3.44 Velocity diagrams for Example 3.6

Substituting in Eq. (I) $\Rightarrow U_2' = \sqrt{19.7^2 + 4.7^2} = 20.2\,\text{m/s}$

$$\lambda = 1 - \frac{V_2}{2u_2} = 1 - \frac{21.6}{2 \times 31.2} = 0.65$$

To determine the vane angles (β_1 and β_2), we proceed as follows:

$$Y_1 = \frac{Q + Q_L}{\pi D_1 b_1 C_c} \quad \text{and} \quad D_1 = 0.5 D_2 = 0.205\,\text{m}$$

Therefore, $Y_1 = \dfrac{0.109}{\pi\,0.205 \times 0.02 \times 0.9} = 9.4\,\text{m/s}$

$$u_1/u_2 = D_1/D_2 \quad \Rightarrow \quad u_1 = 31.2/2 = 15.6\,\text{m/s}$$

$$\beta_1 = \tan^{-1}(Y_1/u_1) = \tan^{-1}(9.4/15.6) \quad \Rightarrow \quad \beta_1 = 31.1°$$

$$\beta_2 = \tan^{-1}\left(\frac{Y_2}{u_2 - V_2}\right) = \tan^{-1}\left(\frac{4.7}{31.2 - 21.6}\right) \quad \Rightarrow \quad \beta_2 = 26°$$

Example 3.7
If in Example 3.4 a similar impeller of diameter 30 cm is used at the same speed of rotation, what will be the flow rate and head developed by the pump?

Solution

Data: $D_{2m} = 0.3$ m, same speed.
    Since prerotation is neglected at inlet, we can use the similarity of the two velocity diagrams to write

$$H = C\frac{u_2 V_2}{g} = K\frac{u_2^2}{g} \quad \text{but since } u_2 \propto D_2, \text{ we can write}$$

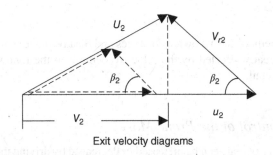

Exit velocity diagrams

Figure 3.45 Diagram for Example 3.7

$$\frac{H_m}{H} = \frac{u_{2m}^2}{u_2^2} = \frac{D_{2m}^2}{D_2^2} \quad \Rightarrow \quad \frac{H_m}{55} = \left(\frac{0.3}{0.41}\right)^2 \quad \Rightarrow \quad H_m = 29.4\,\text{m}$$

$Q = \pi D_2 b_2 Y_2 C_c$ and $b_2 \propto D_2$ (geometric similarity)

$$\therefore \frac{Q_m}{Q} = \frac{D_{2m}^2 Y_{2m}}{D_2^2 Y_2} \quad \text{and} \quad Y_2 \propto u_2 \propto D_2$$

$$\therefore \frac{Q_m}{Q} = \frac{D_{2m}^3}{D_2^3} \quad \text{(same as obtained from dimensional analysis)}$$

$$\therefore \frac{Q_m}{0.109} = \left(\frac{0.3}{0.41}\right)^3 \quad \Rightarrow \quad [Q_m]_{th.} = 0.042\,\text{m}^3/\text{s}$$

$$[Q_m]_{act.} = [Q_m]_{th.} \times \eta_{vol.} = 0.04\,\text{m}^3/\text{s}$$

3.8 Flow Rate Control in Pumping Systems

Energy saving in pumping systems is strongly dependent on the method used for flow rate control. Nowadays, this issue is gaining momentum because of the worldwide thrust towards the reduction of CO_2 emissions to the atmosphere for better environment and for solving the global warming problem. In general, flow rate control in pumping systems is carried out using one or a combination of the following method(s):

a. speed control of the pump driver
b. delivery valve throttling
c. inlet guide vanes
d. use of impellers with adjustable vanes
e. impeller trimming
f. partial circulation of the outflow using a bypass
g. operate the system using more than one pump
h. use of a storage tank

Each of the above methods has its own advantages and disadvantages since the system overall efficiency will be greatly affected by the selected method. In the following, each method is discussed in some detail.

3.8.1 Speed Control of the Prime Mover

The pumping system flow rate can be increased or decreased by driving the pump at a higher or a lower speed. However, operating the pump at speeds higher or lower than its rated speed will

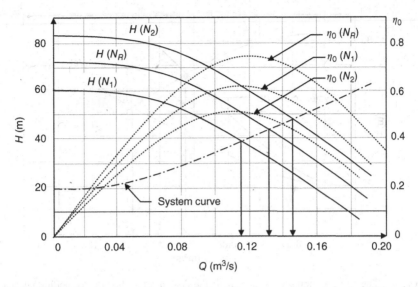

Figure 3.46 Typical $H - Q$ and $\eta_o - Q$ curves for a centrifugal pump at its rated speed (N_R) and at a lower speed (N_1) and a higher speed (N_2)

result in less overall efficiency. Moreover, electric motors with speed control are more expensive. The pump $H–Q$ and $\eta_o–Q$ curves at different speeds can be either provided by the manufacturer (possibly in the form of isoefficiency curves) or predicted with reasonable accuracy from the known $H–Q$ and $\eta_o–Q$ curves (at the pump's rated speed) using the similarity laws presented in Section 2.5. Figure 3.46 shows the $H–Q$ and $\eta_o–Q$ curves at the pump rated speed (N_R) and at a lower speed $(N_1 < N_R)$ and at a higher speed $(N_2 > N_R)$. Knowing the system $H–Q$ curve, we can determine the system flow rate at each speed and can also calculate of the corresponding power consumption.

3.8.2 Delivery Valve Throttling

The system flow rate can be reduced by partial closure of the delivery valve. Although this is the cheapest and most common method for flow rate control, it is accompanied by a large amount of energy loss, which can be as high as 50% or more of the pump output power. The partial closure of the valve tends to change the system $H–Q$ curve from curve (I) to curve (II) as shown in Figure 3.47. The resulting additional friction head loss in the valve is also indicated in the figure. Accordingly, the amount of power loss in the valve can be expressed as

$$P_{loss} = \gamma Q \left(h_L \right)_{Valve} \tag{3.39}$$

where $(h_L)_{Valve}$ is the friction head loss in the valve due to its partial closure. The change in the point of pump operation may also cause additional losses due to a possible decrease in the pump

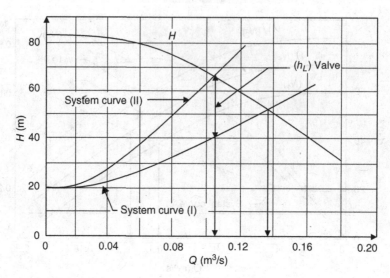

Figure 3.47 The effect of partial valve closure on the system curve showing the additional friction head loss in the valve

overall efficiency. As a result, we may end up getting a lower flow rate at the expense of higher power consumption. Accordingly, the use of a delivery valve for flow rate control should be a temporary solution, not a permanent one.

3.8.3 Using Inlet Guide Vanes for Flow Rate Control

The use of inlet guide vanes for flow rate control is common in fans and compressors and also in axial flow pumps but less common in radial-type centrifugal pumps. The idea is to have a set of guide vanes upstream of the impeller (close to the impeller inlet) to create prewhirl before the vane inlet. This prewhirl (in the direction of vane rotation) creates positive whirl component of the flow velocity (V_1), which tends to reduce the total head developed (see Eqs. (3.9) and (3.19)) and accordingly reduce the system flow rate. In theory, a negative whirl tends to create higher total head; however, the associated reduction in the pump efficiency prevents such practice. Figure 3.48 shows the inlet velocity diagram with no prewhirl (solid lines) and with positive prewhirl (dotted lines). Figure 3.49 shows the effect of the amount of prewhirl on the pump H–Q curve for the three cases of zero prewhirl, 33% prewhirl ($V_1/u_1 = 33\%$), and 67% prewhirl. The pump flow rate decreases from Q_1 to Q_2 and Q_3 as the prewhirl increases from 0 to 33% and 67%.

One of the advantages of this method is the decrease of the pump input power with the increase of inlet prewhirl (+ve prewhirl) as a result of the reduction in the driving torque. This can be understood based on Eq. (3.7), which relates the driving torque to the whirl components (V_1 and V_2). However, one would expect some decrease in the overall pump efficiency in comparison with normal operation. Also, this method is rarely used for increasing the pump flow rate because of the associated reduction in the pump efficiency.

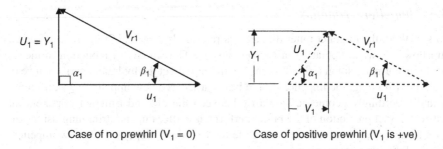

Case of no prewhirl ($V_1 = 0$)   Case of positive prewhirl (V_1 is +ve)

Figure 3.48   Inlet velocity diagrams for the cases of no prewhirl and positive prewhirl

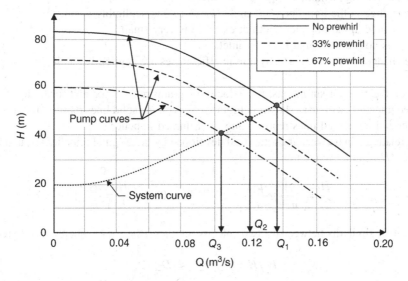

Figure 3.49   Effect of prewhirl on the pump H–Q curve when operating at constant speed

3.8.4   Using Impellers with Adjustable Vanes

The idea of using impellers with adjustable vanes (variable pitch) was first applied to axial flow pumps. The impeller vane angles (β_1 and β_2) can be adjusted by rotating each vane using a special mechanism. By increasing the inlet and exit vane angles, we can increase the head developed and the pump flow rate. This method is normally used in large pumps and it gives flexibility for increasing or decreasing the flow rate at the expense of a possible decrease in the overall efficiency.

One of the problems faced in these pumps is the increase in hydraulic losses in the stationary (exit) guide vanes located downstream of the impeller as a result of changing the impeller vane angles. These guide vanes act as a diffuser and are designed to minimize flow separation and vortex formation when the pump operates at its best efficiency point. Accordingly, the change of impeller vane angles requires a parallel change in the orientation of the exit guide vanes. Zhu *et al.* [6] proposed an axial-flow pump with adjustable guide (diffusion) vanes, in which the inlet angle of guide vanes can be adjusted to coordinate with the change of flow direction resulting from changing the impeller vane angle.

3.8.5 Impeller Trimming

In this method, the impeller outer diameter is reduced by machining in order to decrease the pump flow rate when operating in a given system. The amount of reduction in the impeller diameter is normally small (10–20%) and should not affect the hydraulic performance of other parts. After trimming, the impeller should be regarded as a new impeller with different geometry and accordingly geometric similarity between the old and trimmed impellers may not be correct. Exact prediction of the pump performance after impeller trimming using equations cannot be achieved. In order to estimate the total head developed by the new impeller, let us make the following assumptions:

a. The original and trimmed impellers are driven at the same speed.
b. The change in the impeller vane angle at exit (β_2) is negligibly small.
c. The change in the impeller width at exit (b_2) is very small.
d. There is no prerotation at impeller inlet.
e. The changes in the vane and hydraulic efficiencies are negligibly small.

Based on the above assumptions, the velocity diagrams at the vane exit for both original and trimmed impellers will be as shown in Figure 3.50 with the solid lines representing the original impeller and the dotted lines representing the trimmed impeller.

The relationship between the total head developed by the original impeller (H) and trimmed impeller (H_t) can be written as

$$\frac{H_t}{H} = \frac{u_{2t}V'_{2t} - u_{1t}V'_{1t}}{u_2 V'_2 - u_1 V'_1} \ , \ V'_1 = V'_{1t} = 0 \ \Rightarrow \ \frac{H_t}{H} = \frac{u_{2t}V'_{2t}}{u_2 V'_2}$$

Based on the similarity between the two velocity diagrams, the above equation can be reduced to

$$H_t/H \approx (u_{2t}/u_2)^2 \approx (D_{2t}/D_2)^2 \tag{3.40}$$

The following expressions can also be deduced for the reduction in the flow rate and power consumption:

$$Q_t/Q \approx D_{2t}/D_2 \ \text{ and } \ BP_t/BP \approx (D_{2t}/D_2)^3 \tag{3.41}$$

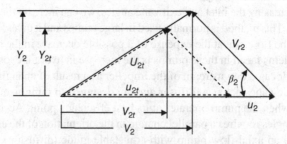

Figure 3.50 Exit velocity diagrams for the original and trimmed impellers (—— original and ------ trimmed)

Equations (3.40) and (3.41) can be used to predict the pump performance curves (H–Q and BP–Q) after impeller trimming. The system flow rate can be determined as explained in Chapter 2. One should keep in mind that impeller trimming may cause changes in the impeller width at exit, vane angle at exit, and the overall efficiency. Such changes differ from one impeller to another based on its design. Accordingly, the reduction in H, Q, and BP may deviate from the predictions of Eqs. (3.40) and (3.41), depending on the pump geometry.

The deviation between the actual performance of the trimmed impeller and that predicted using the above approach may be due to the following:

1. The trimmed impeller will result in more prewhirl at the vane inlet, leading to less total head.
2. The possible mismatch between the trimmed impeller and the casing may result in a reduction in the hydraulic efficiency.
3. The fluid flows through a longer path in the casing, causing more hydraulic losses.
4. Part of the mechanical losses (losses in the bearings and in the mechanical seal) will remain unchanged, whereas the output fluid power is reduced. This results in a reduction in the mechanical efficiency.

3.8.6 Using Bypass for Flow Rate Control

In this method, part of the pump outflow is circulated back to the suction pipe or to the suction reservoir. Figure 3.51 shows a schematic of a typical bypass, where the amount of circulated flow is controlled by the bypass valve. The friction head loss in the bypass (including the valve) is approximately the same as the total head developed by the pump. Accordingly, the power loss in this method is considerable and can be calculated from

$$P_{bypass} = P_{fluid}(Q_c/Q) = \gamma Q_c H \qquad (3.42)$$

where P_{bypass} is the power loss in the bypass and Q_c is the rate of flow circulation. Taking into consideration the power loss in the pump due to hydraulic, mechanical, and leakage losses, the actual power loss in the bypass can be written as

$$P_{actual\,bypass} = \gamma Q_c H / \eta_o \qquad (3.43)$$

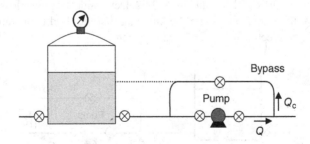

Figure 3.51   Schematic of a bypass with the return pipe connected to the suction line or to the suction reservoir

Knowing that $BP = \gamma QH/\eta_o$, we can write

$$P_{actual\,bypass} = (Q_c/Q)\,BP \qquad (3.44)$$

The main advantages of this method are its low cost and simplicity. However, the disadvantages include the high power loss in fluid friction, and the temperature increase of the circulated flow. This temperature increase may cause temperature buildup that may eventually create cavitation (due to the increase in the fluid vapor pressure). When the bypass return pipe goes back to the suction reservoir (the dotted line in Figure 3.46), the temperature buildup will be much less because of the large heat capacity of the reservoir fluid. We can use the energy equation to determine the temperature increase when the bypass is used in either of the two configurations.

3.8.7   Flow Rate Control by Operating Pumps in Parallel or in Series

In situations requiring considerable flow rate variation (e.g. city water network), the pumping station can be equipped with a number of pumps that can be operated in parallel or in series. This is done by selecting the appropriate number of pumps for supplying the required flow rate at any given time. The combined efficiency of the system was discussed in detail in Section 2.4. The main advantage here is the flexibility of increasing or decreasing the system flow rate while keeping the overall efficiency of the pumping system as high as possible. We can also decide on whether parallel or series operation will produce the best efficiency.

3.8.8   Use of a Storage Tank

Sometimes the system demand temporarily exceeds the normal capacity of the pumping station (e.g. in a city water network). In this case, a water tower can be a good solution. The availability of water from the tower for firefighting, especially in the case of power outage, is considered as additional advantage.

Example 3.8

A centrifugal pump is used to deliver water at a rate of 0.02 m³/s against a total head of 150 m. The pump overall efficiency at the point of operation is 60%. It is required to reduce the flow rate by 20% using a bypass as shown in Figure 3.52. Determine the power loss in the bypass and the temperature at the pump suction nozzle knowing that the fluid temperature in the suction reservoir is 45°C. Neglect heat loss to the atmosphere.

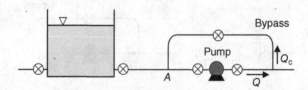

Figure 3.52   Diagram for Example 3.8

Solution

The actual power loss in the bypass can be obtained by using Eq. (3.43):

$$\rightarrow P_{actual\,bypass} = \gamma Q_c H / \eta_o = 9.81\,(0.2 \times 0.02)\,150/0.6 = \underline{9.81\,kW}$$

To determine the temperature at the pump suction nozzle, let us consider the following:

T_s = fluid temperature at the pump suction nozzle
T_d = fluid temperature at the pump delivery nozzle
T_b = fluid temperature in the bypass return pipe
T_r = fluid temperature in the suction reservoir

Using the energy equation, we can determine the temperature increase through the pump as follows:

Rate of heat generation due to various losses = Input power − Output power = $BP(1-\eta_o)$

This amount of heat causes an increase in the fluid temperature through the pump, ΔT. Neglect heat loss to the atmosphere and apply the energy balance,

$$\dot{m}\,C_p\,\Delta T = BP(1-\eta_o) \quad \Rightarrow \quad \rho Q C_p \Delta T = \frac{\gamma Q H}{\eta_o}(1-\eta_o)$$

Simplify to obtain

$$\Delta T = \frac{gH}{C_p}\left(\frac{1}{\eta_o}-1\right) \quad \text{or} \quad T_d - T_s = \frac{gH}{C_p}\left(\frac{1}{\eta_o}-1\right) \tag{I}$$

Also, the power loss in the bypass is equal to the rate of heat generation. Therefore we can use Eq. (3.42) to write

$$\dot{m}_c\,C_p\,\Delta T = \gamma Q_c H \quad \Rightarrow \quad \rho Q_c C_p \Delta T = \rho g Q_c H \quad \Rightarrow \quad \Delta T = gH/C_p$$

Accordingly we can write

$$T_b - T_d = gH/C_p \tag{II}$$

The energy balance at junction A gives

$$\rho Q_c C_p T_b + \rho(Q-Q_c)C_p T_r = \rho Q C_p T_s$$

Simplifying gives

$$T_s = x T_b + (1-x)\,T_r \quad \text{where } x = Q_c/Q \tag{III}$$

The solution of Eqs. (I), (II), and (III) gives

$$T_s = T_r + \frac{x}{(1-x)} \frac{gH}{\eta_o C_p} \qquad \text{(IV)}$$

Substitute from the given data to obtain $T_s = 45 + \dfrac{0.2}{(1-0.2)} \dfrac{9.81 \times 150}{0.6 \times 4200} \approx \underline{45.2^\circ C}$

References

1. Stepanoff, A. (1957) *Centrifugal and Axial Flow Pumps: Theory, Design, and Application*, 2nd edn. John Wiley & Sons, Inc., New York.
2. Pfleiderer, C. (1955) *Die Kreiselpumpen für Flüssigkeiten und Gase: Wasserpumpen, Ventilatoren, Turbogebläse, Turbokompressoren.* Julius Springer, Berlin.
3. Mikhail, S., Khalafallah, M.G., and El-Nady, M. (2001) Disk friction loss in centrifugal and mixed-flow pumps. 7th International Congress on Fluid Dynamics and Propulsion, December 18–20, 2001, Cairo, Egypt.
4. Nemdili, A. and Hellmann, D.H. (2004) Development of an empirical equation to predict the disc friction losses of a centrifugal pump. 6th International Conference on Hydraulic Machinery and Hydrodynamics, Timisoara, Romania; 2004.
5. Nemdili, A. and Hellmann, D.H. (2007) Investigations on fluid friction of rotational disks with and without modified outlet sections in real centrifugal pump casings. *Forsch Ingenieurwes*, 71, 59–67.
6. Zhu, H., Zhang, R., Xi, B., and Hu, D. (2013) Internal flow mechanism of axial-flow pump with adjustable guide vanes. Paper No. FEDSM2013.16613. ASME Fluids Engineering Division Summer Meeting, Incline Village, NV; 2013.

Problems

3.1 A centrifugal water pump is driven at a speed of 1450 rpm. The pump impeller is of radial type and has an average radius of 50 mm at the vane inlet and a radius of 150 mm at the vane exit. The vane width at inlet and exit are 25 and 10 mm respectively. The vane angles are $\beta_1 = 50^\circ$ and $\beta_2 = 30^\circ$. Assuming no prerotation at inlet and neglecting all sources of losses, determine

  a. the volume flow rate
  b. the output fluid power
  c. the static pressure rise through the impeller.

3.2 A centrifugal pump is driven at 1800 rpm and delivers water at a flow rate of 0.12 m³/s. The impeller vanes have the following dimensions:

Dimension	At inlet	At exit
Vane diameter	12 cm	24 cm
Vane width	5 cm	2.7 cm
Vane angle	30°	20°

  a. Determine the Euler head He.
  b. Calculate the absolute velocity of the fluid and the flow angle at the impeller exit.

   c. What would be the total head developed between the impeller inlet and the exit if the fluid motion inside the impeller is approximated by a pure forced vortex.

3.3 The following data is provided for a centrifugal water pump:

Vane inlet diameter	$D_1 = 10$ cm	Vane exit diameter	$D_2 = 25$ cm
Vane width at inlet	$b_1 = 5$ cm	Vane width at exit	$b_2 = 5$ cm
Vane inlet angle	$\beta_1 = 30°$	Vane exit angle	$\beta_2 = 18°$
Pump speed	$N = 1800$ rpm		

The fluid is directed (using guide vanes) at the impeller inlet in order to enter with a small amount of prerotation such that $V_1 = 0.2\, u_1$. Neglecting all types of losses, calculate:

   a. flow rate supplied by the pump, Q
   b. head developed by the pump, H
   c. driving torque, T
   d. impeller degree of reaction, λ.

3.4 A radial-type pump has an impeller with vane inlet radius of 50 mm and an outside radius of 200 mm. The impeller vanes have inlet and exit angles of 18 and 10° respectively and a constant width of 60 mm. The pump is driven at a speed of 1800 rpm and the fluid enters the impeller with no whirl. Neglecting all losses, determine:

   a. pump flow rate
   b. static pressure rise through the impeller
   c. pump fluid power.

3.5 A centrifugal water pump is designed to be driven at 1800 rpm and deliver a flow rate of 0.1 m³/s. The pump impeller has the following dimensions:

Vane radius at inlet = 6 cm	Vane radius at exit = 12 cm
Vane width at inlet = 3.6 cm	Vane width at exit = 2 cm
Vane angle at inlet = 45°	Vane angle at exit = 30°

Neglecting all types of losses, determine:

   a. theoretical head developed by the pump
   b. pump driving torque
   c. degree of impeller reaction.

3.6 A centrifugal pump is to deliver 0.09 m³/s against a head of 24 m when running at 1500 rpm. The flow at inlet is radial, and the radial velocity component through the impeller is to be constant at 3.6 m/s. The outer diameter of the impeller is twice the inner diameter, and the impeller width at exit is to be 12% of the outer diameter. Neglecting impeller losses and the effect of vane thickness, determine:

    a. impeller diameter at inlet and at exit

    b. impeller width at inlet and at exit

    c. inlet and exit vane angles.

3.7 The impeller of a centrifugal water pump rotates at 1260 rpm. The inlet eye radius is 5 cm and the outside radius is 12 cm. The vane width at inlet and exit are 25 and 10 mm respectively. The vane angles are $\beta_1 = 30°$ and $\beta_2 = 20°$. Assuming radial inflow to the impeller vanes, determine:

    a. volume flow rate

    b. output fluid power

    c. stagnation pressure rise through the impeller.

3.8 A radial-type pump has an impeller with vane inlet radius of 50 mm and an outside diameter of 400 mm. The inlet and exit vane angles are 18 and 10°, respectively, and a constant vane width of 60 mm. Knowing that the pump is driven at a speed of 1800 rpm and assuming no whirl at inlet, determine:

    a. pump actual flow rate

    b. static pressure rise through the impeller

    c. input power to the impeller

    d. power loss in leakage.

Note: Consider $\eta_{vol.} = 0.98$, $\eta_{mech.} = 0.94$, $\eta_o = 0.82$, $\eta_{vane} = 0.90$.

3.9 A centrifugal pump is designed for a rated capacity of 0.25 m³/s at a total head of 80 m. The impeller diameter is 50 cm and the vane angle at exit is 140°. Knowing that the vane width at exit is 4 cm and that the flow enters the impeller with no prerotation, determine:

    a. pump rated speed, N

    b. impeller degree of reaction, λ

    c. pump brake power, *BP*.

Note: Consider $\eta_{vane} = 0.90$, $\eta_{hyd.} = 0.78$, $\eta_{mech.} = 0.96$, $\eta_{vol.} = 0.98$.

3.10 Obtain an expression for the disc friction losses for the case when the clearance between the impeller and the casing is small enough to ensure a laminar flow structure. Sketch the variation of the power loss in disc friction versus the clearance width and compare the obtained expression with that reported by Mikhail et al. (2001) and given in Eq. (3.33) for the case of turbulent flow.

3.11 It is required to design a centrifugal pump to deliver water at a rate of 0.2 m³/s against a total head of 25 m when operating at a speed of 1200 rpm. Consider this operating condition to represent the pump design point at which there are no shock losses and no whirl at inlet. In order to obtain the main impeller dimensions the following assumptions are made:

    a. The vane exit angle is 15°.

    b. The radial velocity component through the impeller is kept constant at 3.2 m/s.

c. The hydraulic efficiency is 85% and the vane efficiency is 90%.

d. The ratio between the impeller inlet to exit diameters is 0.24.

You are required to estimate the following:

a. impeller outer diameter

b. impeller width at exit

c. vane inlet angle

d. impeller degree of reaction

3.12 A centrifugal pump lifts water against a total head of 36 m when driven at a speed of 1200 rpm. The impeller outer diameter is 39 cm and the outlet vane angle is 35°. If the impeller width at exit is 3.9 cm, what flow rate would you expect and what is the power of the motor required to drive the pump? Neglect leakage losses and assume a manometric efficiency of 82% and a mechanical efficiency of 90%.

3.13 Considering the data given in Problem 3.3, calculate the leakage flow rate, Q_L, and the actual head developed by the pump, assuming the following efficiencies: $\eta_{hyd.} = 88\%$, $\eta_{vol.} = 97\%$, $\eta_{vane} = 85\%$, $\eta_{mech.} = 90\%$.

3.14 A centrifugal pump is designed to operate at a capacity of 0.24 m³/s against a total head of 25 m when driven at a speed of 1200 rpm. The friction head loss in the impeller can be determined from $h_f = 0.12Y_2^2$ and the shock losses are negligible. The fluid enters with no prerotation when operating at the design point. Initial calculations showed that the impeller outer diameter is 360 mm and the diameter at the vane inlet is 180 mm. Assuming that the vane thickness reduces the area of flow by 6% (i.e. $C_c = 0.94$), determine:

a. width of the impeller at exit

b. vane inlet and exit angles

c. loss of power due to hydraulic losses

d. power required to drive the pump, assuming that the mechanical losses amount to 0.67 kW.

Note: Consider $\eta_{man} = 80\%$, $\eta_{vol} = 98\%$, $\eta_{vane} = 95\%$.

3.15 Figure 3.53 shows the details of a centrifugal pump impeller that has a rated speed of 1450 rpm. The loss of head due to friction and shock losses is approximately 0.15 $U_2'^2/2g$ while the mechanical power loss is 2.4 kW. Assuming no prerotation at inlet, calculate:

a. pump head, H, and flow rate, Q

b. pump brake power, BP

c. power loss in fluid leakage, P_L

Note: If needed consider $\eta_{vane} = 92\%$, $\eta_{vol} = 98\%$.

3.16 A centrifugal water pump is driven at 1500 rpm and has an impeller of dimensions:

$D_1 = 20$ cm, $b_1 = 6$ cm, $\beta_1 = 35°$
$D_2 = 40$ cm, $b_2 = 3$ cm, $\beta_2 = 25°$

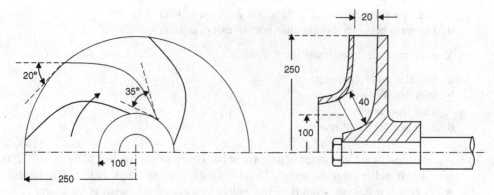

Figure 3.53 Diagram for Problem 3.15 (all dimensions in mm)

The vane thickness reduces the effective area of flow at inlet and at exit by 4%. Knowing that $\eta_{hyd} = 90\%$, $\eta_{vol} = 97\%$, $\eta_{vane} = 88\%$, $\eta_{mech} = 94\%$, and that the flow has no whirl at inlet, determine:

a. actual pump head and flow rate
b. actual fluid velocity at impeller exit
c. power loss due to fluid leakage
d. pump brake power.

3.17 A centrifugal pump has a double-shrouded impeller with nine vanes. The radii at the vane inlet and exit are 90 and 180 mm, while the vane width at exit is 32 mm. The pump delivers 860 L/min when driven at a speed of 1450 rpm and gives a pressure difference of 240 kPa between the inlet and outlet flanges. The vane thickness reduces the area of flow by 5%. Assuming a manometric efficiency of 88%, determine:

a. impeller degree of reaction
b. power required to drive the pump, assuming a mechanical efficiency of 92% and a volumetric efficiency of 97%.

3.18 A centrifugal pump is driven at 1500 rpm and has a double-shrouded impeller with nine vanes. The main dimensions of the impeller are:

Radius at vane inlet = 90 mm	Radius at vane exit = 180 mm
Vane angle at inlet = 45°	Vane angle at exit = 30°
Vane width at inlet = 45 mm	Vane width at exit = 30 mm

The flow enters the impeller at a flow angle $\alpha_1 = 60°$ using inlet guide vanes. The circulatory flow effect increases the whirl component at the vane inlet by 10% while decreasing it at the vane exit by 12%. Determine:

a. flow rate to be delivered by the pump
b. total head developed

   c. power input to the impeller

   d. power required to drive the pump.

Note: If needed consider $\eta_{vol} = 0.98$, $\eta_{hyd.} = 0.9$ and mechanical power losses = 1.8 kW.

3.19 It is required to estimate the dimensions of a single-stage centrifugal pump to be used in delivering 0.32 m³/s of water against a total head of 80 m. The pump is to be driven at a speed of 1500 rpm. Assuming that $D_1/D_2 = 0.4$, $b_1/D_1 = 0.25$, $b_2/D_2 = 0.1$, the flow enters the impeller with no whirl, and the contraction factor due to vane thickness = 0.92, determine:

   a. vane inlet and exit diameters

   b. vane inlet and exit angles

   c. impeller degree of reaction

   d. pump driving torque.

Note: Consider $\eta_{hyd.} = 90\%$, $\eta_{man.} = 80\%$, $\eta_{vol.} = 97\%$, $\eta_{mech.} = 95\%$, and a velocity coefficient, $C_v = u_2/\sqrt{2gH} = 0.9$.

4

Axial and Radial Thrusts in Centrifugal Pumps

4.1 Introduction

The pump shaft is normally deigned to withstand various stresses resulting from several forces and moments acting on it. These include the pump driving torque, the bending moments due to static and dynamic forces, and the unbalanced forces acting on the impeller(s) in both axial and radial directions. If not properly balanced, these forces will be acting on the shaft, causing shaft deflection, and will be transmitted to the bearings, creating overloading. The axial unbalanced thrust in multistage pumps may create a very large force that cannot be supported by the journal or ball bearings. In order to avoid bearing failure, the designer either balances this axial force by hydraulic means or utilizes a thrust bearing. However, adding a thrust bearing leads to an increase in pump size and cost, in addition to increasing the cost of maintenance. Similarly, the radial unbalanced force acting on the impeller will cause shaft deflection that may create rapid wear at surfaces with small clearances, such as the wearing rings, and may also create severe shaft vibration that may eventually cause shaft fatigue failure. In this chapter, we will first determine the magnitude and direction of the unbalanced axial and radial forces acting on the impeller and then discuss various means of balancing it.

4.2 Axial Thrust

Axial thrust represents the axial component of the unbalanced force acting on the pump impeller. This force arises mainly from the non-uniformity of the pressure distributions in the axial direction. The pressure distributions at the front and back shrouds of the centrifugal pump impeller are shown in Figure 4.1. The figure shows lower pressure (suction pressure, p_s) acting at the impeller inlet eye while the corresponding area at the back shroud is acted upon by a

Pumping Machinery Theory and Practice, First Edition. Hassan M. Badr and Wael H. Ahmed.
© 2015 John Wiley & Sons, Ltd. Published 2015 by John Wiley & Sons, Ltd.

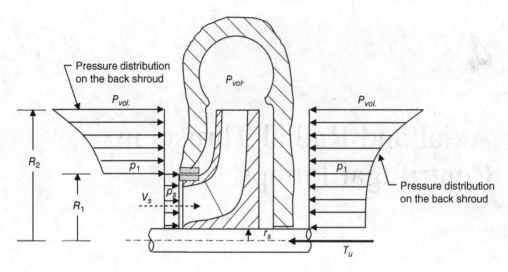

Figure 4.1 The pressure distribution on the impeller back and front shrouds

higher pressure, since it is open to the pressure prevailing in the volute casing ($p_{vol.}$). An additional axial force results also from the change of fluid linear momentum as it flows through the impeller.

4.2.1 Calculation of the Unbalanced Axial Thrust

In order to obtain an expression for the unbalanced axial thrust, let us introduce the following variables:

p_s is the pressure at the suction nozzle
p_1 is the pressure at the wearing rings
$p_{vol.}$ is the pressure in the volute casing
r_s is the shaft radius
R_1 is the radius at the wearing rings
R_2 is the outer radius of the impeller
T_u is the unbalanced axial thrust
ω_s is the shaft angular velocity

The pressure distributions at the back and front shrouds shown in Figure 4.1 are not uniform because of rotation of the fluid in between the impeller and the casing. This fluid rotation prevails due to the shear force arising from the impeller rotation. The resulting pressure distribution is approximately the same as that of a forced vortex (solid body rotation of a fluid). However, the angular velocity will be less than that of the shaft, ω_s. Note that the layer of the fluid in contact with the impeller shroud will rotate with the same impeller angular velocity, ω_s, while the fluid layer in contact with the casing will be at rest. This is mainly due to the no-slip condition at both surfaces. Based on experimental measurements, it was found that using an angular velocity $\omega = \omega_s/2$ in the forced vortex formula will result in a pressure distribution very close to the measured one.

Accordingly, the pressure distributions at the back and front shrouds can be approximated by

$$p_r - p_1 \simeq \frac{\gamma \omega^2}{2g} \left[r^2 - R_1^2 \right] \tag{4.1}$$

where p_r is the pressure at any radius r.

The pressure forces acting on the back and front shrouds at the outer part of the impeller (the annular area between radii R_1 and R_2) are balanced against each other as shown in Figure 4.1. The remaining unbalanced pressure forces are the pressure force acting on the back shroud (between radii r_s and R_1) and the force due to suction pressure at the impeller inlet eye. Let the pressure force acting on the back shroud between r_s and R_1 be T_b, then we can write

$$T_b = \int_A p \, dA = \int_{r_s}^{R_1} \left[p_1 + \frac{\gamma \omega^2}{2g} \left(r^2 - R_1^2 \right) \right] 2\pi r \, dr \tag{4.2}$$

The above integral results in the following expression for T_b

$$T_b = (A_1 - A_s) \left(p_1 - \frac{\gamma}{8} \frac{u_1^2 - u_s^2}{2g} \right) \tag{4.3}$$

where $u_1 = \omega_s R_1$, $u_s = \omega_s r_s$, $A_1 = \pi R_1^2$ and $A_s = \pi r_s^2$.

The total unbalanced force, T_u, resulting from the effects of the change of linear momentum through the impeller as well as the effect of the unbalanced pressure force and can be expressed as

$$T_u = T_b - p_s (A_1 - A_s) - \rho Q V_s \tag{4.4}$$

where V_s is the fluid velocity at the inlet of the impeller suction eye. The last term on the right-hand side of Eq. (4.4) is the momentum term and has a small contribution to the axial thrust.

4.3 Methods of Balancing the Axial Thrust

To completely balance the axial thrust at all operating conditions, a thrust bearing must be used. These bearings are normally used in large pumps and they result in an increase in the cost and size of the pump, and increased maintenance cost. However, in the absence of a thrust bearing, one of the following methods can be used to balance or at least reduce the axial thrust.

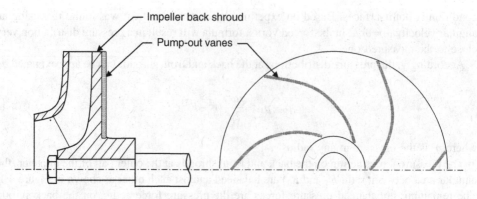

Figure 4.2   A centrifugal pump impeller equipped with pump-out vanes

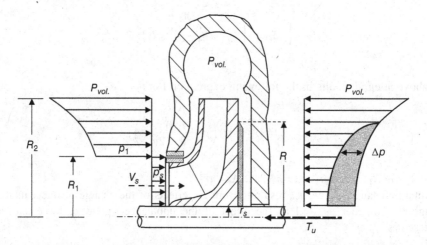

Figure 4.3   The pressure distribution on the impeller back shroud after using the pump-out vanes

4.3.1 Balancing Axial Thrust Using Pump-out Vanes

The *radial ribs* or *pump-out vanes* are simply straight or curved vanes of small width fixed on the back shroud of the impeller (Figure 4.2). The main function of these vanes is to force the fluid in the clearance space between the back shroud and casing to rotate with the same angular velocity of the impeller, ω_s.

The use of pump-out vanes results in a pressure distribution on the back shroud (Figure 4.3) different from the original pressure distribution shown in Figure 4.1. The pressure reduction (Δp) is mainly due to the increase of speed of rotation of the fluid entrained in the clearance space between the impeller and casing (from $0.5\,\omega_s$ to ω_s). The main outcome of this process is a reduction in the axial thrust. The pump-out vanes should be designed in such a way that the reduction in the axial thrust is equal to the unbalanced force (T_u) calculated using Eq. (4.4). The final outcome is a complete balance of the axial thrust.

Let the reduction in the axial thrust be T_R, then

$$T_R = \int_A \Delta p \, dA$$

where Δp is the pressure reduction which can be expressed in the form

$$\Delta p = \left[p_{vol.} - \frac{\gamma(\omega_s/2)^2}{2g} (R_2^2 - r^2) \right] - \left[p_R - \frac{\gamma \omega_s^2}{2g} (R^2 - r^2) \right]$$

and $p_R = p_{vol.} - \dfrac{\gamma(\omega_s/2)^2}{2g} (R_2^2 - R^2)$

where R is the outer radius of the pump-out vanes. The above integral results in the following equation for T_R:

$$T_R = \frac{3}{8} \gamma (A_R - A_s) \left(\frac{u_R^2 - u_s^2}{2g} \right) \tag{4.5}$$

where $u_1 = \omega_s R_1$, $u_s = \omega_s r_s$, $A_R = \pi R^2$.

Now, the axial thrust becomes completely balanced when

$$T_u = T_R \tag{4.6}$$

Equations (4.3)–(4.6) provide enough basis for the determination of the radius of the pump-out vanes that are required to balance the axial thrust. Experimental measurements and theoretical analysis for studying the influence of the flow rate, axial displacement of the impeller, and annular seal clearances on the axial thrust were carried out by Iino et al. [1]. They also considered the effect of leakage flow on the pressure distribution in the space between the impeller and the casing. A detailed investigation of the flow field in the clearance space between the impeller back shroud (including the pump-out vanes) and the casing was carried out by Hong et al. [2]. The effect of the geometrical parameters of the pump-out vanes (including its thickness) on the reduction of the axial thrust and on the pump efficiency was studied for the purpose of obtaining the optimum design of these vanes. It is important here to mention that the pump-out vanes are always designed to balance the axial thrust when the pump operates at its design point. Pump operation away from that point will create a certain amount of unbalanced thrust that can be calculated as demonstrated in the following example.

Example 4.1
Pump-out vanes are usually designed to completely balance the axial thrust when the pump runs at its normal capacity. However, when the pump operates away from its normal capacity the pump-out vanes will only partially balance the axial thrust. The remaining thrust is balanced by the thrust bearing. Determine the axial thrust on the bearing for the pump shown in Figure 4.4.

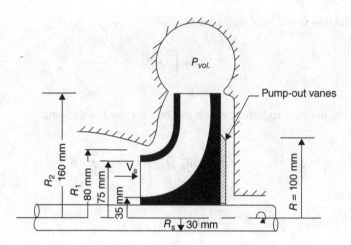

Figure 4.4 Diagram for example 4.1

Data:
  Speed of rotation, $N = 1500$ rpm, flow rate, $Q = 0.12$ m^3/s,
  Suction nozzle pressure, $p_s = 50$ kPa, Volute casing pressure, $p_{vol.} = 350$ kPa.

Preliminary calculations

$$\omega = 2\pi N/60 = 157\, \text{rad/s}, \qquad u_R = \omega R = 15.7\, \text{m/s},$$

$$u_1 = \omega R_1 = 12.6\, \text{m/s}, \qquad V_s = \frac{Q}{A_{inlet}} = \frac{0.12}{\pi\left(0.075^2 - 0.035^2\right)} = 8.68\, \text{m/s},$$

$$u_s = \omega R_s = 4.71\, \text{m/s}, \qquad A_1 - A_s = \pi\left(0.08^2 - 0.03^2\right) = 0.0173\, \text{m}^2$$

Solution
The pressure at the wearing rings (p_1) can be obtained from

$$p_1 = p_{vol.} - \frac{\gamma(\omega_s/2)^2}{2g}\left(R_2^2 - R_1^2\right) = 350 - \frac{9.81(157/2)^2}{2 \times 9.81}\left(0.16^2 - 0.08^2\right) = 291\, \text{kPa}$$

The unbalanced force due to the difference in pressure distributions between the back and front shrouds, T_b, is obtained from

$$\text{Equation (4.3)} \Rightarrow T_b = (A_1 - A_s)\left(p_1 - \frac{\gamma}{8}\frac{u_1^2 - u_s^2}{2g}\right)$$

Substitute to obtain $T_b = 0.0173\left(291 - \frac{9.81}{8}\frac{12.6^2 - 4.71^2}{2 \times 9.81}\right) = \underline{4.89\, \text{kN}}$

The reduction in axial thrust due to the use of pump-out vanes (T_R) is obtained from

$$\text{Equation }(4.5) \Rightarrow T_R = \frac{3}{8}\gamma(A_R-A_s)\left(\frac{u_R^2-u_s^2}{2g}\right)$$

Substitute to obtain $T_R = \frac{3}{8}\times 9.81\left[\pi\left(0.1^2-0.03^2\right)\right]\left(\frac{15.7^2-4.71^2}{2\times 9.81}\right) = \underline{1.2\,\text{kN}}$

To determine the unbalanced axial thrust (T_u), we can use Eq. (4.4) to write

$$T_u = T_b - p_s(A_1-A_s) - \rho Q V_s - T_R$$

Substitute to obtain

$$T_u = 4.89 - 50(0.0173) - \frac{10^3 \times 0.12 \times 8.68}{10^3} - 1.2 = \underline{1.78\,\text{kN}}$$

4.3.2   Balancing Axial Thrust Using Balancing Chambers

A balancing chamber is a chamber located at the back of the impeller which is subjected to the lowest pressure in the system (by connecting it to the pump suction nozzle). If this chamber is properly sized, an axial force will be developed for balancing the axial thrust. Various designs have been used; here are some examples.

a. *Balancing chamber with balancing holes*
   Figure 4.5 shows a balancing chamber located at the back of the impeller which is connected by balancing holes to the impeller suction eye. The low pressure created in the chamber

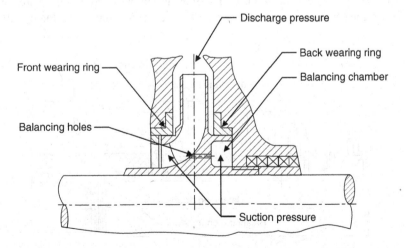

Figure 4.5   A pump impeller equipped with balancing chamber and balancing holes [3]

reduces the unbalanced axial thrust. Proper design of the chamber may lead to a complete balance of the axial thrust. The disadvantages of this design are:

1. The cost of impeller is higher because of higher cost of manufacturing.
2. The addition of one more set of wearing rings increases the initial cost of the pump and also the cost of maintenance.
3. The pump overall efficiency decreases because of the additional leakage occurring through the back wearing rings and the balancing holes.

b. *Balancing chamber with balancing drum*

Figure 4.6 shows a typical balancing drum used to balance the axial thrust in a multistage centrifugal pump. The drum has two parts: one stationary (fixed with the casing) and the other mounted on the shaft, as shown in the figure. A balancing chamber is located on the right side of the drum and is subjected to the suction pressure. The left side of the drum is subjected to the high pressure prevailing in the volute of the last stage of the pump. The drum can be sized so that the pressure difference between the two sides creates a pressure force equal and opposite to the axial thrust. The main advantage of the balancing drum is the reduction of the balancing devices/components for multistage pumps to only one device. The disadvantages are almost the same as those mentioned in using impellers with balancing holes. Also, using a balancing drum makes the pump operation sensitive to any vibrations/misalignment. The labyrinth balancing drum shown in Figure 4.7 is designed to minimize the rate of leakage through the clearance space between the stationary and rotating parts of the drum. The addition of two pressure relief chambers helps to increase the resistance to the fluid motion and accordingly reduces the rate of leakage. Using the labyrinth balancing drum tends to increase the initial cost of the pump as well as the cost of maintenance. In addition, the pump becomes more sensitive to both radial and axial misalignment.

c. *Balancing chamber with balancing disk*

Figures 4.8 and 4.9 show two designs of balancing disks used to balance the axial thrust in multistage pumps. In the first design (Figure 4.8), a balancing disk is again subjected on the right side to the lowest pressure in the system (suction pressure) while being subjected on the left side to the pressure in the volute casing of the last pump stage. The large area of the disk creates a large force that may be needed for balancing the axial thrust. Figure 4.9 shows a

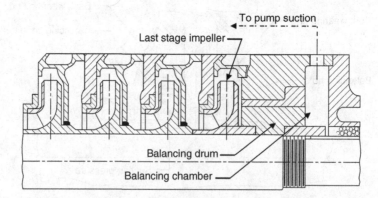

Figure 4.6 A typical balancing drum of a multistage pump (Adapted from Ref. [3])

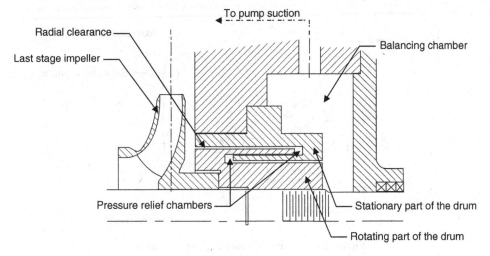

Figure 4.7   Configuration of a labyrinth balancing drum

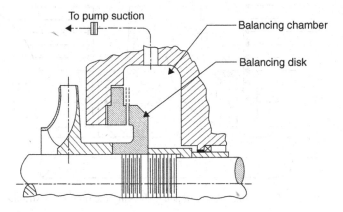

Figure 4.8   A typical balancing disk [3]

different design in which a disk–drum combination is used with a long radial clearance to reduce fluid leakage from the high-pressure side to the low-pressure side. The added relief chamber creates higher resistance to the flow of the leaking fluid.

4.3.3   *Balancing Axial Thrust in Double-Suction and Multistage Pumps*

Figure 4.10 shows the pressure distribution on both sides of a double-suction impeller. Because of symmetry, the pressure forces are balanced, and the linear momentum forces cancel each other. Ideally, the axial thrust in this case is completely balanced, but there may be a small amount of unbalanced thrust when the two inlets are not identical. This small amount can easily be taken by the bearings.

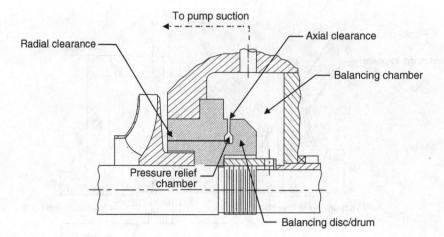

Figure 4.9   Combined balancing disk and drum [3]

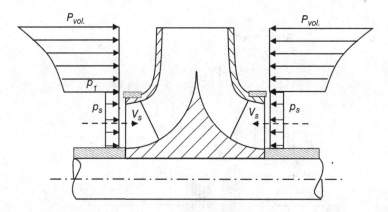

Figure 4.10   Schematic of a double-suction impeller showing the pressure distribution on both sides

In multistage pumps, the impellers are sometimes arranged such that the axial thrust from one cancels the other. Figure 4.11 shows various combinations of impeller arrangements for pumps with an even number of impellers. In designing the crossovers (flow passages from one impeller to next one), it is important to keep the size of the pump as compact as possible and at the same time minimize the internal losses, in addition to other factors such as minimizing the cost of manufacturing and ease of maintenance. It is important to mention that there are more losses associated with every change of flow direction and every change in the cross-sectional area. Such losses will definitely reduce the overall pump efficiency.

In some cases, with an odd number of stages, a combination of single-suction and double-suction impellers are used in order to balance the axial thrust as shown in Figure 4.12.

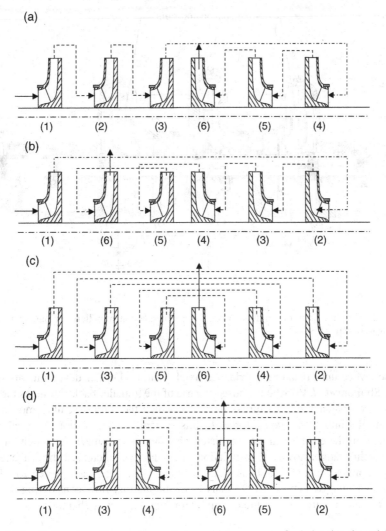

Figure 4.11   Various impeller arrangements in multistage pumps for balancing the axial thrust

4.3.4   Other Balancing Systems

Other balancing systems were developed with the objective of balancing the axial thrust at different operating conditions and at the same time reduce the power loss due to leakage and disc friction. Fairman and Mabe [5] developed a thrust balancing device that integrates a balancing piston with the impeller shrouds. The device was proven to achieve self-compensating thrust balancing at different flow rates and to minimize leakage and disc friction losses. The main idea is to control the pressure at the back shroud using a variable-constriction high-pressure orifice. The pressure control is achieved through a small axial movement of the impeller (resulting from the unbalanced axial force) that controls (regulates) the pressure drop

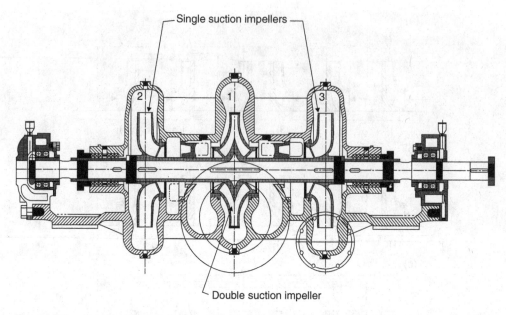

Single suction impellers

Double suction impeller

Figure 4.12 Combinations of single-suction and double-suction impellers for balancing axial thrust (from Stepanoff [4])

across the orifice until reaching a balanced axial position. Further developments were introduced by Shimura *et al.* [6] who proposed the use of the impeller back shroud as the balancing piston of a self-balancing axial thrust system in which the rotor assembly moves axially to compensate the unbalanced axial force. In this system, a combination of a balance piston and grooves on the casing wall of the piston chamber were utilized for extending the axial thrust balancing range; however, installation of the grooves increased the leakage flow rate and the frictional torque. Detailed calculations of the balancing system were carried out using computational fluid dynamics (CFD) analysis of the internal flow considering the boundary layer effects and the grooves on the casing wall of the balancing chamber.

4.4 Radial Thrust

4.4.1 Source of the Radial Thrust

Radial thrust arises from the non-uniformity of the pressure distribution around the periphery of the impeller. The volute collector is normally designed to keep a constant average velocity around the volute, which results in equal pressures around the pump casing when operating at the design point. A uniform pressure distribution in the casing results in zero radial thrust on the impeller. Figure 4.13 shows the effect of flow rate on the pressure at different points around the circumference of the impeller when using a standard volute. It is clear from the figure that the pressure distribution is uniform when the pump operates at its rated capacity. However, when the pump flow rate is much less than, or exceeds, the rated capacity, radial thrust develops as a result of the non-uniform pressure distribution in the volute collector.

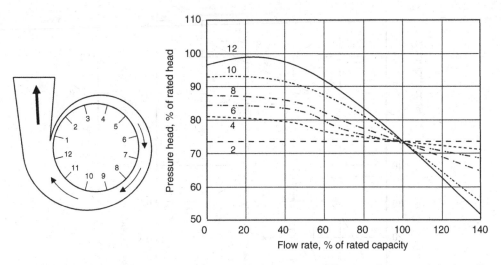

Figure 4.13 Approximate pressure variation with flow rate at different points around the impeller (adapted from Stepanoff [4])

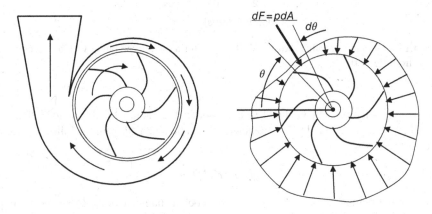

Figure 4.14 Schematic of pressure variation around the impeller periphery (from Stepanoff [4])

This unbalanced thrust, if allowed to persist, could result in shaft deflection that may damage the bearing, mechanical seal (or packing rings), or the wearing rings and may also cause severe pump vibration. Any of these problems (or their combined effect) may result in pump failure. Experimental measurements were carried out to obtain an expression for the radial thrust in terms of the impeller main dimensions and the point of pump operation [5–7].

In general, for a given pressure distribution, such as that shown in Figure 4.14, we can write

$$T_{rad} = \int_0^{2\pi} dF = \int_0^{2\pi} \frac{1}{2} p(\theta) b_2 D_2 \, d\theta$$

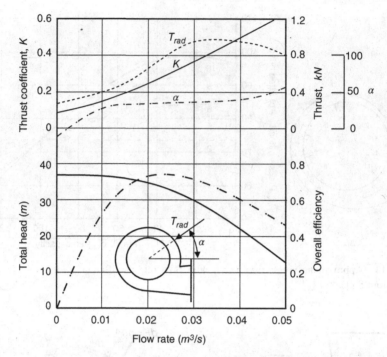

Figure 4.15 Variation of radial thrust with flow rate for a radial-type pump with a concentric casing (adapted from Ref. [4])

The above equation can be integrated to obtain the radial thrust provided that $p(\theta)$ is known. In practice, the radial thrust (in newtons) can be obtained from the following empirical formula

$$T_{rad} = KHD_2b_2 \tag{4.7}$$

where H is the pump head (m), D_2 and is the impeller diameter (cm), b_2 and is the impeller width including the back and front shrouds (cm), and K is a thrust coefficient that depends on the departure from the rated capacity, and is given by

$$K = C\left[1 - (Q/Q_r)^2\right] \tag{4.8}$$

where Q_r is the pump rated capacity and C is a numerical constant that depends on the pump geometry/specific speed [8–9].

The variation of the radial thrust with the pump flow rate depends on the shape of the pump casing. Figure 4.15 shows the variation of radial thrust (magnitude and direction) with the flow rate for a circular concentric casing. Pumps equipped with diffuser vanes surrounding or downstream of the impeller (as in vertical-axis mixed-flow turbine pumps) have uniform pressure distribution around the impeller for a wide range of flow rate. In such pumps, the radial thrust is insignificant thus resulting in lower vibration levels and wider operating range.

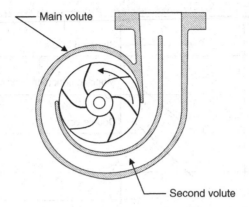

Figure 4.16   A radial-type pump with double-volute casing

4.4.2   Methods for Balancing Radial Thrust

4.4.2.1   Use of Double-Volute Casing

The use of double-volute casing in a radial-type centrifugal pump (shown in Figure 4.16) provides a method for substantial reduction of the radial thrust over the entire range of flow rates. This is achieved by splitting the lower part of the volute into two compartments, one in the immediate neighborhood of the impeller and the other on the outside. The two flow passages at the top and bottom parts of the impeller are asymmetric, resulting in approximately equal and opposite elementary radial forces balancing each other. This gives the double-volute pump an advantage of operating over a wide range of flow rates with negligible radial force, resulting in less shaft deflection and less vibration. On the other hand, the double-volute casing is difficult to manufacture and it results in slightly lower pump efficiency (due to additional hydraulic losses).

The effect of casing design on the resulting radial thrust is shown in Figure 4.17 for the cases of a standard volute, a concentric casing, and a double-volute casing [10]. It is clear that a double-volute casing results in a very small thrust over the entire range of Q, while the standard volute casing is the worst at low flow rates.

4.4.2.2   Use of Diffuser Vanes

Another method for balancing the radial thrust is the use of diffuser vanes in the pump volute casing as shown in Figure 4.18. The diffuser vanes are located in an annular space (ring) surrounding the impeller. The vanes may have an airfoil shape or simply be curved plates forming a number of channels with area increasing in the direction of flow creating the diffuser effect. Also, the diffuser streamlines the flow entering the volute casing and thus reducing the hydraulic losses. A photograph of typical diffuser vanes is shown in Figure 4.19. The axially symmetric geometry surrounding the impeller creates a more uniform pressure distribution, resulting in less radial thrust and lower vibration levels. The diffuser vanes improve the pump hydraulic efficiency when operating at the design point. However, operation away from this point creates more hydraulic losses than that in the case of a vaneless casing. This is mainly because the flow direction at the impeller exit will not be aligned with the diffuser vane angle, resulting in flow separation and higher level of turbulence.

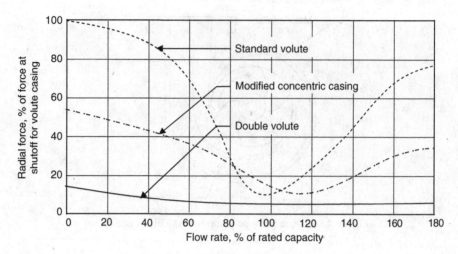

Figure 4.17 Typical variation of radial thrust with flow rate for three casing designs (adapted from Ref. [7])

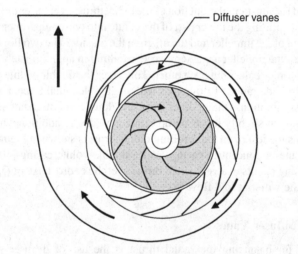

Figure 4.18 A volute casing with diffuser vanes for balancing the radial thrust

4.4.2.3   Use of Turbine Pump Casing

A third method for balancing the radial thrust is the use of a turbine pump casing. This type of casing is commonly used in multistage vertical axis submersible pumps that are normally equipped with mixed-flow impellers, and the casing contains diffuser vanes as shown in Figure 4.20. The vanes not only function as a diffuser but also remove the swirl component of flow velocity (at impeller exit) and direct the fluid to enter the next stage axially (with no prewhirl). Because of axi-symmetry, the radial thrust vanishes at all flow rates.

The diffuser vanes are designed so that the direction and magnitude of flow velocity at the impeller exit (characterized by high swirl velocity) changes smoothly until reaching the

Figure 4.19 A photograph showing typical pump diffuser vanes

(a)   (b)   (c)

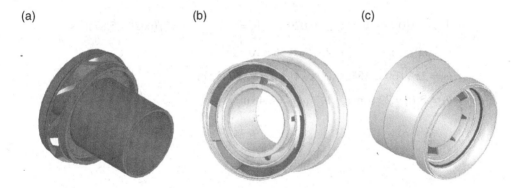

Figure 4.20 A three-dimensional view of impeller and casing of one stage of an electric submersible pump: (a) impeller, (b) front view of the casing, (c) side view of the casing

suction eye of the next impeller. However, pump operation away from the design point (design speed and/or flow rate) will create flow separation (shock losses) and a high level of turbulence in the diffuser channel, causing a considerable increase in hydraulic losses.

4.4.3 Problems Arising from the Unbalanced Radial Thrust

Several problems can occur as a result of the unbalanced radial thrust. The first one is the shaft deflection, which depends on the magnitude of the unbalanced thrust. This deflection may damage mechanical parts having small clearances such as the wearing rings and the mechanical seal. Shaft deflection can also create excessive friction in the packing rings (when using a stuffing box) resulting in high temperature. Excessive radial loads on the bearings produce high temperature, which shortens the bearing life. Mechanical vibrations resulting from shaft deflection create additional dynamic load on the shaft and other pump components and may lead to shaft failure. The location of shaft failure depends on the pump geometry. For a through shaft, supported by bearings on both sides of the impeller, the shaft failure usually occurs at the middle section of the shaft span. On the other hand, the failure occurs close to the bearing (or shaft sleeve) for an overhung impeller.

Example 4.2
Figure 4.21 shows the basic dimensions of a centrifugal pump impeller that is designed to operate at a speed of 1800 rpm and deliver water at a rate of 0.17 m³/s against a total head of 75 m. Based on the pumping system design, the predicted pressures at the impeller inlet and exit sections are 50 and 520 kPa respectively.

a. Determine the magnitude and direction of the total unbalanced axial thrust for this impeller.
b. If it is required to balance the axial thrust using pump-out vanes, determine the radius of these vanes.

Solution

$$A_{suc.} = \pi\left(0.1^2 - 0.028^2\right) = 0.029\,\text{m}^2, \quad V_s = Q/A_{suc.} = 0.17/0.029 = 5.87\,\text{m/s},$$

$$r_s = 0.045\,\text{m}, \quad A_s = \pi(0.045)^2 = 6.36 \times 10^{-3}\,\text{m}^2,$$

$$\omega_s = 2\pi N/60 = 188.5\,\text{rad/s}, \quad A_1 = \pi R_1^2 = 0.0452\,\text{m}^2,$$

$$u_1 = \omega_s R_1 = 22.6\,\text{m/s}, \quad u_s = \omega_s r_s = 8.48\,\text{m/s}.$$

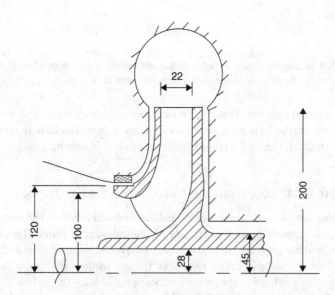

Figure 4.21   Diagram for Example 4.2 (all dimensions in mm)

a.
$$p_1 = p_{vol} - \frac{\gamma(\omega_s/2)^2}{2g}\left[R_2^2 - R_1^2\right]$$

$$= 520 - \frac{9.81(188.5/2)^2}{2 \times 9.81}\left[0.2^2 - 0.12^2\right] = 406\,\text{kPa}$$

$$T_b = (A_1 - A_s)\left[p_1 - \frac{\gamma}{8}\frac{u_1^2 - u_s^2}{2g}\right]$$

$$= (0.0452 - 0.00636)\left[406 - \frac{9.81}{8}\frac{22.6^2 - 8.48^2}{2}\right] = 14.7\,\text{kN}$$

$$T_u = T_b - p_s(A_1 - A_s) - \rho Q V_s$$

$$= 14.7 - 50\left[0.0452 - \pi(0.028)^2\right] - \frac{10^3 \times 0.17 \times 5.87}{10^3} = 11.56\,\text{kN}$$

b. For complete balance of axial thrust, we must have $T_u = T_R$

$$T_u = \frac{3}{8}\gamma(A_R - A_s)\left[\frac{u_R^2 - u_s^2}{2g}\right] = \frac{3}{16}\frac{\pi\gamma\omega_s^2}{g}\left[R^2 - r_s^2\right]^2$$

$$11.56 = \frac{3}{16}\frac{\pi 9.81(188.5)^2}{9.81}\left[R^2 - 0.045^2\right]^2$$

$$\Rightarrow R = 0.16\,\text{m} = 160\,\text{mm}$$

Example 4.3

Figure 4.22 shows the basic dimensions of a centrifugal pump. Under actual working conditions, the variations of the suction and delivery pressures p_s and p_d with the flow rate Q are given when the pump runs at a constant speed of 1850 rpm. It is required to replace the impeller with a new one having pump-out vanes, in order to reduce the axial thrust. You are required to determine the radius of the pump-out vanes in order to completely balance the axial thrust when operating at the best efficiency point. Assume that the volute casing pressure (p_v) is equal to the discharge pressure (p_d).

Data:

At design point, $Q = 1.2\,\text{m}^3/\text{s}$, $p_d = p_{vol.} = 3\,\text{MPa}$, $p_s = 95\,\text{kPa}$

Pump data: $R_2 = 0.38\,\text{m}$, $R_1 = 0.2\,\text{m}$, $R_s = 0.05\,\text{m}$,

Preliminary calculations

$$A_{suc.} = \pi(0.15^2) = 0.0707\,\text{m}^2, \quad V_s = Q/A_{suc.} = 1.2/0.0707 = 17\,\text{m/s},$$

$$A_s = \pi(0.05)^2 = 7.85 \times 10^{-3}\,\text{m}^2, \quad \omega_s = 2\pi N/60 = 193.7\,\text{rad/s}, \quad A_1 = \pi R_1^2 = 0.126\,\text{m}^2,$$

$$u_1 = \omega_s R_1 = 38.7\,\text{m/s}, \quad u_s = \omega_s r_s = 9.7\,\text{m/s}.$$

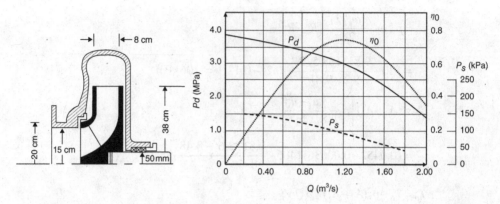

Figure 4.22 Diagram for Example 4.4

Solution

$$p_1 = p_{vol} - \frac{\gamma (\omega_s/2)^2}{2g} \left[R_2^2 - R_1^2 \right]$$

$$= 3000 - \frac{9.81 (193.7/2)^2}{2 \times 9.81} \left[0.38^2 - 0.2^2 \right] = 2510 \text{kPa}$$

$$T_b = (A_1 - A_s) \left[p_1 - \frac{\gamma}{8} \frac{u_1^2 - u_s^2}{2g} \right]$$

$$= (0.126 - 0.00785) \left[2510 - \frac{9.81}{8} \frac{38.7^2 - 9.7^2}{2 \times 9.81} \right] = \underline{286 \, \text{kN}}$$

$$T_u = T_b - p_s (A_1 - A_s) - \rho Q V_s$$

$$= 286 - 95 \left[0.126 - 0.00785 \right] - \frac{10^3 \times 1.2 \times 17}{10^3} = \underline{254 \, \text{kN}}$$

For complete balance of axial thrust, we must have $T_u = T_R$

$$T_u = \frac{3}{8} \gamma (A_R - A_s) \left[\frac{u_R^2 - u_s^2}{2g} \right] = \frac{3}{16} \frac{\pi \gamma \omega_s^2}{g} \left[R^2 - r_s^2 \right]^2$$

$$254 = \frac{3}{16} \frac{\pi 9.81 (193.7)^2}{9.81} \left[R^2 - 0.05^2 \right]^2$$

$$\Rightarrow R = 0.33 \, m = 330 \, \text{mm}$$

References

1. Iino, T., Sato, H., and Miyashiro, H. (1980) Hydraulic axial thrust in multistage centrifugal pumps. *ASME J Fluid Eng*, 102 (1), 64–69.
2. Hong, F., Yuan, J., Heng, Y., Fu, Y., Zhou, B., and Zong, W. (2013) Numerical optimal design of impeller back pump-out vanes on axial thrust in centrifugal pumps. Paper No. FEDSM2013-16598. ASME 2013 Fluids Engineering Division Summer Meeting, Incline Village, NV.
3. Karassik, I.J. (1986) *Pump Handbook*, 2nd edn, McGraw-Hill, New York.
4. Stepanoff, A.J. (1957) *Centrifugal and Axial Flow Pumps*, 2nd edn, John Wiley & Sons, Inc., New York.
5. Fairman, K. and Mabe, W. (1991) An integral balance piston for centrifugal pump impellers. Proceedings of the Eighth International Pump Users Symposium. March 5–7, 1991, Turbomachinery Laboratory, Department of Mechanical Engineering, Texas A&M University, Texas, USA.
6. Shimura, T., Matsui, J., Kawasaki, S., Uchiumi, M., and Kimura, T. (2012) Internal flow and axial thrust balancing of a rocket pump. *ASME J Fluid Eng*, 134 (4), 041103.
7. Agostinelli, A., Mockridge, C.R., and Nobles, D. (1960) An experimental investigation of radial thrust in centrifugal pumps. *ASME J Eng Power*, 82 (2), 120–126.
8. Biheller, H.J. (1965) Radial force on the impeller of centrifugal pumps with volute, semi-volute, and fully concentric casings. *ASME J Eng Power*, 87 (3), 319–322.
9. Hergt, P. and Krieger, P. (1969) Radial forces in centrifugal pumps with guide vanes. *Proc Inst Mech Eng Conf Proc*, 184 (14), 101–107.
10. Sulzer Pumps. (2010) *Centrifugal Pump Handbook*. 3rd edn. Elsevier, Amsterdam.

Problems

4.1 Figure 4.23 shows the basic dimensions of a centrifugal pump impeller that is equipped with radial ribs for balancing the axial thrust. The pump normal operating conditions are as follows:

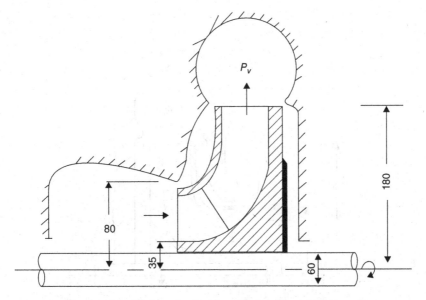

Figure 4.23 Diagram for Problem 4.1 (all dimensions in mm)

Pump speed = 1450 rpm, pump flow rate = 0.1 m³/s, pump total head = 40 m, pressure at the suction nozzle = 60 kPa abs, pressure in the volute casing = 380 kPa abs.

You are required to determine:

a. total unbalanced axial thrust in the absence of the radial ribs
b. radius of the radial ribs needed to completely balance the axial thrust.

4.2 A centrifugal pump impeller is equipped with pump-out vanes for reducing the axial thrust. The impeller is mounted on a through shaft that is supported at both sides by ball bearings. The following data provides the pump operating conditions as well as the main dimensions of the impeller:

pump speed = 1800 rpm	impeller outer diameter = 0.4 m
pump flow rate = 0.1 m³/s	radius of pump-out vanes = 12 cm
suction nozzle pressure = 60 kPa abs	impeller inlet eye diameter = 0.18 m
volute casing pressure = 320 kPa abs	driving shaft diameter = 50 mm

a. Determine the pressure in the clearance space between the back shroud and the casing at a radius $r = 80$ mm
b. Calculate the reduction in axial thrust due to the use of pump-out vanes.
c. Determine the magnitude and direction of the remaining unbalanced axial thrust.

4.3 The axial thrust of the single-stage single-suction centrifugal pump shown in Figure 4.24 is to be balanced using a thrust bearing. At normal running condition, the pump delivers a flow rate of 0.12 m³/s, while driven at a speed of 1500 rpm. The suction pressure is 50 kPa abs. while the pressure in the volute casing is 540 kPa abs. Knowing that the driving shaft diameter is 60 mm, the impeller outer diameter is 400 mm, and the diameter at the wearing

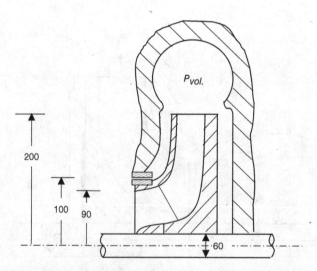

Figure 4.24 Diagram for Problem 4.3 (all dimensions in mm)

rings is 200 mm, determine the unbalanced axial thrust, taking into consideration all the forces acting on the impeller.

4.4 The multistage centrifugal pump shown in Figure 4.25 has four identical impellers, each one exactly the same as that given in Problem 4.3. The suction pressure is 50 kPa abs. and the pressure in the volute casing of the last stage is 2.0 MPa abs. The axial thrust is to be balanced by using a simple balancing drum, as shown in the figure. The fluid pumped is water and the pump is driven at 1500 rpm. Determine the outer diameter of the drum if the axial thrust is to be completely balanced.

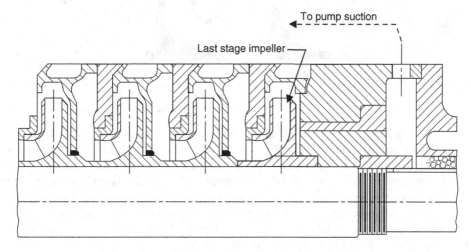

Figure 4.25   A multistage pump with a balancing drum

Hint: Consider the unbalanced *thrust/stage* to be the same as the axial thrust in Problem 4.3.

4.5 Consider the pump in Example 4.3 that utilizes the impeller with pump-out vanes designed to completely balance the axial thrust at the best efficiency point as calculated therein. You are required to determine the residual axial thrust when the pump operates away from its design point (say at $Q = 0.4$ m^3/s, 0.8 m^3/s, and 1.6 m^3/s).

4.6 The two-stage centrifugal pump shown in Figure 4.26 has two identical impellers and is designed to pump water at a rate of 0.1 m^3/s when driven at 1500 rpm. The suction pressure is atmospheric and the pressures in the volute casing of the first and second stages are 0.8 and 1.6 MPa gage respectively. The axial thrust is to be balanced by using a simple balancing disk as shown in Figure 4.26.

a. Calculate the axial thrust resulting from each impeller taking all forces into consideration.

b. Determine the radius R_d of the balancing disk.

Note: State clearly any assumption that you may find necessary.

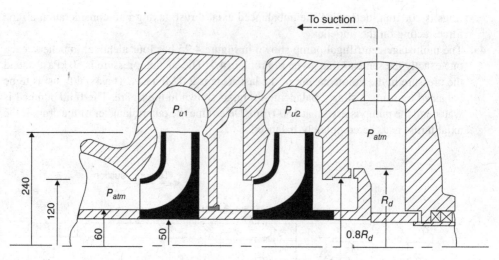

Figure 4.26 Diagram for Problem 4.6 (all dimensions in mm)

4.7 The single-stage centrifugal pump shown in Figure 4.27 is to be driven at a speed of 1500 rpm to pump a fluid ($\gamma = 8.5$ kN/m^3) at a rate of 0.28 m^3/s in a given system. At this flow rate, the pump develops a total head of 70 m and the pressures in the suction nozzle and volute casing are 80 and 620 kPa respectively. The impeller is equipped with pump-out vanes on the back shroud, as shown in the figure. Carry out the necessary calculations for all axial forces acting on the impeller in order to determine the magnitude and direction of the unbalanced axial thrust at this operating condition. *Main dimensions*: $D_S = 10$ cm, $D_E = 21$ cm, $D_1 = 25$ cm, $D_R = 42$ cm, $D_2 = 45$ cm.

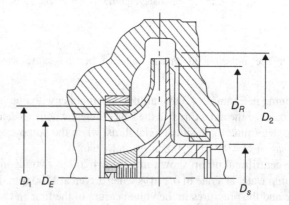

Figure 4.27 Diagram for Problem 4.7

4.8 The unbalanced force due to the pressure distribution on the back and front shrouds (T_b) and the reduction in axial thrust (T_R) due to the use of pump-out vanes are given by the following equations:

$$T_b = (A_1 - A_s)\left(p_1 - \frac{\gamma}{8}\frac{u_1^2 - u_s^2}{2g}\right)$$

$$T_R = \frac{3}{8}\gamma(A_R - A_s)\left(\frac{u_R^2 - u_s^2}{2g}\right)$$

Verify these two equations by using a rigorous mathematical proof.

4.9　A radial-type centrifugal pump is equipped with an impeller that is mounted on a through shaft and is supported at both sides by ball bearings. The pump is to be driven at 1500 rpm and will deliver water at a rate of 0.25 m^3/s. This operating condition is slightly away from the design point. The main dimensions of the impeller and the driving shaft are as follows:

impeller outer diameter	=45 cm
impeller inlet eye diameter	=25 cm
Impeller width at inlet	=5 cm
Impeller width at exit	=2.5 cm
driving shaft diameter	=8 cm

The suction pressure at the normal operating condition is 40 kPa abs. and the corresponding pressure in the volute casing is 580 kPa abs. Knowing that the impeller is equipped with pump-out vanes having an outer radius, $R = 18$ cm, determine the magnitude of the remaining axial thrust transmitted to the ball bearings.

4.10　Figure 4.28 shows the basic dimensions of the first stage of a multistage centrifugal pump. Because of a problem in the sealing system between the first and second stages, fluid at high pressure leaks from the second stage back to the first stage, as shown in the figure. This leakage caused the fluid pressure at the back shroud to have approximately a uniform pressure distribution with $p = p_{vol}$. The pump operating data is given below.

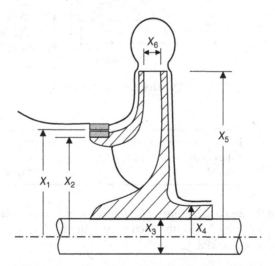

Dimensions: $X_1 = 60$ mm, $X_2 = 50$ mm, $X_3 = 14$ mm, $X_4 = 23$ mm, $X_5 = 100$ mm, $X_6 = 11$ mm

Figure 4.28　Diagram for Problem 4.10

pump speed, N	=1800 rpm
pressure at the impeller inlet eye, p_s	=50 kPa abs
pressure in the volute casing, $p_{vol.}$	=180 kPa abs
pump flow rate, Q	=0.02 m³/s
pump total head, H	=16 m

You are required to determine the unbalanced axial thrust.

4.11 A centrifugal pump has a double-shrouded impeller mounted on an overhung shaft, as shown in Figure 4.29. The impeller is equipped with pump-out vanes ($R = 97$ mm) for completely balancing the axial thrust when operating at its design point ($N = 2900$ rpm and $Q = 0.06$ m³/s). The main dimensions of the impeller are as follows:

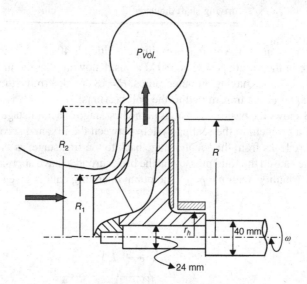

Figure 4.29   Diagram for Problem 4.11

impeller outer radius (R_2)	=105 mm
radius at the wearing ring (R_1)	=60 mm
hub radius (r_h)	=25 mm

Determine the unbalanced axial thrust (T_u) when the pump operates at partial capacity at which the flow rate is 0.04 m³/s, the suction pressure is 60 kPa, and the pressure in the volute casing, $p_{vol.} = 720$ kPa.

5

Common Problems in Centrifugal Pumps

5.1   Introduction

Problems that may be encountered during centrifugal pump operation are numerous and many of them are interrelated. These problems may be attributed to one or more of the following:

1. inadequate pump selection
2. improper pump operation
3. poor design of the pump and/or piping system
4. type of pumped fluid
5. lack of regular maintenance

   The selection of the proper pump for any given pumping system is very important in order to avoid problems that may arise when the pump operates away from its design point. Improper pump operation may also lead to problems such as cavitation, flow discontinuity, pump surge, and many others. The cleanliness of the pumped fluid helps to avoid problems such as erosion by solid particles (as in the case of pumping sea water) that may cause damage to the pump components (impeller, wearing rings, and casing). Other problems may be due to the corrosive nature of the pumped fluid that may cause similar failure of pump components. A third problem related to the type of fluid is the loss of pump efficiency due to the high viscosity of the pumped liquid. Fluid leakage through sealing systems represents another problem, not only because of the loss of fluid leaking outside the pump but also due to the fire and toxicity hazards (depending on the type of fluid). Also, leakage of air through the pump seal into the casing may cause considerable reduction in the pump flow rate, in addition to possible loss of pump prime. The pump operation at low efficiency may arise as a result of erosion/corrosion problems and may also arise from operation far away from the design point. Pump surge is another problem in which the pump operation suffers severe fluctuations. Mechanical vibration is another of

Pumping Machinery Theory and Practice, First Edition. Hassan M. Badr and Wael H. Ahmed.

the serious problems that may even cause complete failure of one or more of the pump components (shaft, bearings, wearing rings, or mechanical seals) in a short time. These vibrations can originate from different sources such as unbalanced rotor(s), bearing failure, clogged impeller, and flow-induced forces. Cavitation is another serious problem, not only because of the accompanying reduction in the pump total head and flow rate but also because of the erosion damage of pump components and the resulting vibration. Cavitation may also lead to loss of pump prime which may cause damage of pump components due to overheating. Operational-type problems may also arise from poor design of the piping system. For example, a suction vortex occurring as a result of poor design of pump intake may cause flow discontinuity. Air locks represent another problem that can cause the pump to lose its prime. It occurs when an air pocket is trapped at a high spot in the suction pipe. Also, poor selection of the location of the suction reservoir may lead to cavitation because of the resulting low available net positive suction head (*NPSHA*). In this chapter, some of the common operational-type problems will be discussed in some detail.

5.2 Cavitation

Cavitation is a serious operational-type problem that arises from the low pressure in the pump suction side. This low suction pressure can cause the pressure inside the pump to fall below the fluid vapor pressure, thus creating a very large number of vapor cavities (bubbles). Once a cavity is formed, it grows in the direction of flow so long as the static pressure is less than the vapor pressure (i.e. $p < p_v$). As the fluid moves to high-pressure zones, some of these cavities collapse near the impeller inner surfaces (and/or vane surfaces). The collapse of these cavities creates very high localized pressure that may reach as high as 10^4 atm. Cavitation causes material erosion in the pump components (mainly the impeller) and can result in complete failure of the pump. Figure 5.1 shows cavitation erosion in a mixed-flow impeller.

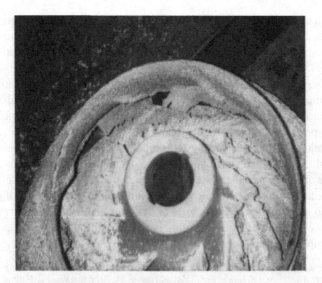

Figure 5.1   Cavitation erosion in a mixed flow impeller

5.2.1   Origin of Cavitation

Cavitation starts when the fluid static pressure falls below its vapor pressure. In order to study cavitation in pumps, the pressure variation starting from the suction vessel up to the minimum pressure point (inside the pump) must be investigated. Consider the simple system shown in Figure 5.2.

Apply the energy equation between the fluid free surface (a) and section s-s (inlet of the suction nozzle) to obtain,

$$\frac{p_s}{\gamma} = \frac{p_a}{\gamma} + (z_a - z_s) - \sum h_{Ls} - \frac{V_s^2}{2g} \tag{5.1}$$

where p_s is the pressure at section s-s, p_a is the pressure at the fluid free surface, and $\sum h_{Ls}$ is the total head loss in the pump suction side starting from the suction tank up to the pump suction nozzle (major and minor losses).

Similarly, the energy equation can be used to show that the pressure at the entrance of the impeller vane p_1 can be written as

$$\frac{p_1}{\gamma} = \frac{p_a}{\gamma} + (z_a - z_s) - \sum h_{Ls} - h_{fsn} - \frac{U_1^2}{2g} \tag{5.2}$$

where U_1 is the absolute velocity at vane inlet and the term h_{fsn} represents the additional friction losses occurring in the suction nozzle (between the pump suction flange and the impeller inlet eye).

In order to determine the origin of the cavitation, we have to investigate the pressure variation in the flow passage between the vanes. An accurate determination of the pressure distribution on both sides of the vane is a difficult task, either theoretically or experimentally. Although the front and back sides of the vane have almost the same geometrical shape, the pressure distributions on the two sides are different. Considering an impeller with backward-curved vanes, the velocity distribution on both vane sides will be approximately as shown in Figure 5.3. The high- and low-pressure zones on the impelling and suction sides of the vane are shown in Figure 5.4.

Following the vane inlet, the pressure is further reduced on the back side (suction side) while increasing on the front side (impelling side) as shown in Figure 5.5. The maximum local pressure drop on the suction side is labeled p_{ld}. Now, cavitation first occurs when

$$p_v = p_{min} = p_1 - p_{ld} \tag{5.3}$$

Figure 5.2   The suction side of a simple pumping system

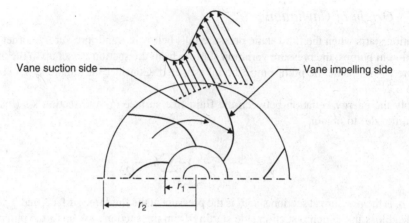

Figure 5.3 Velocity distribution on front (impelling) and back (suction) sides of the impeller vane

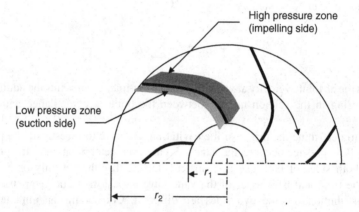

Figure 5.4 Schematic of an impeller showing the high- and low-pressure zones on the impelling and suction sides of the vane

Using Eq. (5.2) for p_1, we can write

$$\frac{p_a}{\gamma} + h_{ss} - h_{Ls} - h_{fsn} - \frac{U_1^2}{2g} - \frac{p_{ld}}{\gamma} = \frac{p_v}{\gamma} \tag{5.4}$$

Note that Eq. (5.4) is applicable only at cavitation inception, since it is deduced based on Eq. (5.3).

The local pressure drop, p_{ld}, depends on the fluid properties, the flow rate, the pump speed, and the impeller geometry (shape, size, and surface roughness). In general, it is difficult to determine p_{ld} at all flow rates either theoretically or experimentally. However, we can argue that the minimum pressure must be equal to the fluid vapor pressure when cavitation starts. This concept can be used to estimate the local pressure drop experimentally only at cavitation inception using Eq. (5.3). In order to prevent cavitation, the minimum pressure must

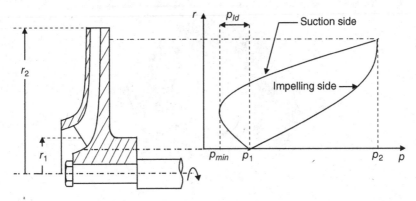

Figure 5.5   Pressure variation along the suction and impelling sides of a backward-curved vane

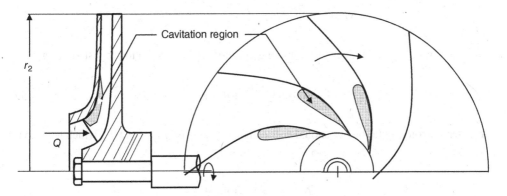

Figure 5.6   Location of the origin of cavitation

be maintained higher than the fluid vapor pressure at the normal operating temperature (i.e. $p_{\min} > p_v$). The minimum pressure point is always inside the impeller and is located close to the inlet of the impeller vanes and adjacent to the front shroud side, as shown in Figure 5.6. Figure 5.7 shows exactly the same location of cavitation erosion (suction side of the impeller vane close to the vane inlet) in a mixed-flow impeller.

5.2.2   Basic System Requirements to Avoid Cavitation

Consider section s-s in Figure 5.2 to represent the pump suction flange (suction nozzle inlet) and apply the energy equation to obtain

$$\frac{p_{sn}}{\gamma} + \frac{V_s^2}{2g} = \frac{p_a}{\gamma} + h_{ss} - \sum h_{Ls}$$

But since $NPSHA = h_{sn} + \frac{V_{sn}^2}{2g} - \frac{p_v}{\gamma}$ we can use the above equations to write

Figure 5.7   A photograph showing cavitation damage in a mixed-flow impeller

$$NPSHA + \frac{p_v}{\gamma} = h_{sn} + \frac{V_{sn}^2}{2g} = \frac{p_a}{\gamma} + h_{ss} - \sum h_{Ls} \tag{5.5}$$

Substituting in Eq. (5.4) gives the following condition when cavitation is just starting,

$$NPSHA = \frac{p_{ld}}{\gamma} + h_{fsn} + \frac{U_1^2}{2g} \tag{5.6}$$

In order to avoid cavitation, we should ensure that $p_{min.} > p_v$ or $(p_1 - p_{ld}) > p_v$, which leads to the condition

$$NPSHA > \frac{p_{ld}}{\gamma} + h_{fsn} + \frac{U_1^2}{2g} \tag{5.7}$$

The term $\left(\frac{p_{ld}}{\gamma} + h_{fsn} + \frac{U_1^2}{2g}\right)$ in the above inequality represents the minimum value of $NPSHA$ to avoid cavitation, and this value is known as the required net positive suction head ($NPSHR$).

$$\text{Therefore, } NPSHR = \frac{p_{ld}}{\gamma} + h_{fsn} + \frac{U_1^2}{2g} \tag{5.8}$$

The above equation specifies the physical significance of $NPSHR$ and removes any ambiguity about the definition of that term. Based on the above analysis, we can conclude that the term $NPSHR$ represents the pressure drop occurring inside the pump due to the friction head loss in the suction nozzle, the local pressure drop inside the impeller, and the change of the velocity head. Based on the above equation, the factors affecting $NPSHR$ are:

1. shape and size of the suction nozzle including surface roughness
2. geometry of the impeller and impeller vanes
3. speed of rotation
4. fluid properties (density and viscosity)
5. flow rate.

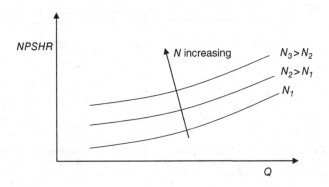

Figure 5.8   Variation of *NPSHR* with Q at different pump speeds

For a better pump performance with respect to cavitation *NPSHR* must have as low a value as possible. The determination of *NPSHR* at different flow rates and different speeds is normally carried out by the pump manufacturer, using water as the fluid medium. Figure 5.8 shows the variation of *NPSHR* with Q at different pump speeds.

The value of net positive suction head (NPSH) is sometimes expressed in terms of a dimensionless quantity called the Thoma's cavitation factor, σ, which is defined as,

$$\sigma = NPSH/H \tag{5.9}$$

When cavitation starts, $NPSH = NPSHR$, and the value of σ is called the critical value σ_c, where

$$\sigma_c = NPSHR/H \tag{5.10}$$

The critical value, σ_c, can be used to determine the minimum fluid level in the suction reservoir required to avoid cavitation. For example, the minimum value of the static suction head, h_{ss}, can be obtained by equating *NPSHA* to *NPSHR*, resulting in

$$NPSHA = NPSHR \quad \Rightarrow \quad \frac{p_a}{\gamma} + [h_{ss}]_{min} - \sum h_{Ls} - \frac{p_v}{\gamma} = \sigma_c H$$

$$\text{Therefore } (h_{ss})_{min.} = \sigma_c H + \sum h_{Ls} + \frac{p_v}{\gamma} - \frac{p_a}{\gamma} \tag{5.11}$$

If $(h_{ss})_{min}$ obtained from the above equation is positive, then the fluid free surface must be higher than the pump level. Equation (5.11) may be used for the selection of a suitable pump location in order to avoid cavitation. The variation of *NPSHR* or σ_c with capacity Q is usually supplied by the pump manufacturer, as shown in Figure 5.9.

Once the details of the piping system are known, we can study the system characteristics in order to determine the variation of *NPSHA* with Q or the variation of σ with Q. By plotting these curves on the same graphs (Figure 5.9), we can determine the range of flow rates for which cavitation is unlikely to occur.

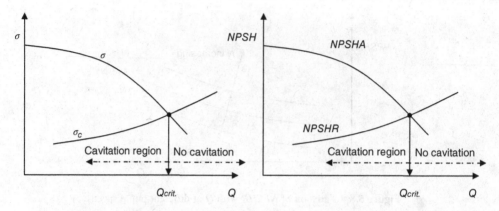

Figure 5.9   Variation of *NPSHR* and σ_c with flow rate *Q* at constant pump speed

The relationship between *NPSHA* and *Q* can be easily deduced from Eq. (5.5) as follows:

$$\text{Equation (5.5)} \Rightarrow NPSHA = \frac{p_a - p_v}{\gamma} + h_{ss} - \sum h_{Ls}$$

Considering that p_a, p_v, and h_{ss} are constants and assuming that the friction coefficient is independent of *Q* (this is reasonably accurate in a wide range of applications in which rough pipe are used and *Re* is high), the above equation can be expressed as

$$NPSHA = C_1 - C_2 Q^2 \tag{5.12}$$

where C_1 and C_2 are easily identifiable constants. A similar equation can be deduced for the relationship between σ and *Q*.

5.2.3   Effect of Cavitation on Pump Performance

Experimental measurements have shown that the pump total head, *H*, and overall efficiency, η_o, drop off suddenly when cavitation occurs in pumps having low specific speed ($N_s < 1500$). The reason for this behavior is that the low-pressure zone on the suction side of the impeller vane allows the cavity to grow rapidly, resulting in flow discontinuity. It was also found that the pump head, *H*, drops off at lower flow rate as the pump speed, *N*, increases, as shown in Figure 5.10a. Figure 5.10b explains this phenomenon since higher pump speed tends to increase the local pressure drop and thus to increase *NPSHR*. Accordingly, the *NPSHR–Q* curve changes with the change of the pump speed while the *NPSHA–Q* curve is completely independent of the pump speed.

Figure 5.11 shows the behavior of *H*, η_o, and *BP* with *NPSHA* for a typical low N_s pump operating at constant speed and constant flow rate. The decrease of *NPSHA* here can be achieved in a laboratory experiment by gradual throttling of the suction valve and opening the delivery valve to maintain a constant flow rate. The process continues until cavitation starts

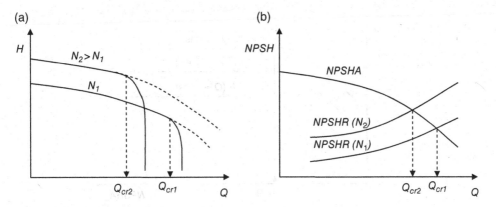

Figure 5.10 (a) Effect of pump speed on the start of cavitation. (b) Effect of pump speed on the *NPSHR–Q* curve

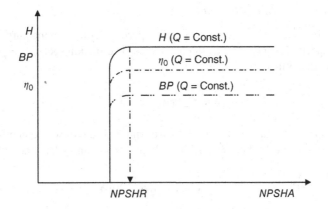

Figure 5.11 Effect of *NPSHA* on H, η_o, and *BP* at constant flow rate for low specific speed pumps

and then it reaches complete flow discontinuity. The suction valve throttling tends to increase $\sum h_{Ls}$ and thus decrease *NPSHA* (see Eq. (5.5)).

The effect of cavitation on the performance of high N_s pumps (such as axial flow pumps) is different. These pumps can tolerate slight cavitation without severe drop in their performance. The drop in H, η_o, and *BP* occurs gradually, as shown in Figure 5.12.

As well as its effect on pump hydraulic performance, cavitation can also cause severe damage to the pump because of the associated material erosion. The implosion of cavitation bubbles at the impeller surface or surfaces of other pump components generates very high localized pressure that may exceed the fatigue strength of the material, causing surface pitting which is known as *cavitation erosion*. The rate of erosion depends on many factors such as material hardness, size of the cavitation bubble zone and the gas content of the pumped fluid. The material damage will be much higher if cavitation erosion is coupled with corrosion effects.

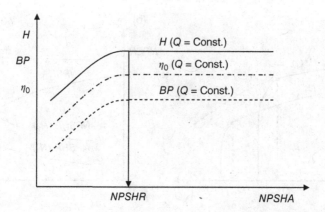

Figure 5.12 Effect of *NPSHA* on *H*, η_o, and *BP* at constant flow rate for high specific speed pumps (axial-flow pumps)

5.2.4 The Suction Specific Speed, S

Similar to specific speed, N_s, the suction specific speed is defined for the purpose of evaluating pumps with respect to their cavitation characteristics. The suction specific speed, S, is a dimensionless parameter defined as,

$$s = n\sqrt{Q}/(g \times NPSHR)^{3/4} \tag{5.13a}$$

where n is the speed of rotation in revolutions/second, Q is the flow rate in m³/s, g is the gravitational acceleration in m/s², and *NPSHR* in m. In the American system, the following equation is used

$$S = N\sqrt{Q}/(NPSHR)^{3/4} \tag{5.13b}$$

where N is in rpm, Q is in gallons/min, and *NPSHR* is in ft. The values of n (or N), Q, and *NPSHR* used in Eq. (5.13a and b) correspond to the pump design point (best efficiency point). Accordingly, every pump has only one value of the suction specific speed.

The relationship between S and the Thoma's cavitation factor, σ_c, is obtained as follows:

$$S = \left(N\sqrt{Q}/H^{3/4}\right)\left(H^{3/4}/NPSHR^{3/4}\right) = N_s/\sigma_c^{3/4}$$

$$\text{Or } \sigma_c = (N_s/S)^{4/3} \tag{5.14}$$

It is clear from Eq. (5.13) that S increases as *NPSHR* decreases. Accordingly, the higher the value of S the better the pump design with respect to cavitation. The suction specific speed is independent of the pump size and speed of rotation. In normal applications, it is not advisable to select pumps with specific speeds higher than 10 000, in order to have a stable and reasonably wide window of operation.

5.2.5 Cavitation Prevention Measures

There are several measures that can be taken to reduce or prevent pump malfunction and damage of its components due to cavitation.

5.2.5.1 Measures Related to the Piping System

- Install the pump as close as possible to the suction reservoir.
- Choose the pump location that maximizes the static suction head (h_{ss}).
- Use large diameter pipes in the suction side.
- Reduce the number of fittings (such as bends, elbows, valves) in the suction side to a minimum.

5.2.5.2 Measures Related to Pump Design

- Reduce surface roughness for all inner surfaces of pump components.
- Avoid sharp changes in direction of flow in the suction nozzle and at impeller inlet.
- Use the optimum value of the impeller inlet vane angle (β_1) to reduce the local pressure drop (p_{ld}).
- Avoid using pumps having an unstable range of operation.
- Introducing a small amount of prerotation (using inlet guide vanes) to improve the performance with respect to cavitation.

5.2.5.3 Measures Related to Pump Operation

- Avoid operating the pump much lower or much higher than its rated capacity.
- Operate the pump at or near its shockless flow rate, which may be slightly away from its best efficiency point.
- Reduce vibrations as much as possible.
- Reinvestigate the operation conditions (such as maximum suction lift and maximum flow rate) when changing the pumped fluid or when there is a significant change in fluid temperature.

5.2.5.4 Measures Related to the Use of Auxiliary Devices

- In some applications, auxiliary units (booster pumps) are used to pump the fluid from the large reservoir in order to reach the main pumping station with sufficient suction head.
- Use an axial flow stage or an inducer immediately before the first stage (Figure 5.13). This may result in a small reduction in the overall pump efficiency.
- A jet pump can be used in combination with a centrifugal pump, as shown in Figure 5.14 for lifting underground water from a deep well to a level suitable for pump operation. In other words, the jet pump is used for increasing *NPSH* available to the centrifugal pump. The jet pump utilizes part of the pumped fluid as a motive fluid for the jet pump.

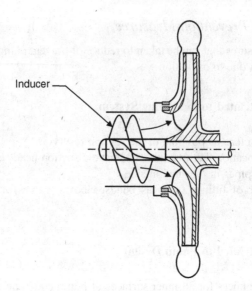

Inducer

Figure 5.13   Schematic of a centrifugal pump fitted with an inducer in the suction nozzle

Example 5.1
The table below gives the variation of *NPSHR* with Q for the pump shown in Figure 5.15. Assume a coefficient of pipe friction of 0.012 and neglect minor losses.

a. If the pump is required to deliver 0.16 m³/s of water, what will be the minimum depth (y) at which the pump should be installed?
b. If the pump is installed at a depth of 3 m, what will be the maximum flow rate if cavitation is to be avoided?

Q (m³/s)	0.04	0.08	0.12	0.16	0.20
NPSHR (m)	0.30	0.50	0.80	1.25	1.80

Solution

a. For Q = 0.16 m³/s, $V_s = \dfrac{Q}{A_s} = \dfrac{0.16}{\pi(0.15)^2/4} = 9.05 \, \text{m/s}$ and *NPSHR* = 1.22 m.

The minimum depth (y) is the depth at which cavitation will be about to start or when *NPSHA = NPSHR*.

Or $\dfrac{p_a - p_v}{\gamma} + h_1 - \sum h_{Ls} = NPSHR$

$$\frac{101 - 5.6}{9.81} - (10 - y) - \frac{0.012(16 - y)(9.05)^2}{2 \times 9.81 \times 0.15} = 1.22$$

Solve to obtain $y = 5.14 \, \text{m}$

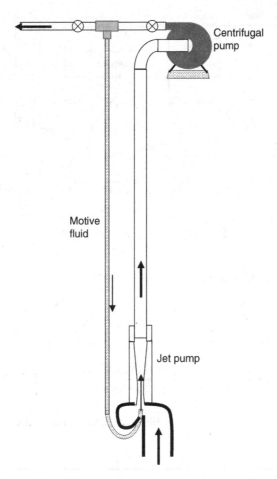

Figure 5.14   Schematic of centrifugal-jet pump combination

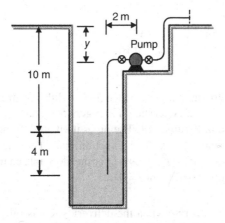

Figure 5.15   Diagram for Example 5.1

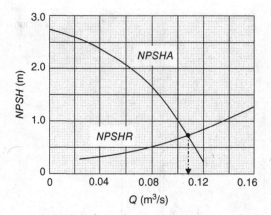

Figure 5.16   Diagram for Example 5.1

b. For y = 3 m, the *NPSHA* will be

$$NPSHA = \frac{p_a - p_v}{\gamma} + h_1 - \sum h_{Ls}$$

$$NPSHA = \frac{101 - 5.6}{9.81} - 7 - \frac{0.012 \times 13 \times Q^2}{2 \times 9.81(\pi/4)^2(0.15)^5} = 2.72 - 170 \, Q^2 \tag{I}$$

Based on Eq. (I), we can generate the table below and then draw the *NPSHA* curve as shown in Figure 5.16.

Q (m³/s)	0.04	0.08	0.12	0.16	0.20
NPSHR (m)	0.30	0.50	0.80	1.25	1.80

The point of intersection of *NPSHA* and *NPSHR* curves shown in the figure gives the maximum (critical) flow rate $\Rightarrow Q_{max} = 0.108$ m³/s

Example 5.2

An experimental set up to circulate hot water ($p_v = 75$ kPa) from reservoir A through a heat exchanger and back to reservoir A, as shown in Figure 5.17, has a small centrifugal pump with the characteristics shown in Figure 5.18. The pipes used are 2.5 cm diameter and have a total length of 10 m out of which 3 m is in the suction side. The friction head loss in the heat exchanger is given by $1.4 \times 10^6 \, Q^2$, where Q is the flow rate in m³/s. Neglect minor losses and consider a pipe friction coefficient $f = 0.02$.

a. Determine the system flow rate when the delivery valve is fully open.
b. Investigate the system with respect to cavitation, and determine the critical flow rate, Q_{crit}.

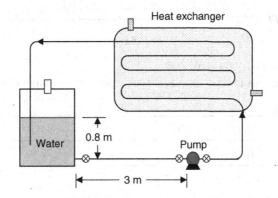

Figure 5.17 Diagram for Example 5.2

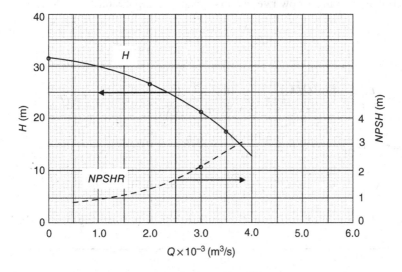

Figure 5.18 Diagram for Example 5.2

c. In order to operate the system at a flow rate of 2.4 L/s with no cavitation, it is suggested to pressurize reservoir A by connecting it to a compressed air tank. Determine the minimum pressure required to avoid cavitation.

Solution

Applying the energy equation, we can deduce that

$$H = h_{st} + \sum h_L = h_{st} + \sum (h_L)_{major} + \sum (h_L)_{minor} + (h_L)_{HE}$$

where the subscript *HE* is used for heat exchanger. Since the suction and delivery reservoirs are the same, therefore $h_{st} = 0$, and by neglecting minor losses, we can write

$$H = \frac{fL(Q/A)^2}{2gD} + (h_L)_{HE} = \frac{0.02 \times 10 \times Q^2}{2 \times 9.81(\pi/4)^2(0.025)^5} + 1.4 \times 10^6 Q^2$$

Simplify to obtain $H = 3.09 \times 10^6 Q^2$ (I)

Equation (I) represents the system curve from which the following table can be constructed.

Q (m³/s)	0.0	1×10^{-3}	2×10^{-3}	3×10^{-3}	4×10^{-3}
H (m)	0.0	3.1	12.4	27.8	49.4

The point of intersection of the pump and system H–Q curves gives the flow rate, $Q = 2.72 \times 10^{-3}$ m³/s

To determine the critical flow rate, we first obtain an expression for NPSHA as

$$NPSHA = \frac{p_a - p_v}{\gamma} + h_1 - \sum h_{Ls} \tag{II}$$

and $h_{Ls} = \dfrac{fL_s V_s^2}{2gD_s} = \dfrac{fL_s(Q/A_s)^2}{2gD_s} = \dfrac{0.02 \times 3 \times Q^2}{2 \times 9.81(\pi/4)^2(0.025)^5} = 0.51 \times 10^6 Q^2$

Substitute in Eq. (II) $NPSHA = \dfrac{101 - 75}{9.81} + 0.8 - 0.51 \times 10^6 Q^2$

Simplify $\Rightarrow NPSHA = 3.45 - 0.51 \times 10^6 Q^2$ (III)

Equation (III) represents the NPSHA–Q curve from which the following table can be constructed.

Q (m³/s)	0.0	1×10^{-3}	2×10^{-3}	3×10^{-3}	4×10^{-3}
NPSHA (m)	3.45	3.1	1.41	−1.14	−4.71

As shown in Figure 5.19, the point of intersection of the NPSHA and NPSHR curves gives the critical flow rate $\Rightarrow Q_{crit} = 2.0 \times 10^{-3}$ m³/s.

Since $Q > Q_{crit}$, cavitation will occur.

To operate the system at a flow rate of $Q = 2.4$ L/s without cavitation, the minimum pressure $(p_a)_{min.}$, will be the value at which NPSHA = NPSHR for this flow rate. Accordingly we can write

$$NPSHA = \frac{p_a - p_v}{\gamma} + h_1 - \sum h_{Ls} = 1.65$$

Therefore, $\dfrac{p_a - 75}{9.81} + 0.8 - 0.51 \times 10^6 (2.4 \times 10^{-3})^2 = 1.65$

Finally, $(p_a)_{min.} = 112$ kPa abs.   or $(p_a)_{min.} = \underline{11 \text{ kPa gage}}$

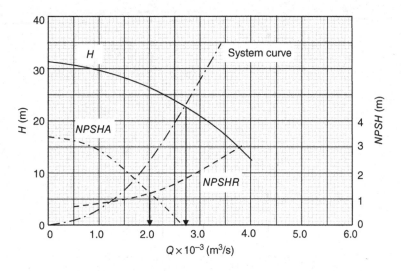

Figure 5.19 Diagram for Example 5.2

Example 5.3

A centrifugal pump is used to lift water from reservoir A to reservoir B through a 200 mm diameter pipe that has a total length of 200 m, as shown in Figure 5.20. The characteristic curves of the pump at 1800 rpm are as shown in Figure 5.21. The total static lift is 20 m and the pump is located 5 m above the water free surface in reservoir A. The total length of the suction pipe is 10 m. The minor losses in the suction pipe are given by $1.5 \ V_s^2/2g$, where V_s is the flow velocity in the pipe.

a. Investigate the system with respect to cavitation.
b. A young engineer suggested reducing the flow rate to 80% of its original value by partially closing the delivery valve in order to avoid cavitation. Show whether this could solve the problem.
c. What would be the power loss (in kW) resulting from the solution suggested in part (b).

Data: Consider that the friction coefficient for all pipes $f = 0.02$, sp. wt. of water $= 9.81 \ kN/m^3$, water vapor pressure $= 4.5 \ kPa$ abs., and $p_{atm} = 101 \ kPa$.

Solution

In order to investigate the system with respect to cavitation, we first determine the variation of NPSHA–Q and compare with NPSHR. Use the energy equation to write

$$NPSHA = \frac{p_a - p_v}{\gamma} + h_1 - \sum h_{Ls} \tag{I}$$

and $\sum h_{Ls} = \dfrac{f L_s V_s^2}{2g D_s} + 1.5 \dfrac{V_s^2}{2g} = \dfrac{Q^2}{2g A_s^2}[(f L_s/D_s) + 1.5] = 129.2 Q^2$

Substitute in Eq. (I) $NPSHA = \dfrac{(101 - 4.5)}{9.81} - 5.0 - 129 Q^2$

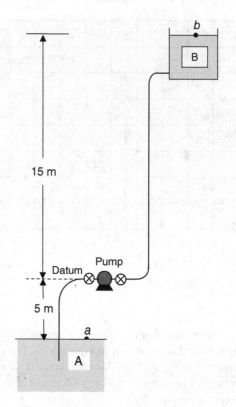

Figure 5.20 Diagram for Example 5.3

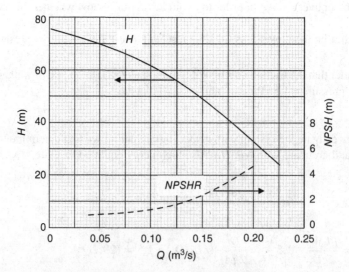

Figure 5.21 Diagram for Example 5.3

$$\text{Simplify} \Rightarrow NPSHA = 4.84 - 129 Q^2 \tag{II}$$

To determine the system flow rate, let us draw the system H–Q curve. First, we apply the energy equation to obtain

$$H = h_{st} + \sum h_L = h_{st} + \sum (h_L)_{major} + \sum (h_L)_{minor}$$

$$H = h_{st} + \left(\frac{f L Q^2}{2 g A^2 D} \right) + \left(1.5 \frac{Q^2}{2 g A^2} + \frac{Q^2}{2 g A^2} \right)$$

$$\text{Simplify } H = 20 + 1163 \, Q^2 \tag{III}$$

a. Draw each of $NPSHA$ and H–Q curves using Eqs (II) and (III), as shown in Figure 5.22. From the figure, we can see that the system curve and the pump curve intersect at a flow rate $Q = 0.152 \, \text{m}^3/\text{s}$. At this flow rate we find that $NPSHA < NPSHR$ and so cavitation will occur and the pump will fail to supply the predicted flow rate.
b. When the flow rate is reduced by partial closure of the delivery valve, the new flow rate becomes

$$Q_2 = 0.8 \times 0.152 = 0.12 \, \text{m}^3/\text{s}$$

At this flow rate, it is clear from the figure that $NPSHA > NPSHR$ and accordingly no cavitation will take place.

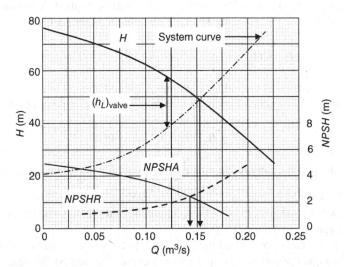

Figure 5.22 Diagram for Example 5.3

c. From the figure, we can see that the head developed by the pump is approximately 57 m
while the useful head used by the system is approximately 38 m. Accordingly, the head loss
in the valve becomes $(h_L)_{valve} = 57 - 38 = 19$ m. The power loss in the valve, $(P_L)_{valve} = \gamma Q$
$(h_L)_{valve} = 9.81 \times 0.12 \times 19 = 22.4$ kW

Example 5.4
If you want to suggest a better solution to stop cavitation in the problem given in Example 5.3
by regulating the pump speed, at what speed should the pump operate?

Solution

Preliminary Analysis
Before we proceed to solve this problem, let us first emphasize the following:

1. The previously obtained *NPSHA–Q* curve is totally independent of the pump operating
speed and so will continue to be the same.
2. By contrast, the *NPSHR–Q* curve depends on the pump speed, and it is expected that
NPSHR values become lower at any given flow rate.
3. There is more than one solution to this problem (i.e. more than one speed), keeping in mind
that the difference between the predicted flow rate (0.152 m³/s) and the critical flow rate
(0.142 m³/s) is small.
4. We may assume that the *NPSHR–Q* curve will not change with the change of pump speed as
a rough approximation. This assumption will be discussed after obtaining a solution.

At the normal operating speed, we found (in Example 5.3) that the critical flow rate at which
cavitation will start is $Q_2 = 0.142$ m³/s at which the head required by the system is $H_2 = 43.5$ m.
We will consider this point to be the new (required) point of operation at which we wish to
determine the new pump speed. The similarity rules will be applied as follows:
The locus of similarity is given by $H = C Q^2$, where the constant C can be obtained from

$$H_2 = CQ_2^2 \quad \Rightarrow \quad C = H_2/Q_2^2 = 43.5/(0.142)^2 = 2157$$

Now, we can draw the locus of similarity (Figure 5.23) and determine the dynamically similar
point on the original *H–Q* curve (at speed 1800 rpm) and label it point (1), then

$$Q_1 = 0.149 \, \text{m}^3/\text{s}, \quad H_1 = 49 \, \text{m}$$

Therefore, applying the similarity criterion $C_{Q1} = C_{Q2}$ gives, $\frac{Q_1}{N_1 D_1^3} = \frac{Q_2}{N_2 D_2^3}$ $\Rightarrow$ and since $D_1 = D_2$,
therefore $N_2 = N_1(Q_2/Q_1)$. Finally, $N_2 = 1800(0.142/0.149) = \underline{1715 \, \text{rpm}}$.

Accordingly, when the delivery valve is fully open, the pump will operate with no cavitation at
any operating speed less than 1715 rpm.
In fact, the pump can operate at a little bit higher speed with no cavitation since the *NPSHR*
gets less (at the same flow rate) as the speed decreases (see Figure 5.8). So the assumption made
is a valid assumption and the obtained solution is on the safe side.

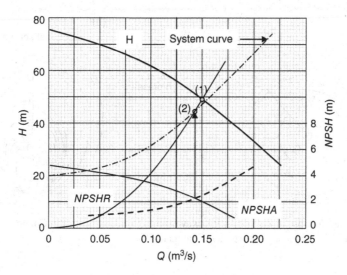

Figure 5.23   Diagram for Example 5.4

5.3   Mechanism of Cavitation Erosion

The term "cavitation erosion" refers to the process of material removal from solid surfaces adjacent to a flowing fluid (liquid) due to the formation and collapse of vapor cavities. The formation of a vapor cavity starts when the fluid pressure falls below the vapor pressure. The cavity continues to grow as long as $p < p_v$. When the cavity moves to a region where fluid pressure increases above the vapor pressure, it collapses (due to vapor condensation). The collapse of the cavity occurs in a very short time, resulting in a very high localized pressure. The majority of the formed cavities collapse away from the solid surface of the impeller. In theory, the collapse of a single cavity away from the solid boundary occurs spherically with the highest pressure at the geometric center of the cavity as shown in Figure 5.24a.

On the other hand, the collapse of the cavity near a solid boundary is not a perfectly spherical collapse (Figure 5.24b). Instead, a liquid jet is formed which strikes the solid surface (acting as a water knife) causing a very high localized pressure. The pressure wave resulting from the collapse of the cavity, together with the high pressure created by the impinging of the liquid jet are believed by most researchers to be the main causes of surface damage (Van Terwisga et al. 2009). Due to the presence of a very large number of cavities, the phenomenon becomes more like high pressure pulsation at a very high frequency. The effect on the solid surface depends on the material properties. For ductile materials, the high frequency pressure pulsation causes local strain hardening that eventually leads to fatigue failure. Long-term exposure to this phenomenon causes surface cracking and loss of material (Berchiche et al. 2001). The main difficulty in achieving a full understanding of the cavitation erosion mechanism is mainly due to the very short time of collapse of the cavity, the presence of two phases (liquid and vapor) and to the very large number of cavities formed (Schmidt et al. 2007). The pitting generated from the interaction between the strong pressure waves emitted during bubble collapse and the solid boundary depends on the characteristic of that boundary (Fortes-Patella et al. 2001). Experimental and computational erosion models

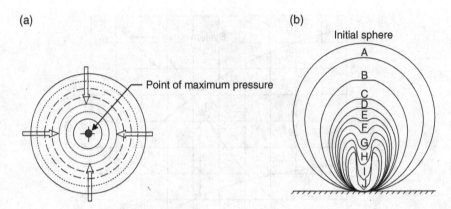

Figure 5.24   (a) Collapse of a single cavity away from the solid boundary. (b) Collapse of a single cavity adjacent to the solid boundary (Plesset and Chapman, 1971)

have been developed aiming at prediction of cavitation intensity and erosion area in centrifugal pumps (Gülich and Pace 1986; Dular 2006; Hattori *et al.* 2008; Fukaya *et al.* 2010).

5.4   Solid Particle Erosion

Erosion caused by the presence of solid particles in the pumped fluid is one of the major sources of surface damage to the impeller and casing materials. This damage tends to reduce the overall pump efficiency due to the increase in hydraulic losses (because of the increase of surface roughness) and increase the rate of fluid leakage (due to the increase in the gap between the wearing rings). In general, the process of material erosion by solid particles depends on the solid particle characteristics (particle type, shape, and size), particle intensity, velocity of impact (magnitude and direction), and properties of the impacted surface. In many industrial applications, the presence of solid particles is unavoidable despite the filtration process. One example is the pumping of sea water for cooling purposes in petroleum refineries, power stations, and petrochemical plants. Other examples include oil production using submersible pumps and water injection in oil wells. In some cases, erosion and corrosion occur simultaneously exacerbating each other. Erosion removes the corroded layer (protective layer of metal compound) thus exposing the metal to more corrosion and, at the same time, the removed corroded solid particles impact the walls of the flow passage causing more erosion. Figure 5.25 shows severe sand erosion at the back shroud of an impeller equipped with a balancing chamber and balancing holes.

5.5   Pump Surge

In pumping systems, the term "surge" refers to pressure variation resulting from sudden (or fast) changes in fluid velocity. This may occur, for example, due to rapid valve closure, pump

Figure 5.25   Severe solid particle erosion at the back shroud of a radial-type impeller

startup, sudden pump stoppage due to power failure, removal of air pockets in the piping system, and water hammer. Uncontrolled surge may result in failure of the pump and other components in the piping system (pipe joints, valves, etc.). Also, the vacuum that can occur as a result of this fluid transient can cause fluid contamination from groundwater.

The unstable pump performance can be the origin of the fluid transient in the pumping system and is called "pump surge." It describes a continuous increase and decrease (fluctuation) of the pump flow rate in an oscillatory fashion. To explain this process, consider the typical pumping system shown in Figure 5.26, in which a fluid is pumped from reservoir A to a pressurized chemical reactor B. Suppose also that the pressure in the reactor is regulated by using a control valve at C. The system is also equipped with a non-return valve (check valve) at D. Consider that the pump has an impeller with forward-curved vanes with the H–Q characteristic EFG shown in Figure 5.27.

Now, let us assume that the system characteristic curve is IJK, shown in Figure 5.27, where the point of normal operation is point J. The flow rate unsteadiness in this system may result from different sources such as the malfunctioning of the non-return valve at D or the partial opening or closing of the control valve at C. Suppose that the partial valve closure at C causes the point of operation to move to point J' and the system curve changes momentarily to IJ'K'. The pump response to the higher pressure in the delivery side (or additional resistance of the system) is an increase in the pump head and a reduction in the flow rate Q. As soon as the system returns to normal characteristic IK, the flow rate increases (since ΔH is positive) and the point of operation moves back to J.

Let us now suppose that the system curve is ILM, and the point of normal operation is L. A similar unsteadiness will cause the system curve to change to AL'M'. The pump response is now a reduction of Q and H (ΔH is now negative). This results in having the head developed by the pump less than the delivery pressure which causes the flow rate to decrease further. The process continues until reaching point E, and the non-return valve closes to prevent reverse

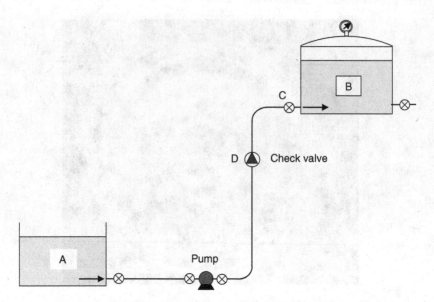

Figure 5.26   Schematic of a typical pumping system

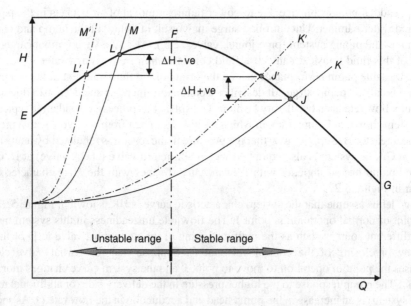

Figure 5.27   Pump response to system unsteadiness

flow. Now the back pressure clears off and the difference between H_E and H_I causes the flow rate to increase again and the process to be repeated.

In the above explanation, the surge process is simplified since we assumed that the speed of rotation of the pump is unchanged and also assumed that the pump and system H–Q curves are unaffected by the flow unsteadiness (inertia effects).

In conclusion, the range between points E and F on the pump curve represents an unstable operation zone and should be avoided. Accordingly, it is always better to use pumps having a full range of stable operation (i.e. pumps having H–Q curves with a continuous downward slope are self-protected against surge). Knowing that impellers with backward-curved vanes do not have humps in their H–Q characteristic, it becomes clear why such impellers are widely used. This should be recorded as a third advantage for using impellers with backward-curved vanes.

5.6 Operation at Other Than the Normal Capacity

Operating pumps far from their design capacity (best efficiency point) may create pressure pulsation, resulting in noise and vibration, and also cavitation problems. As explained earlier, when the flow rate is higher or lower than the normal capacity, Q_n, the fluid enters the impeller at an angle different from the inlet vane angle, causing severe hydraulic losses. These hydraulic losses lead to a severe drop in the overall pump efficiency. Other problems associated with operation away from the design point are the secondary flows—usually called recirculating flows, back flows, or flow reversal—occurring at the impeller inlet eye or at the impeller discharge section in radial-type pumps. Such recirculating (vortical) flows may cause localized pressure drop, resulting in cavitation damage. These phenomena were investigated experimentally by Fraser (1981a,b), who obtained the associated recirculation flow patterns and studied the root cause of the problem and the extent of impeller damage due to the resulting cavitation. The two phenomena are presented in the following sections.

5.6.1 Impeller Suction Recirculation

Impeller suction recirculation is a phenomenon that takes place at the impeller inlet eye at low flow rates and is shown schematically in Figure 5.28a for radial-type pumps and Figure 5.28b for axial-flow pumps. Fraser (1981b) argued that the reverse flow in radial-type impellers is attributed to the asymmetrical pressure field, resulting in a vortex that is locked to the vane

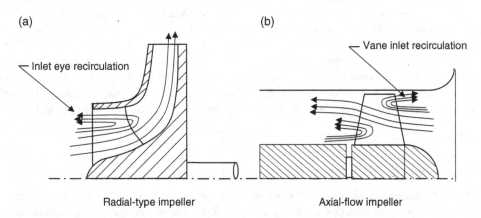

Figure 5.28   Flow pattern at partial capacities for radial- and axial-flow impellers showing regions of recirculation

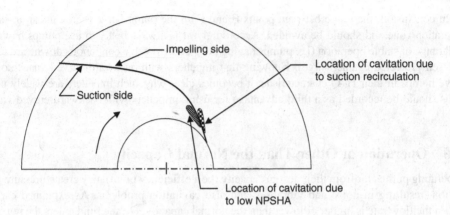

Figure 5.29 Location of cavitation due to suction recirculation

system and rotates with it. This reverse flow, which acquires a large amount of angular momentum, creates a rotating annulus of fluid at the outskirts of the suction nozzle, leaving an axial stream entering the impeller inlet eye in the core region. The shear layer in between these two streams (the rotating fluid annulus and the axial flow at the core) creates vortex streets that also cavitate and produce a sharp crackling noise. On the other hand, the fluid motion created by the reverse rotating flow in the outer region of the suction nozzle and the inflow in the core region results in a stationary vortex that rotates with the impeller vanes. The cavitation occurring at the center of this vortex because of the low pressure attacks the metal on the pressure side of the impeller vanes, causing surface pitting.

The occurrence of cavitation due to suction recirculation is completely independent of the value of *NPSHA*. In other words, it occurs even when *NPSHA* is greater than *NPSHR*. Figure 5.29 shows the location of cavitation caused by suction recirculation which is on the impelling side (pressure side) of the vane near the vane inlet. This contrasts with low *NPSHA* cavitation, which originates on the suction side of the impeller vane, as shown in the figure.

Fraser (1981b) reported that the extent of damage resulting from suction and discharge recirculation depends on the pump shape and size, power rating, head developed, properties of the pumped fluid, and the materials of construction. He also found a relationship between the suction specific speed and the points of suction and discharge recirculation. Pumps with suction specific speeds over 9000 and specific speeds over 2550 should be carefully tested for recirculation within the flow rate operating range. He also reported that pumps designed for higher suction specific speed will have suction recirculation at flow rates close to their rated capacity. Although such pumps have high efficiency, their operating range of flow rate becomes very limited.

5.6.2 Impeller Discharge Recirculation

The same phenomenon may take place at the impeller exit section, where the outflow is reversed back to the impeller, as shown in Figure 5.30. The resulting discharge vortex rotates with the impeller vane and causes pressure pulsation (Fraser 1981a,b). Cavitation at the impelling (pressure) side of the impeller vane, near the vane tip, starts when the velocity of the reverse flow is high enough to create a sufficiently low pressure at the vortex core. Figure 5.31 shows

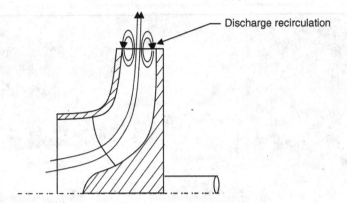

Figure 5.30 Schematic of a radial-type impeller showing discharge recirculation(adapted from Nelson 1980)

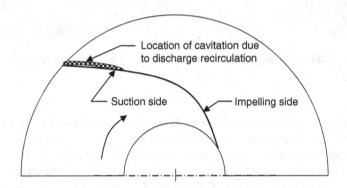

Figure 5.31 Location of cavitation damage due to discharge recirculation

the location of the damage (cavitation pitting) resulting from discharge recirculation and Figure 5.32 shows a photograph of this effect. Similar to suction recirculation, cavitation due to discharge recirculation is completely independent of *NPSHA*.

Aside from the above cavitation damage, the pump operation away from its normal capacity (near shut off) results in severe increase in hydraulic losses, which accounts for approximately 30% of the rated power in low specific speed pumps and reaches 300% of the rated power in high specific speed pumps. Furthermore, operating the pump at low flow rates may tend to increase the fluid temperature, as discussed in Section 5.7.

5.6.3 Effects on Axial and Radial Thrusts

Another problem resulting from operating the pump away from its rated capacity is the increase in the unbalanced axial and radial thrusts. The axial thrust balancing components (e.g. pump-out vanes, balancing chambers, etc.) are normally designed to completely balance the axial thrust when the pump operates at its design point. Operation at flow rates much lower or much higher than the rated capacity will result in unbalanced axial thrust acting on the impeller and

Figure 5.32   A photograph showing cavitation pitting due to discharge recirculation

transmitted to the shaft and to the bearings. Long-term operation under these conditions may lead to bearing failure, and may also result in mechanical seal problems.

With respect to radial thrust, some pumps are equipped with diffuser vanes (such as mixed-flow turbine pumps) that create uniform pressure distribution around the impeller and result in a completely balanced radial thrust at all flow rates. On the other hand, radial-type pumps with single volute casing are designed to produce uniform pressure distribution around the impeller when operating at the pump design point. However, operation at flow rates much lower or much higher than the rated pump capacity will produce unbalanced radial thrust. This force represents an additional load on the shaft and bearings which may lead to pump failure, especially for high-head pumps that are equipped with large diameter impellers. The use of double-volute casing (see details in Chapter 4) in radial-type pumps reduces the radial thrust to small values over a wide range of the flow rate.

5.7   Temperature Rise of Pumped Fluid

When pumps are operated at low capacities (either using the delivery valve for controlling the flow rate or by using a bypass), the fluid temperature increases due to the power loss (in hydraulic, leakage, and mechanical losses). The temperature rise ΔT can be obtained by applying the energy conservation equation as follows:

Total power loss = Rate of heat generation

Or Brake power − Fluid power = $\dot{m}\, C_p \Delta T \;\Rightarrow\; (1-\eta_o) BP = \rho Q C_p \Delta T$

But $BP = \dfrac{\gamma Q H}{\eta_o} = \dfrac{\rho g Q H}{\eta_o}$

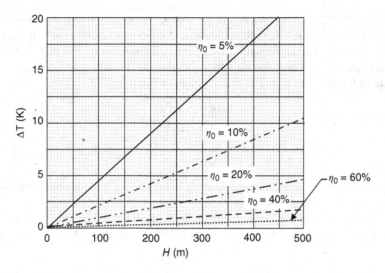

Figure 5.33 Approximate temperature rise in centrifugal pumps handling water

Substitute in the above equation and simplify to obtain

$$\Delta T = \left(\frac{1}{\eta_o} - 1\right)\frac{gH}{C_p} \tag{5.15}$$

In the above analysis, we assumed that all of the heat generated (as a result of mechanical, hydraulic, and leakage losses) is carried by the fluid, neglecting the heat loss from the pump casing to the atmosphere by convection and radiation. Figure 5.33 shows the temperature rise for water ($C_p = 4.2$ kJ/kg·K) when the pump operates at various heads and efficiencies based on Eq. (5.15).

Note that increasing the fluid temperature tends to decrease *NPSHA* which may, in some cases, cause cavitation problems. When using a bypass to reduce the net delivered flow rate while operating the pump near Q_n, it is recommended to have the bypass flow returned to the suction vessel rather than to the suction pipe. This will lead to a smaller increase in the fluid temperature.

The temperature rise may cause serious consequences if the pump operates at very low flow rate or in shut-off condition. For operation at shut-off, all of the input power is transformed into heat, causing the fluid temperature to continuously increase with time. If we neglect heat loss to the atmosphere, the rate of temperature rise can be calculated from the energy balance as

$$\text{Brake power at shut-off} = BP_0 = \text{Rate of heat generation} = (m_L C_L + m_c C_c)\frac{dT}{dt}$$

where m_L and C_L represent the mass of liquid contained in the casing and its specific heat, and m_c and C_c represent the mass of the casing and its specific heat, respectively.

Since the amount of fluid contained in the casing is normally small, the rate of temperature increase at shut-off is relatively high. In this time-dependent process, most of the heat generated will be absorbed by the liquid with a small amount absorbed by the casing. Karassik 1986 suggested neglecting the heat absorbed by the casing, as this will give a conservative estimate of the rate of temperature rise. The above equation becomes

$$\frac{dT}{dt} \approx \frac{BP_0}{m_L C_L} \qquad\qquad (5.16)$$

Example 5.5
The following data belongs to the operating condition of a multistage centrifugal pump. Estimate the overall pump efficiency.

$N = 1800$ rpm, $Q = 0.01$ m³/s, $H = 250$ m, fluid sp. wt. $= 8.5$ kN/m³, fluid sp. heat $C_p = 2.4$ kJ/kg.K, fluid temperature at pump inlet $= 35\,^\circ$C, and fluid temperature at pump exit $= 38\,^\circ$C

Solution
Neglecting the heat loss from the pump casing, the fluid temperature increase is related to the pump overall efficiency by the equation

$$\Delta T = \frac{gH}{C_p}\left(\frac{1}{\eta_o}-1\right)$$

Substituting from the above data,

$$(38-35) = \frac{9.81 \times 250}{2.4 \times 10^3}\left(\frac{1}{\eta_o}-1\right)$$

Therefore, $\eta_o \approx 25\%$

Example 5.6
The power consumption at shut-off for the multistage pump mentioned in Example 5.5 is 56 kW. The volume of fluid in the casing is approximately 120 L. The pump casing is made of carbon steel (sp. heat $= 0.49$ kJ/kg.K) and has a total weight of 1.8 kN. Estimate the rate of fluid temperature increase considering the following two assumptions:

a. The temperature distribution is uniform in the fluid and pump casing.
b. Most of the heat is absorbed by the fluid and negligible amount is absorbed by the casing.

Solution

a. At shut-off, all the input power is transformed into heat. Neglecting heat loss from the pump casing to the atmosphere by convection and radiation, we can use the energy balance to write

$$BP_0 \approx (m_L C_L + m_c C_c) \frac{dT}{dt}$$

Therefore, $\dfrac{dT}{dt} \approx \dfrac{BP_0}{m_L C_L + m_c C_c}$

$$= \frac{56 \times 10^3 \times 60}{(8.5 \times 0.12 \times 10^3 / 9.81) 2.4 + (1.8 \times 10^3 / 9.81) 0.49} = 9.9\,\mathrm{K/min.}$$

b. If we neglect the heat absorbed by the casing, we get

$$\frac{dT}{dt} \approx \frac{BP_0}{m_L C_L} = \frac{56 \times 10^3 \times 60}{(8.5 \times 0.12 \times 10^3 / 9.81) 2.4} = 13.5\,\mathrm{K/min.}$$

5.8 Change of Pump Performance with Fluid Viscosity

The performance of dynamic pumps is determined when handling clean cold water in almost all cases (even if the pump is manufactured to handle a special fluid). This performance may change considerably when handling liquids other than water. However, the performance of displacement pumps is less affected by fluid viscosity in comparison with dynamic pumps.

In general, fluid properties (density, viscosity, chemical activity, presence of solid particles in the pumped fluid, etc.) play an important role in the selection of the type of pump to be used for a specific application. For aqueous fluids (fluids having approximately the same density and viscosity as water), the pump performance is almost the same as quoted by the manufacturer. For dense aqueous solutions, such as brine, the density change affects the pump performance. With an increase of fluid viscosity, both the head developed by the pump and its efficiency decrease. This is accompanied by an increase in the pump power consumption as shown in Figure 5.34.

Experiments have shown that for fluids with $\nu < 20$ centistokes, the centrifugal pump performance is almost the same as when handling water. For fluids with $20 < \nu < 100$ centistokes, there is an increasing reduction in the head developed, but the reduction in efficiency is appreciable. For $\nu > 100$ centistokes, the reduction in head and loss of efficiency are considerable.

Special centrifugal pumps are sometimes designed to handle viscous fluids. These pumps have large impeller diameters and run at high speeds in order to increase the Reynolds number ($\omega D^2 / \nu$) and accordingly decrease the relative effect of viscous forces.

In order to determine the pump characteristics when used to pump viscous liquids, correction factors are used to obtain the head, efficiency, and power curves. These correction factors, which are based on experimental evidence, are well explained in the *Pump Handbook* by Karassik (1986).

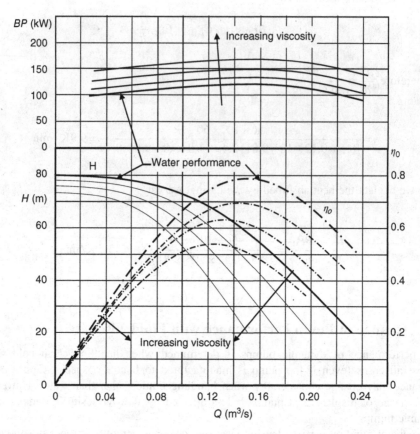

Figure 5.34   Effect of viscosity on the performance of a centrifugal pump at constant speed

5.9   Rotating Stall in Centrifugal Pumps

The rotating stall phenomenon is very common in centrifugal and axial-flow compressors and may also occur in pumps when the flow rate is much lower than the normal capacity. Due to the mismatch between the inflow relative velocity and the vane inlet angle at low flow rates, flow separation occurs downstream of the vane inlet, causing blade stalling. The phenomenon is similar to an airfoil stalling due to high angle of incidence. The rotating stall is characterized by flow instabilities and accompanied by pressure pulsation at a frequency of about 60% of the impeller speed with amplitudes equivalent to approximately 50% of the pump total head. The development of a large region of circulation (vortices) in the flow passage between adjacent impeller vanes occurs due to the detachment of the main stream away from the vane surface. It has been found experimentally that the vortices start appearing in one of the impeller flow passages at $0.5\,Q_n$, where Q_n is the normal capacity. If the flow rate is reduced further, these stalling cells propagate around the circumference. Figure 5.35a shows typical flow pattern when $Q = Q_n$, while Figure 5.35b shows the flow pattern when the flow rate is reduced to $0.35\,Q_n$, where the vortical motion is clearly visible.

(a)   (b)

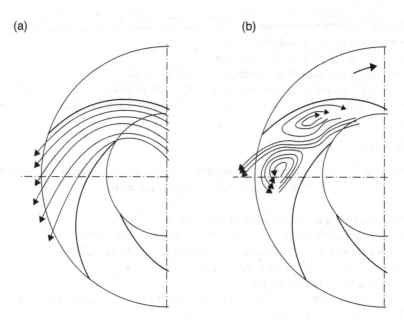

Figure 5.35   Representative of flow patterns in the flow passage between adjacent vanes for the cases (a) $Q = Q_n$ and (b) $Q = 0.35 \, Q_n$

5.10   Pump Vibration

Pump vibration is one of the serious problems that can result in complete failure of one or more of the pump components. These components may include (but not limited to) the stuffing box or mechanical seal, bearings, wearing rings, impeller(s), and pump/motor coupling. The radial and axial movement of the pump shaft arising from mechanical vibrations will definitely have a detrimental effect on the packing material in the stuffing box and may also cause excessive wear in the shaft or shaft sleeve. Mechanical seals are very sensitive to shaft vibration, and the seal life may be drastically shortened as a result of severe vibrations. Critical dimensions and tolerances such as wearing ring clearance, impeller setting, and the clearances of axial thrust balancing devices (balancing drum and balancing disc) will be affected by vibration. Another detrimental effect of mechanical vibration is the possibility of developing cavitation near the vibrating solid surface (e.g. pump casing). This cavitation may create material erosion (pitting of metal parts) as a result of the very high localized pressure associated with this phenomenon. The causes of pump vibration are numerous and can be categorized under mechanical causes, hydraulic causes, and others.

5.10.1   Mechanical Causes

The unbalanced periodic forces originating from the mechanical components of the pump represent the main source of pump vibration. These include

a. mechanical unbalance originating from rotating parts (e.g. impellers, shaft, shaft sleeves, couplings, mechanical seal, balancing drum, and balancing disc)
b. impeller unbalance due to erosion/corrosion of impeller material; this may also originate from a clogged impeller

c. operation near shaft critical speed
d. misalignment of pump and driver shafts
e. faulty bearings (e.g. worn or loose bearings)
f. loose of damaged parts (e.g. impellers or wearing rings)
g. strained piping (due to design problems or thermal growth)
h. vibrations inflected from pump driver (e.g. gasoline or diesel engines).

5.10.2   Hydraulic Causes

In some cases, pump vibration can originate either from fluid forces or from interaction between the fluid and the solid parts of the pump. These include

a. formation and implosion of vapor bubbles due to cavitation
b. vane passing excitations resulting from small clearance between impeller vanes and cutwater or between impeller and diffuser vanes (Figures 5.36 and 5.37)
c. strong pressure waves due to water hammer
d. flow unsteadiness caused by pump surge
e. internal recirculation (e.g. recirculation at the inlet eye and/or at impeller exit)
f. air leakage at the suction side
g. pump operation away from its best efficiency point (BEP).

5.10.3   Other Causes

In some cases, external sources may cause pump vibration such as

a. vibrations from nearby equipment
b. vibrations originating in the gear box (or pulleys if the pump is belt-driven)

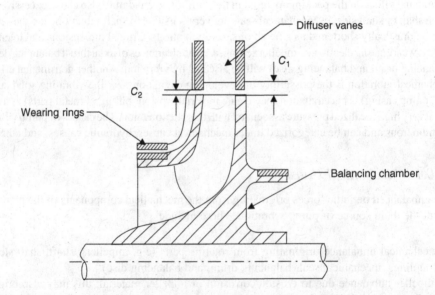

Figure 5.36   Schematic of pump impeller showing the gaps (C_1 and C_2) that may cause vibration

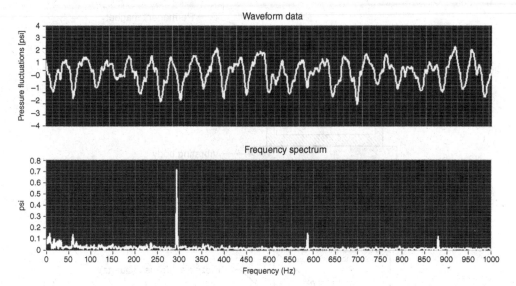

Figure 5.37 Pressure fluctuation inside a pump due to vane passing excitations in both time and frequency domains

c. pump installation problems (e.g. too small foundation or loose hold-down bolts)
d. critical vibrations may result from the interaction between two or more of the above sources, such as the vane passing excitations of the same frequency as the natural frequency of the suction or discharge lines
e. vortex shedding in the suction line due to the presence of one or more ill-designed fittings (such as strainer or check valve) may also be a source of mechanical vibration.

5.11   Vibration Measurements

The severity of mechanical vibrations can be determined by making standard vibration measurements. Such measurements can be used to schedule regular maintenance. It can also be used, with reasonable accuracy, for identifying the origin of vibration excitation (problem diagnosis tool). These standard measurements include frequency of oscillations, amplitude, velocity, and acceleration and may also include acoustic emissions. Vibration measuring transducers are either low- or high-frequency types. A typical low-frequency transducer is shown in Figure 5.38 and is made of a mass m surrounded by a coil and supported by a spring and a dashpot. The casing is equipped with a permanent magnet, as shown in the figure. In some transducers, the moving mass is the permanent magnet, and the coil is fixed in the casing. The output voltage is proportional to the relative velocity between the magnet and the coil. That relative velocity changes periodically during the oscillating motion, and the frequency of the output signal is accordingly the same as the frequency of vibrating body.

The most widely used transducers for measuring high-frequency vibrations are piezoelectric accelerometers. These are made of a mass m mounted on a piezoelectric material (Figure 5.39) that generates an electric charge when stressed (subjected to tension or compression). In principle, the accelerometer measures the acceleration of motion of the vibrating body.

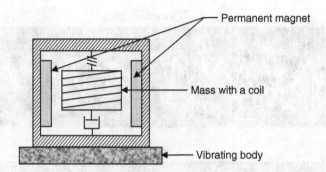

Figure 5.38   Main components of a typical low-frequency vibration transducer

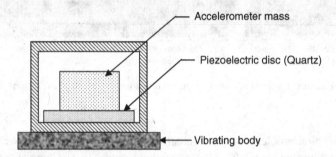

Figure 5.39   Main components of a typical piezoelectric accelerometer

The inertia force of the accelerometer mass causes compression (or elongation) of the piezoelectric material thus producing an electrical charge proportional to the exerted force. Knowing that the inertia force is proportional to the acceleration, the voltage generated will consequently be proportional to the acceleration.

Piezoelectric accelerometers are also designed based on shear deformation of the piezoelectric material. The produced electric charge is also proportional to the exerted shear force. Both compression and shear types transducers are very widely used because of their small size, reliability, and stable characteristics over long time periods. These transducers do not require a power supply and can be used to measure vibrations at high frequencies since their natural frequency is very high (>30 kHz).

5.12   Vibration Signal Analysis

Let us consider a body vibrating in a simple harmonic motion in the form of Eq. (I), below, due to a harmonic excitation having the same frequency. If another exciting force with different frequency is applied, and it results in a harmonic motion in the form of Eq. (II), the resultant motion will be in the form of Eq. (III) and the three motions are presented graphically in Figure 5.40. Now, if we carry out the fast Fourier transform (FFT) for the third signal (Eq. (III)), we can convert the vibration–time data into vibration–frequency data as shown in Figure 5.41. The vibration–time signal can be in the form of displacement, velocity, or

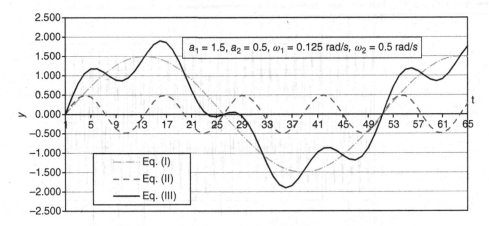

Figure 5.40 Graphical representation of Eqs (I), (II), and (III)

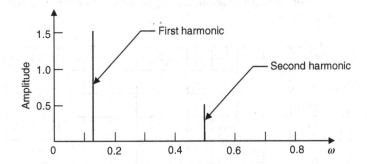

Figure 5.41 Fast Fourier transform of vibration signal (III)

acceleration, and the ordinate of Figure 5.41 can represent the amplitude of any of the three signals. The magnitude of this amplitude is indicative of the severity of the excitation.

$$y_1 = a_1 \sin \omega_1 t \qquad \text{(I)}$$

$$y_2 = a_2 \sin \omega_2 t \qquad \text{(II)}$$

$$y = a_1 \sin \omega_1 t + a_2 \sin \omega_2 t \qquad \text{(III)}$$

The pump vibration analysis starts with the selection of the vibration transducer and the required system components such as filters (high-pass and/or low-pass filters), FFT processor (splitting time-varying signal to individual frequency components). Samples of vibration–time measurements will then be taken and analyzed. Figure 5.42 shows a typical vibration–time sample and Figure 5.43 shows the same sample in the amplitude–frequency domain.

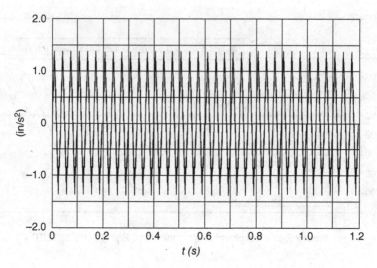

Figure 5.42   Schematic of a typical acceleration-time signal obtained from low-frequency accelerometer

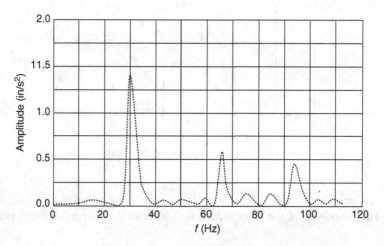

Figure 5.43   FFT of the vibration-time signal

The magnitude of the vibration amplitude (amplitude of acceleration) at different frequencies can be understood as decomposing the vibration signal into a large number of harmonics with different amplitudes for different frequencies. By comparing the frequencies with the pump characteristics and looking at the large amplitudes, we can find out accurately the source of the severe vibration excitation. For example, excitations at frequencies equal to the shaft speed indicate unbalance of a rotor (e.g. pump shaft, impeller, coupling, thrust balancing device). On the other hand, excitations at high frequencies (1–20 kHz) indicate the presence of cavitation. Excitations at frequencies equal to N times the shaft speed (where N is the number of impeller vanes) may indicate blade passing pressure pulsation. Table 5.1 gives the probable cause of vibration for each frequency range.

Table 5.1 Probable cause of vibration at different frequency ranges (n is the pump speed of rotation in revs/s, m is the number of impeller vanes, k is the number of diffuser vanes)

Probable cause of vibration	Vibration frequency (Hz)
Surge	Very low frequency ($\ll n$)
Diffuser stall	$0.05\,n$–$0.4\,n$
Rotating stall at low capacity due to shock at vane inlet (relative velocity not tangential to vane angle) or due to inlet eye recirculation	$0.6\,n$–$0.9\,n$
Mechanical unbalance of different rotors which includes one or more of the following:	$1\,n$
• Eroded or clogged impeller • Eroded or clogged impeller • Unbalance of shaft and rotors assembly • Misalignment of pump-driver coupling • Excessive impeller-volute eccentricity • Volute tongue very close to impeller • Unbalance of shaft sleeves or any of the thrust balancing devices	
a) Pressure pulsation due to damaged or clogged impeller passages b) Loose bearing retainer or shaft cracks c) Pump-driver coupling misalignment	$2\,n$
a. Vane passing pressure pulsation due to small clearance between the vane tip and volute tongue b. Discharge pressure pulsation c. Suction pressure pulsation d. Acoustic resonance in suction or discharge pipes	$m\,n$
Vane passing pressure pulsation due to small clearance between the vane tip and the diffuser vanes.	$k\,m\,n$
Bearing misalignment, loose or poorly lubricated gear drive, loose bearing housing, or loose pump casing	Several multiples of n
Vortex shedding in the suction side	Frequency of vortex shedding (depends on the source)
Cavitation	High frequency (1–20 kHz)

Example 5.7

A single-stage centrifugal pump has an impeller equipped with six vanes and is driven at a speed of 1500 rpm. The pump casing is equipped with a diffuser having 15 vanes equally spaced around the impeller. Estimate the frequency (in kHz) of the vane-passing pressure pulsation.

Solution

Frequency = N (revs/s) × Number of impeller vanes × Number of diffuser vanes

$$= (1500/60) \times 6 \times 15 = 2250\,\text{Hz} = 2.25\,\text{kHz}$$

Example 5.8

Consider the same data as Example 5.7 and assume that the casing has no diffuser vanes but is equipped with a double-volute. Estimate the frequency of the vane-passing pressure pulsation.

Solution

Frequency = N (revs/s) × Number of impeller vanes × Number of volute tongues

$$= (1500/60) \times 6 \times 2 = 300\,\text{Hz} = 0.3\,\text{kHz}$$

References

Berchiche, N., Franc, J.P., and Michel, J.M. (2001) A cavitation erosion model for ductile materials. Fourth International Symposium on Cavitation, June 20–23, 2001, California Institute of Technology, Pasadena, CA.

Dular, M.Z., Stoffel, B., and Sirok, B. (2006) Development of a cavitation erosion model. *Wear*, 261, 642–655.

Fortes-Patella, R., Challier, G., Reboud, J.L., and Archer, A. (2001) Cavitation erosion mechanism: numerical simulations of the interaction between pressure waves and solid boundaries. Fourth International Symposium on Cavitation, *California Institute of Technology*, June20–23, 2001, Pasadena, CA.

Fraser, W.H. (1981a) Recirculation in centrifugal pumps. ASME Winter Annual Meeting, November 15–20, 1981, Washington, DC.

Fraser, W.H. (1981b) Flow recirculation in centrifugal pumps. Proceedings of the Tenth Turbomachinery Symposium, Turbomachinery Laboratories, Department of Mechanical Engineering, Texas A&M University, College Station, TX.

Fukaya, M., Tamura, Y., and Matsumoto, Y. (2010) Prediction of cavitation intensity and erosion area in centrifugal pump by using cavitating flow simulation with bubble flow model. *J Fluid Sci Technol*, 5 (2), 305–316.

Gülich, J.F. and Pace, S. (1986) Quantitative prediction of cavitation erosion in centrifugal pumps. Proceedings of the 13th IAHR Symposium, Montreal, Canada.

Hattori, S. and Kishimoto, M. (2008) Prediction of cavitation erosion on stainless steel components in centrifugal pumps. *Wear*, 265 (11–12), 1870–1874.

Karassik, I.J. (1986) *Pump Handbook*, 2nd edn, McGraw-Hill, New York.

Nelson, W.E. (1980) Pump curves can be deceptive. Refinery and Petrochemical Plant Maintenance Conference, January 22–25, 1980, Convention Center, San Antonio, TX.

Plesset, M.S. and Chapman, R.B. (1971) Collapse of an initially spherical vapour cavity in the neighborhood of a solid boundary. *J Fluid Mech*, 47 (2), 283–290.

Schmidt, S.J., Sezal, I.H., Schnerr, G.H., and Thalhamer, M. (2007) Shock waves as driving mechanism for cavitation erosion. Proceedings of the 8th International Symposium on Experimental and Computational Aerothermodynamics of Internal Flows, July 2007, Lyon.

Van Terwisga, T.J.C., Fitzsimmons, P.A., Ziru, L., and Foeth, E.J. (2009) Cavitation erosion: a review of physical mechanisms and erosion risk models. Proceedings of the 7th International Symposium on Cavitation, August 17–22, Ann Arbor, MI.

Problems

5.1 A boiler feed pump delivers water at 90 °C from an open hot well with a total friction loss of 1.2 m in the suction side of the pump. The critical value of the Thoma's cavitation factor at the normal operating flow rate is 0.08. Knowing that the total head developed by the pump is 80 m, what must be the minimum static suction head in order to avoid cavitation?

5.2 A centrifugal pump draws oil (sp. gr. = 0.75 and vapor pressure = 64.0 kPa) from a closed reservoir at a rate of 0.5 m³/s. The reservoir is partially filled with oil, and the pressure gage located at the top of the reservoir reads 60.8 kPa gage. The pump is driven at 1450 rpm and located 3.0 m above the oil level in the reservoir.

 a. Assuming that the friction losses in the suction pipe to be 1.0 m, determine *NPSHA*.
 b. If the pressure gage at the delivery side of the pump reads 150 kPa gage, determine the Thoma's cavitation factor.
 c. Knowing that $\sigma_c = 0.05$, show whether cavitation will occur or not.

Data: Consider $D_s = D_d = 40$ cm.

5.3 A centrifugal pump has the characteristics shown in Figure 5.44 when operating at a speed of 1500 rpm. Identify the point of operation at which the rate of heat generation is

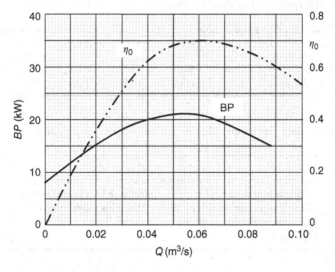

Figure 5.44 Diagram for Problem 5.3

maximum. What is the rate of heat generation at that point? If the fluid pumped is water and the temperature at the pump suction nozzle is 32 °C, calculate the temperature at the discharge nozzle, neglecting heat losses.

5.4 The table below gives the H–Q and σ_c–Q characteristics for a mixed flow pump operating at a speed of 1800 rpm in a certain pumping system.

Q (m³/s)	0.0	0.10	0.20	0.30	0.40	0.50
H (m)	85.0	81.5	74.0	65.0	53.5	39.0
σ_c	—	0.009	0.011	0.017	0.026	0.039

The fluid pumped is water at 60 °C ($\rho = 983$ kg/m³ and $p_v = 20$ kPa abs.). Other details of the pumping system are:

Length of suction pipe = 12 m	Length of delivery pipe = 200 m
Static suction head = −3 m	Static delivery head = 20 m
Diameter of all pipes = 25 cm	Pipe friction coefficient = 0.02

Neglecting minor losses, calculate the following:

a. The *NPSHA* at a flow rate of 0.1 m³/s
b. The pump critical flow rate

5.5 A centrifugal pump is used to move a fluid ($\gamma = 8.8$ kN/m³, $p_v = 36.5$ kPa abs.) from a suction tank A to a delivery tank B, where both tanks are open to the atmosphere. The pump is driven at 1500 rpm and has the characteristics given in the table below.

Q (m³/s)	0.0	0.01	0.02	0.03	0.04	0.05
H (m)	95.0	92.5	87.5	80.5	71.5	59.0
NPSHR (m)	—	0.85	1.10	1.55	2.25	3.30

The following data represents other details of the pumping system:

Length of suction pipe = 12 m	Length of delivery pipe = 150 m
Static suction head = −4 m	Static delivery head = 16 m
Minor losses (suction) = 1.2 $V^2/2g$	Minor losses (delivery) = 5.6 $V^2/2g$
Diameter of all pipes = 10 cm	
Pipe friction coeff. = 0.01	

Determine the critical flow rate for this pumping system.

5.6 A centrifugal pump sucks water at 80 °C ($p_v = 47.3$ kPa) from reservoir A as shown in Figure 5.45. The suction pipe is 10 cm diameter and 6 m long. The pump H–Q and *NPSHR*–Q curves when running at speeds N_1 and N_2 are shown in Figure 5.46. The secondary losses in the suction side of the pump amount to $0.8V_s^2/2g$, where V_s is the velocity in the suction pipe. The system H–Q curve is also plotted in Figure 5.46. The friction coefficient for all pipes can be assumed to be 0.015.

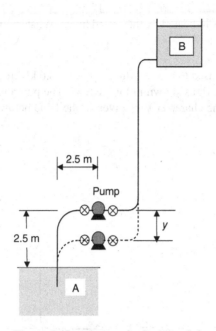

Figure 5.45   Diagram for Problem 5.6

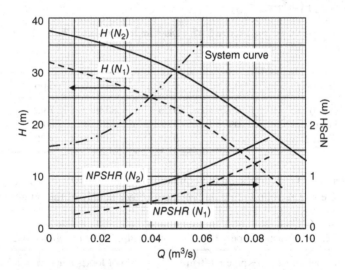

Figure 5.46   Diagram for Problem 5.6

a. Investigate the system with respect to cavitation considering pump speed N_1.
b. What is the critical flow rate at that speed?
c. It is required to increase the system flow rate by 25% by increasing the pump speed to N_2. In order to achieve that with no cavitation, a young engineer suggested changing the

pump location (as indicated by the dotted lines). What should be the minimum value of *y* to avoid cavitation?

5.7 A dynamic pump is used to move a fluid (sp. wt. = 8.6 kN/m³, p_v = 32 kPa abs.) from reservoir A to reservoir B, as shown in Figure 5.47. The pump is driven at its rated speed of 1500 rpm and has the characteristics given in the table below.

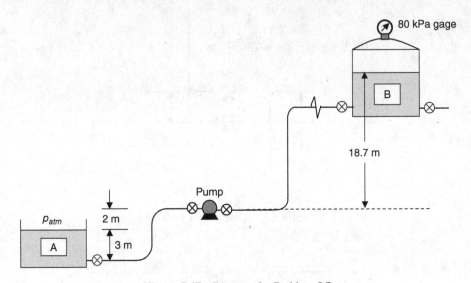

Figure 5.47   Diagram for Problem 5.7

Q (m³/s)	0.0	0.02	0.04	0.06	0.08
H (m)	95.0	92.0	85.0	73.5	50.0
Overall eff.	0%	53%	74%	80%	67%
NPSHR (m)	—	0.65	1.00	1.70	2.80

The system H–Q curve follows the equation $H = 30 + 2 \times 10^4 Q^2$, where Q is in m³/s and H is in meters. The following data represents other details of the pumping system:

Length of suction pipe = 12 m	Pipe friction coeff. = 0.012
Length of delivery pipe = 140 m	Minor losses (suction) = 1.2 $V^2/2g$
Diameter of all pipes = 10 cm	Minor losses (delivery) = 4.8 $V^2/2g$

a. Calculate the pump suction specific speed.
b. Determine the pump critical flow rate.
c. Will cavitation occur? If yes, suggest a method to solve this problem taking into consideration that the conditions at reservoirs A and B cannot be altered.

5.8 A centrifugal pump is used to move a fluid (sp. wt. $= 8.2\,kN/m^3$, $p_v = 15\,kPa$ abs.) from reservoir A to reservoir B, as shown in Figure 5.48. The pump is driven at its rated speed of 1500 rpm and the variation of *NPSHR* with Q is as tabulated below.

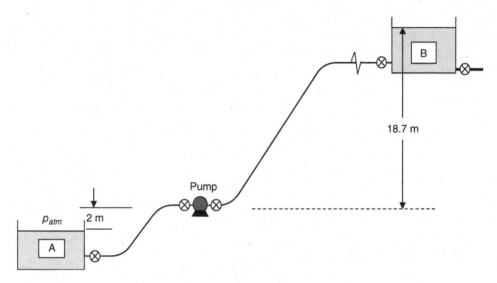

Figure 5.48   Diagram for Problem 5.8

Q (m^3/s)	0.02	0.04	0.08	0.12	0.16	0.20
NPSHR (m)	0.5	0.6	0.8	1.1	1.5	2.2

The following data represents other details of the pumping system:

Length of suction pipe = 15 m	Pipe friction coeff. = 0.01
Length of delivery pipe = 160 m	Minor losses (suction) = 0.8 $V^2/2\,g$
Diameter of all pipes = 15 cm	Minor losses (delivery) = 5 $V^2/2\,g$

a. Determine the maximum flow rate if cavitation is to be avoided.
b. If the rated capacity for this pump is 0.12 m^3/s, calculate the suction specific speed.

6

Axial Flow Pumps

6.1 Introduction

Axial flow pumps are commonly used when there are high flow rate and low head requirements. This is mainly because of the very large area of flow and the comparatively small impeller diameter. The best efficiency point for these pumps exists at the aforementioned operating condition. The specific speed for these pumps ranges from 9000 to 15 000 (based on the American system). At the lower specific speeds range (5000–9000), mixed flow pumps have higher efficiency than axial flow pumps and will produce better performance. One of the main disadvantages of axial flow pumps is their very low efficiency and very high power consumption at low flow rate (near shut-off). Although mixed flow impellers are sometimes used in multistage pumps (especially submersible pumps), axial flow impellers were hardly used in multistage pumps except in research [1–3]. However, multistage axial flow compressors are widely used in the industry and in jet engines. Also blowers designed for specific speeds of around 5000 were used in multistage units to compete with centrifugal compressors in size and efficiency [4,5].

6.2 Definitions and General Considerations

In axial flow pumps, the fluid approaches the impeller axially and leaves it also axially but with swirling motion resulting from the impeller rotation. A typical pump impeller (sometimes called propeller) is shown in Figure 6.1. Figure 6.2 shows a general view of an elbow-type axial flow pump while Figure 6.3 shows a horizontal-axis pump equipped with diffusion vanes. Axial flow impellers normally have small number of vanes (three to six vanes) and some of these impellers are designed to have adjustable vane angles. Axial flow pumps can also be equipped with movable inlet guide vanes for flow rate control.

Pumping Machinery Theory and Practice, First Edition. Hassan M. Badr and Wael H. Ahmed.
© 2015 John Wiley & Sons, Ltd. Published 2015 by John Wiley & Sons, Ltd.

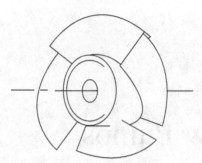

Figure 6.1   Axial flow impeller

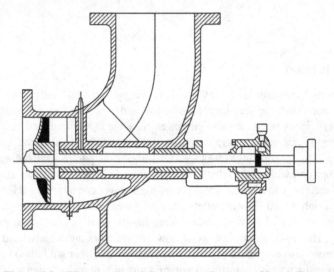

Figure 6.2   Elbow-type axial flow pump (Courtesy of Lewis Pumps)

Axial flow pumps are usually single stage and characterized by low head and high flow rate. The impeller vanes are equally spaced around the hub and are followed downstream by a system of fixed diffusion vanes which are used to straighten the flow (i.e., to remove the whirl component of the velocity) and transform the rotational kinetic energy of the fluid into pressure (act as a diffuser). The pitch t at any radius r is given by $t = 2\pi r/n$, where n is the number of vanes. Figure 6.4a shows a schematic of an axial flow pump in which no guide vanes are used at inlet, and Figure 6.4b shows a blown-up view of the impeller and guide vanes, assuming that the hub is unfolded. The rotational motion of the impeller vanes in Figure 6.4a is transformed to linear motion in Figure 6.4b.

The impeller vanes are designed such that the vane angle at exit, β_2, is slightly greater than the vane angle at inlet, β_1. The velocity diagrams (at any radius r) at the vane inlet and exit sections are shown in Figure 6.5.

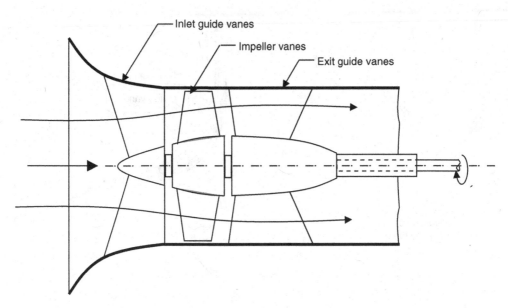

Figure 6.3   A horizontal-axis pump equipped with inlet and exit guide vanes

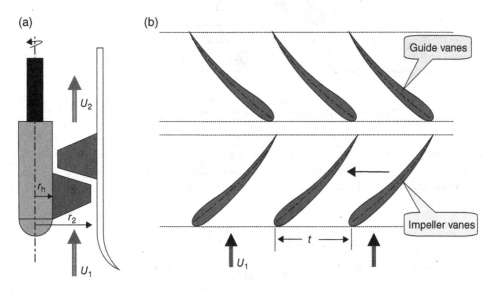

Figure 6.4   A sketch showing the impeller and diffusing vanes of an axial flow pump

Normally, the vanes have a cross-section similar to that of an airfoil. For the airfoil shown in Figure 6.6, the relative velocity makes an angle α with the chord line (called the angle of attack). The relative velocity changes its direction slightly, resulting in a force on the vane due to the change of fluid momentum. This force is represented by the lift and drag components as shown in Figure 6.6. The drag force, D, is in the direction of the approaching stream while the lift force

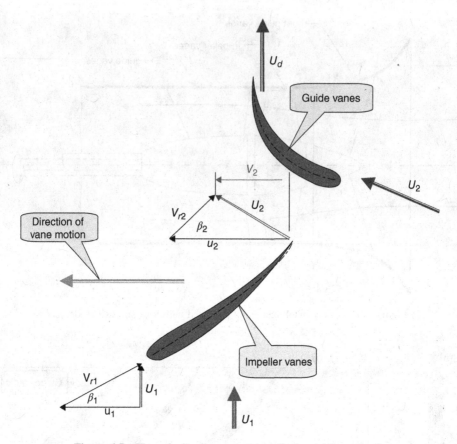

Figure 6.5 The velocity diagrams at impeller vane inlet and exit

L is normal to D. The resultant of the lift and drag forces, R, represents the summation of the elementary pressure and shear forces acting on the entire surface of the airfoil.

The lift and drag coefficients (C_L and C_D) are defined as,

$$C_L = \frac{L}{0.5\rho V_r^2 A} \tag{6.1a}$$

$$C_D = \frac{D}{0.5\rho V_r^2 A} \tag{6.1b}$$

where A is a characteristic area. The effect of the angle of attack, α, on C_L and C_D is to be determined experimentally. Figure 6.7 shows a typical variation of C_L and C_D with α. The energy transfer between the impeller vane and the fluid is highly dependent on the lift force L which increases with the increase of α up to a certain limit after which L decreases (stalling phenomenon). Note that L is much greater than D.

Let the average value of the vane angle be β_m, as shown in Figure 6.8. The rate of doing work by the vane can be expressed as

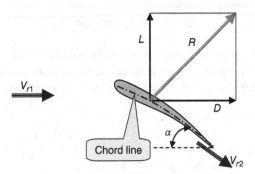

Figure 6.6 Flow over an airfoil showing the lift and drag forces

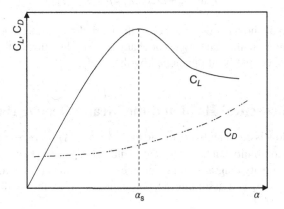

Figure 6.7 Typical variation of C_L and C_D with α

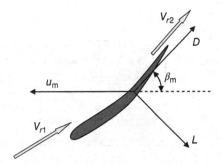

Figure 6.8 The impeller vane showing the lift and drag forces and the average vane angle

$$\dot{W} = (L\sin\beta_m + D\cos\beta_m)u_m \tag{6.2}$$

where u_m is the vane velocity at some average radius (mean effective radius).

Using the angular momentum conservation equation, we know that the torque exerted by the vanes on the fluid is equal to the rate of change of the fluid angular momentum. Therefore, we can write

$$T = \rho Q(V_2 r_m - V_1 r_m) = \rho Q(V_2 - V_1) r_m \tag{6.3}$$

So the rate of doing work can be expressed as

$$\dot{W} = T\omega = \rho Q(V_2 - V_1)\omega r_m = \rho Q(V_2 - V_1)u_m \tag{6.4}$$

Using Eqs. (6.2) and (6.4), we can write

$$L\sin\beta_m + D\cos\beta_m = \rho Q(V_2 - V_1) \tag{6.5}$$

Equations (6.2)–(6.5) show the dependence of the work done by the vanes on the lift and drag forces. So any changes in the vane angles or changes in the direction of the approaching stream (V_r) will directly affect the head developed by the pump.

6.3   Pump Theoretical Head and the Mean Effective Radius

Since the area of flow is constant from inlet to exit $\left[A = A_1 = A_2 = \pi\left(r_o^2 - r_h^2\right)\right]$ (Figure 6.9), therefore, by using the continuity equation, we can write $Q = A_1 Y_1 = A_2 Y_2$ and accordingly, $Y_1 = Y_2$.

If we draw the velocity diagrams at the mean effective radius (r_m) taking into consideration that $u_1 = u_2$ and $Y_1 = Y_2$ and assuming no prerotation at inlet, they will appear as shown in Figure 6.10.

Assuming ideal performance (no losses of any type), we can use Eq. (6.4) and equate input and output powers to write

$$\dot{W} = \rho Q(V_{2m} - V_{1m})u_m = \gamma Q H_e \quad \Rightarrow \quad H_e = \frac{(V_{2m} - V_{1m})u_m}{g} \tag{6.6}$$

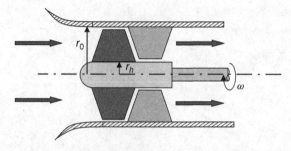

Figure 6.9   A schematic of an axial flow pump showing the main dimensions

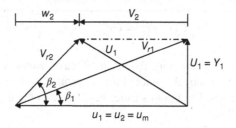

Figure 6.10   Inlet and exit velocity diagrams

In case of no whirl at inlet ($V_{1m} = 0$) Eq. (6.6) can be reduced to

$$H_e = \frac{u_m V_{2m}}{g} = \frac{u_m (u_m - w_{2m})}{g}$$

or

$$H_e = \frac{u_m^2}{g} - \frac{u_m w_{2m}}{g} \qquad (6.7)$$

The first term on the right side of Eq. (6.7) represents the total head developed by forced vortex (pressure and velocity heads) and the second term ($u_m w_{2m}/g$) represents the turbine reaction of the impeller.

As presented in Chapter 3, the fluid reaches the vane inlet of a radial-flow impeller at a constant radius r_1, where the velocity of the vane is u_1 and leaves the vane at radius r_2 where the velocity of the vane is u_2. In that case, the work is being done by the vane between the two radii r_1 and r_2 and all fluid particles gain almost the same amount of energy. However, in the case of axial flow impeller, the fluid particles approaching the impeller vane near the hub where the velocity of the vane is $\omega\, r_h$ receive less energy than the fluid particles approaching the impeller near the outer edge of the vane where the vane tip speed is $\omega\, r_o$. So the velocity diagrams vary from one radius to another and the head developed will follow suit. Downstream of the impeller, the flow passes through the diffusing vanes, where the swirling motion is removed and the energy added to the fluid will reach an average value.

Now, we wish to determine the mean effective radius (r_m) that should be used in Eq. (6.6) or (6.7) to obtain the average theoretical head developed (H_e). By examining the two terms on the right-hand side of Eq. (6.7), we can assume that the head developed at any radius r can be approximated by $H_r = C r^2$. Therefore, the average head can be obtained from

$$H_{av.} = \frac{1}{A} \int_A H\, dA = \frac{1}{\pi\left(r_o^2 - r_h^2\right)} \int_{r_h}^{r_o} C r^2\, 2\pi r\, dr = \frac{2\pi C}{4\pi\left(r_o^2 - r_h^2\right)} \left(r_o^4 - r_h^4\right)$$

which simplifies to $H_{av.} = \frac{1}{2} C\left(r_o^2 + r_h^2\right) = \frac{1}{2}\left(H_o + H_h\right)$.

To find out the mean effective radius r_m at which $H_{av.}$ occurs, we write

$$H_{av.} = \frac{1}{2}C\left(r_o^2 + r_h^2\right) = C r_m^2 \;\Rightarrow\; r_m = \sqrt{\frac{r_o^2 + r_h^2}{2}} \tag{6.8}$$

Equation (6.8) can be used to determine the mean effective radius that can be used for the calculation of the head developed by the pump. From now on, all velocity diagrams will be drawn at the mean effective radius. The pump's Euler head can be obtained using Eq. (6.6) and the theoretical flow rate will be

$$Q_{th} = A Y_m = \pi\left(r_o^2 - r_h^2\right) Y_m \tag{6.9}$$

All efficiencies are defined in exactly the same way as in the radial-type pump (see Chapter 3) and the types of losses (hydraulic, leakage, and mechanical losses) are also the same.

6.4 Performance Characteristics of Axial-Flow Pumps

The performance curves of axial-flow pumps are different from those of radial-type pumps. Figure 6.11 shows typical performance curves for an axial flow pump. In these pumps, the head developed and power consumption are very high at shut off and the best efficiency point is located at high flow rate. Alpan and Peng [6] reported that fluid recirculation in the suction nozzle (suction recirculation) is the main source of power loss at low flow rates. Experimental measurements have shown that the power consumption at shut-off may reach three times the rated brake power. For that reason it is recommended to start these pumps with the delivery valve wide open. Also both the H–Q and P–Q curves have steep negative slope at low flow rates. One of the common features in the performance characteristics of these pumps is the presence of an unstable operation zone at flow rates far below their rated capacities. The dotted lines in Figure 6.11 mark this unstable region. The H–Q curve in this region has a positive slope

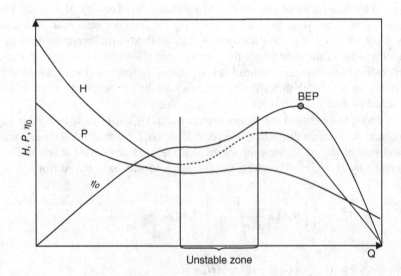

Figure 6.11 Typical performance curves for an axial flow pump

($dH/dQ > 0$), a condition similar to the surge region of radial-type pumps equipped with forward-curved vanes (see Section 5.5). Cheng *et al.* [7] showed that the performance instability is accompanied by vibration and noise, and occurs because of two factors: rotating stall and recirculation at vane inlet. The exact location of the unstable zone depends on the pump geometry and starts at approximately 50% of the normal capacity.

To prove that the H–Q curve has a steep negative slope, let us assume axial flow at inlet (no whirl), then Eq. (6.6) can be written as

$$H_e = \frac{u_m V_{2m}}{g} = \frac{u_m\left(u_m - w_{2m}\right)}{g} = \frac{u_m\left(u_m - Y_m \cot\beta_2\right)}{g}$$

But since $Y_m = Q/A$, therefore

$$H_e = \frac{u_m\left[u_m - (Q/A)\cot\beta_2\right]}{g}$$

From this, we see that

$$dH_e/dQ = -\left(u_m/gA\right)\cot\beta_2 \tag{6.10}$$

For a constant speed of rotation, u_m is constant. Therefore, the variation of the Euler head, H_e, with the flow rate Q depends on the vane angle β_2 which is usually a small angle. This leads to a steep negative slope of the H-Q curve as shown in Figure 6.11.

6.5 Axial Thrust in Axial Flow Pumps

The application of the linear momentum equation in the axial direction for the control volume shown in Figure 6.12 gives the following expression for axial thrust T:

$$\text{Axial thrust} = (p_2 - p_1)A_e \tag{6.11}$$

where A_e is the area of the annulus between the hub and the casing $\left[A_e = \pi\left(r_o^2 - r_h^2\right)\right]$. The contribution due to the change in linear momentum vanishes because the axial component of the

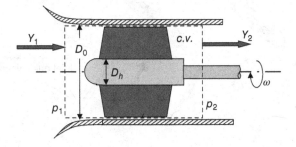

Figure 6.12 A control volume showing the inlet and exit axial velocities and pressures

flow velocity is the same at inlet and exit. Now since $(p_2 - p_1) \approx \gamma H_i$, we can therefore write Eq. (6.11) in the form

$$Axial \ thrust \approx \gamma HA_e / \eta_{hyd} \tag{6.12}$$

Unlike radial-type pumps, there is no radial thrust in axial-flow pumps because of the axisymmetric velocity and pressure fields.

6.6   Flow Rate Control in Axial Flow Pumps

The flow rate control in axial-flow pump systems is carried out using any of the following methods:

a. Speed control of the pump driver
b. Using adjustable inlet guide vanes
c. Changing the impeller vane angle (available in some pumps)
d. Delivery valve throttling

Each of the above methods has its own advantages and disadvantages since the overall system efficiency will be greatly affected by the selected method. In the first method, the flow rate can be reduced or increased by reducing or increasing the pump speed. This method requires a variable-speed prime mover, and we expect a reduction in the overall pump efficiency due to operation at a speed higher or lower than its rated speed. The pump performance curves at a higher (or lower) speed can be predicted using the affinity laws as explained in Chapter 2.

Some pumps are equipped with movable inlet guide vanes (Figure 6.3) that can be used for flow rate control. The set of inlet guide vanes can be adjusted (through a special mechanism) to direct the fluid to enter the impeller with some whirl, as shown in Figure 6.13. The head

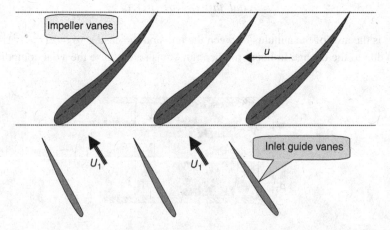

Figure 6.13   Inlet guide vanes adjusted to direct the fluid to enter the impeller with prerotation

developed will be decreased, resulting in a reduction in the pump flow rate. The driving torque will also be decreased (see Eq. (6.3)) and similarly the input power. However, the use of inlet guide vanes for flow rate control will result in a reduction in the overall pump efficiency.

Some pumps are equipped with adjustable impeller vanes. The vane orientation (vane angles) can be adjusted through a special mechanism by allowing each vane to rotate within allowable limits. By changing the vane angle, the result is a new pump with different performance characteristics. For example, reducing the vane angles (β_1 and β_2) will result in less total head, and vice versa. The effect of changing the impeller vane angles on the performance of the axial-flow pump was investigated by Zhu et al. [8] who focused on the increase in hydraulic losses occurring in the diffuser vanes and they proposed the use of adjustable vanes.

The use of delivery valve throttling for flow rate control may result in excessive energy loss and should be avoided. Reducing the flow rate by partial closure of the delivery valve leads to high total head and high power consumption and lower efficiency according to the performance curves shown in Figure 6.11.

Example 6.1

An axial flow pump is driven at 1750 rpm. The impeller has an outer diameter of 50 cm and a hub diameter of 20 cm. The vane inlet and outlet angles at the mean effective radius are 15 and 25° respectively. The hydraulic and manometric efficiencies are 90 and 85% respectively. Knowing that the flow approaches the impeller in the axial direction, determine

a. the mean effective radius
b. the flow rate
c. the actual head developed by the pump
d. the degree of pump reaction
e. the unbalanced axial thrust.

Solution

a. $r_m = \sqrt{\frac{r_h^2 + r_o^2}{2}} = \sqrt{\frac{(0.1)^2 + (0.25)^2}{2}} = 0.19$ m

b. $Q = Y_m A = Y_m \pi \left(r_o^2 - r_h^2 \right)$ (I)

$$\omega = \frac{2\pi N}{60} = \cdots = 183.3 \, \text{rad/s} \quad \text{and} \quad u_m = \omega r_m = 183.3 \times 0.19 = 34.9 \, \text{m/s}$$

From the inlet velocity diagram shown in Figure 6.14, we can write

$$Y_{1m} = u_{1m} \tan \beta_1 = 34.9 \tan 15° = 9.35 \, \text{m/s} \quad \Rightarrow \quad Y_{1m} = Y_{2m} = 9.35 \, \text{m/s}$$

Substituting in Eq. (I)

$$Q = 9.35 \pi \left[(0.25)^2 - (0.1)^2 \right] = 1.54 \, \text{m}^3/\text{s}$$

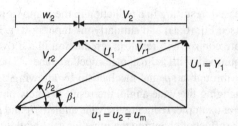

Figure 6.14 Inlet and exit velocity diagrams

c. $H_e = \dfrac{u_m (V_{2m} - V_{1m})}{g}$ (II)

But $V_{1m} = 0$ since the flow approaches the impeller axially and

$$V_{2m} = u_m - Y_m \cot \beta_2 = 34.9 - 9.35 \cot 25° = 14.8 \, \text{m/s}$$

Substituting in Eq. (II)

$$H_e = \frac{u_m (V_{2m} - V_{1m})}{g} = \frac{34.9 (14.8 - 0)}{9.81} = 52.8 \, \text{m}$$

$$H = \eta_{man.} H_e = 0.85 \times 52.8 = 44.9 \, \text{m}$$

d. Degree of reaction, $\lambda \approx 1 - \dfrac{V_2}{2u_2} = 1 - \dfrac{14.8}{2 \times 34.9} = 0.79$

e. Axial thrust $\approx \gamma H A_e / \eta_{hyd.}$ and $A_e = \pi(0.25^2 - 0.1^2) = 0.165 \, \text{m}^2$

Therefore, Axial thrust $\approx \gamma H A_e / \eta_{hyd.} = 9.81 \times 44.9 \times 0.165 / 0.9 = 80.8 \, \text{kN}$

Example 6.2
An axial flow water pump is designed to operate at a speed of 1450 rpm. The pump impeller has an outer diameter of 40 cm and a hub diameter of 20 cm. The inlet and exit vane angles are 15 and 30° respectively. Assuming that the flow enters the impeller axially (with no whirl), sketch the inlet and exit velocity diagrams and determine the following:

a. the actual head developed and flow rate supplied by the pump
b. the pump power consumption.
   The pump is equipped with guide vanes at inlet (upstream of the rotor) to be used for flow rate control. The vanes cause the flow to enter with some prewhirl.
c. Determine the new head developed by the pump if the guide vanes are used to decrease the flow rate by 20%. If all efficiencies are unchanged except for $\eta_{hyd.}$ which is reduced to 90%, what will be the percentage reduction in the pump input power.

Note: If needed consider $\eta_{vane} = 88\%$, $\eta_{vol.} = 98\%$, $\eta_{hyd.} = 94\%$, $\eta_{mech.} = 96\%$.

Solution

a. $r_m = \sqrt{\dfrac{r_o^2 + r_h^2}{2}} = \sqrt{\dfrac{0.2^2 + 0.1^2}{2}} = 0.158\,\text{m}$

$$A_e = \frac{\pi}{4}\left(D_o^2 - D_h^2\right) = \frac{\pi}{4}\left(0.4^2 - 0.2^2\right) = 0.0943\,\text{m}^2$$

$$u_m = \omega\, r_m = \cdots = 24\,\text{m/s}$$

From the velocity diagrams shown in Figure 6.15, we can write

$$Y_1 = u_m \tan\beta_1 = \cdots = 6.43\,\text{m/s} = Y_2 \quad \text{and} \quad V_1 = 0,$$

$$V_2 = u_m - Y_2 \cot\beta_2 = 24 - 6.43 \cot 30° = 12.9\,\text{m/s}$$

$$H_e = \frac{u_m(V_2 - V_1)}{g} = \frac{24 \times 12.9 - 0}{9.81} = 31.5\,\text{m}$$

$$H = \eta_{vane}\,\eta_{hyd}\,H_e = 0.88 \times 0.94 \times 31.5 = 26.1\,\text{m}$$

$$Q = Y_1 A_e\,\eta_{vol} = 6.43 \times 0.0943 \times 0.98 = 0.594\,\text{m}^3/\text{s}$$

b. $BP = \gamma QH/\eta_o$ and $\eta_o = \eta_{vol}\,\eta_{hyd}\,\eta_{mech} = \cdots = 0.884$

$$BP = \frac{9.81 \times 0.594 \times 26.1}{0.884} = 172\,\text{kW}$$

c. The velocity diagram for inlet prerotation is shown in Figure 6.16.

$$Q_{new} = 0.8\,Q_{old} = 0.475\,\text{m}^3/\text{s} = Y_1 A_e\,\eta_{vol} \Rightarrow Y_1 = 5.14\,\text{m/s}$$

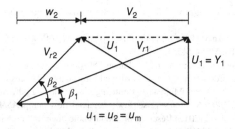

Figure 6.15 Inlet and exit velocity diagrams with no whirl at inlet

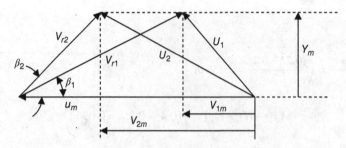

Figure 6.16   Inlet and exit velocity diagrams with prerotation at inlet using guide vanes

$$V_1 = u_m - Y_1 \cot \beta_1 = 24 - 5.14 \cot 15° = 4.82 \, \text{m/s}$$

$$V_2 = u_2 - Y_2 \cot \beta_2 = 24 - 5.14 \cot 30° = 15.1 \, \text{m/s}$$

$$H_e = \frac{u_m (V_2 - V_1)}{g} = \frac{24(15.1 - 4.82)}{9.81} = 25.1 \, \text{m}$$

$$H = \eta_{vane} \, \eta_{hyd} \, H_e = 0.88 \times 0.90 \times 25.1 = 19.9 \, \text{m}$$

$$(\eta_o)_{new} = \eta_{vol} \eta_{hyd} \, \eta_{mech} = 0.98 \times 0.9 \times 0.96 = 0.847$$

$$(BP)_{new} = \frac{9.81 \times 0.475 \times 19.9}{0.847} = 109.5 \, \text{kW}$$

$$\% \, reduction \, in \, input \, power = \frac{172 - 109.5}{172} \approx 36\%$$

Note: The above example shows how much energy can be saved by using the inlet guide vanes for flow rate control.

References

1. http://www.envita.si/index.php/eng/projekti/crpalke/vecstopenjske_crpalke/vecstopenjske_crpalke_1. Accessed June 28, 2014.
2. Wakisaka, Y., Nakatani, T., Anai, H., *et al.* (1995). A two-stage axial flow pump: new approach to reduction of hemolysis. *ASAIO J,* **41** (III), M584-M587.
3. Multistage axial flow pumps and compressors, US Patent 5562405 A.
4. http://www.hsiblowers.com/products/multistage-centrifugal-blowers.html. Accessed June 28, 2014.
5. http://www.trijayasatya.com/Hoffman_Multistage_Centrifugal%20_Catalog.pdf. Accessed June 28, 2014.
6. Alpan, K. and Peng, W.W. (1991) Suction reverse flow in an axial-flow pump. *ASME J Fluid Eng,* **113** (I), 90–97.
7. Cheng, L., Liu, C., Luo, C., *et al.* (2012) Research on the unstable operating region of axial-flow and mixed flow pump. Proceedings of the 26th IAHR Symposium on Hydraulic Machinery and Systems, August 19–23, 2012, Tsinghua University, Beijing, China.

8. Zhu, H., Zhang, R., Xi, B., and Hu, D. (2013) Internal flow mechanism of axial-flow pump with adjustable guide vanes, Paper No. FEDSM2013.16613, ASME Fluids Engineering Division Summer Meeting, July 7–11, 2013, Incline Village, NV.

Problems

6.1 An axial flow pump is driven at 1500 rpm. The impeller hub and outer diameters are 0.14 and 0.32 m respectively. The inlet and exit angles of the impeller vane at the mean effective radius are 18 and 25° respectively. The pump is provided with guide vanes on the suction side of the impeller in order to be used for flow rate control.

Note: Assume a vane efficiency of 92% and a hydraulic efficiency of 88% for all cases.

a. Calculate the flow rate Q and the head developed H when the flow enters the impeller axially.

b. Calculate the axial thrust.

c. If the flow rate is to be reduced by 15% using the inlet guide vanes, determine the whirl component at inlet V_1 and the head H developed by the pump.

6.2 An axial flow pump is designed to deliver a flow rate of 0.5 m³/s when running at a speed of 1200 rpm. The inlet and exit vane angles at the mean effective radius are 12 and 15°, respectively. Assuming that the ratio between impeller outer diameter and hub diameter is 3.0, determine

Note: Neglect fluid leakage and consider $\eta_{hyd} = 92\%$, $\eta_{mech} = 95\%$, $\eta_{vane} = 90\%$.

a. The impeller outer diameter, Do

b. The net head developed by the pump, H

c. The axial thrust acting on the driving shaft, T

6.3 An axial flow water pump is designed to operate at a constant speed of 1200 rpm. The pump impeller has an outer diameter of 30 cm and a hub diameter of 12 cm. The inlet and exit vane angles are 18 and 28° respectively. Assume that the flow enters the impeller axially (with no whirl), and consider $\eta_{vane} = 90\%$, $\eta_{vol.} = 97\%$, $\eta_{hyd.} = 93\%$, $\eta_{mech.} = 95\%$.

a. What are the actual head developed and flow rate supplied by the pump, and the pump power consumption?

b. The pump is equipped with movable impeller vanes to be used for flow rate control. Determine the new flow rate and power consumption if each of the impeller vane angles is reduced by 3°. Assume no change in all efficiencies except for the hydraulic efficiency which will be reduced to 89%.

6.4 An axial flow pump is driven at 1500 rpm. The impeller has an outer diameter of 25 cm
     and a hub diameter of 10 cm. The vane inlet and outlet angles at the mean effective
     radius are 15 and 25° respectively. The vane, hydraulic and volumetric efficiencies are
     92%, 88% and 96% respectively. Knowing that the flow approaches the impeller in the
     axial direction, determine:

     a. mean effective radius
     b. flow rate and actual head developed by the pump
     c. degree of impeller reaction.

6.5 An axial flow pump is driven at 1200 rpm. The impeller hub and outer diameters are 0.16
     and 0.36 m respectively. The inlet and exit angles of the impeller vane at the mean effective
     radius are 18 and 25° respectively. The pump is provided with guide vanes on the suction
     side of the impeller in order to be used for flow rate control.

     Note: Assume a vane efficiency of 92% and a hydraulic efficiency of 88% for all cases.

     a. Calculate the flow rate, Q, and the head developed, H, when the flow enters the impeller
         axially.
     b. Calculate the axial thrust.
     c. If the flow rate is to be reduced by 15% using the inlet guide vanes, determine the inlet
         flow angle, α_1, and the new head developed by the pump, H.

7

Displacement Pumps

7.1 Introduction

Displacement pumps represent a large category of pumps in which a fixed volume of the pumped liquid is delivered every revolution of the pump shaft (or every cycle if the pump is pneumatically operated). That volume depends on the pump geometry and size. In these pumps, energy is added to the fluid by the direct application of a force that moves the fluid from the low pressure side (suction side) to the high pressure side (delivery side). For applications in which a small flow rate is required to be supplied at high pressure, the use of displacement pumps becomes unavoidable. This is mainly because the total head per stage developed by a centrifugal pump is limited by the impeller diameter, speed of rotation, and impeller vane shape. As the impeller diameter increases, the size of the pump increases and the mechanical losses become very high. On the other hand, the pump speed has a limit because of the prime mover speed limitation and the severe increase in hydraulic losses. Moreover, the head developed by centrifugal pumps is drastically reduced when pumping liquids of high viscosity.

Displacement pumps are designed to handle a wide variety of liquids (or mixtures) for a wide range of capacity and pressure. Liquids (or mixtures) handled range from very clean liquids of low viscosity such as diesel fuel in fuel injection pumps (as in diesel engines) to concrete mixtures in concrete handling pumps. The delivery pressure may be very small as in rotary pumps used in food processing or gasoline pumps (in gas stations) and may reach 40 000 psi or higher as in reciprocating pumps utilized in water jetting equipment and also in the petroleum and petrochemical industries. Displacement pumps (such as plunger or diaphragm pumps) are also used for the dual purpose of pumping and metering whenever accurate and reliable metering of liquids is required. This is mainly because the number of revolutions of the pump shaft is an accurate indicator of the volume of fluid supplied. Unlike centrifugal pumps, displacement pumps can be used for handling liquids of high viscosity while maintaining high overall efficiency. They also maintain sufficiently high efficiency when operating at flow rates lower or

Pumping Machinery Theory and Practice, First Edition. Hassan M. Badr and Wael H. Ahmed.
© 2015 John Wiley & Sons, Ltd. Published 2015 by John Wiley & Sons, Ltd.

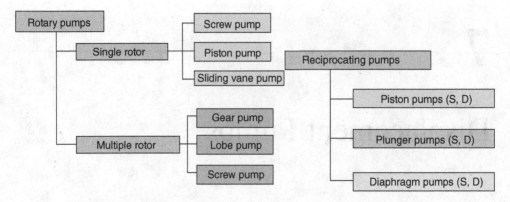

Figure 7.1 Classification of reciprocating and rotary displacement pumps

higher than their normal capacities. The specific speed range for displacement pumps ($N_s < 500$) is the lowest of all types of pumps because of the low flow rate and high head characteristics. Displacement pumps are also used in special applications where centrifugal pumps are avoided. One such application is the pumping of shear-sensitive liquids (or mixtures) as in food processing industries.

Displacement pumps are categorized under reciprocating or rotary pumps, depending on the motion of the moving element within the pump. Reciprocating pumps are classified into piston, plunger, and diaphragm pumps that can be either single- or double-acting (see examples in Chapter 2). The rotary pumps are also classified as having single or multiple rotors with each category having its sub-classification as shown in Figure 7.1. The following sections include detailed analysis of reciprocating and rotary pumps.

7.2 Reciprocating Pumps

Reciprocating pumps are normally used for high head and low flow rate applications. Because of their common use in fluid power systems, they are sometimes called power pumps. Unlike centrifugal pumps, reciprocating pumps can be used for pumping viscous liquids while maintaining sufficiently high efficiency. Piston pumps are used to develop pressures up to 140 bar while plunger pumps can develop pressure up to 2000 bar. These pumps are used in many industrial applications such as pumping heavy petroleum products in pipelines, water injection in boiler feed pumps, chemical industries, hydraulic presses, and other fluid power systems.

7.2.1 General Characteristics of Reciprocating Pumps

Reciprocating pumps usually operate at speeds lower than 500 rpm. The two commonly known types of reciprocating pumps are the piston type and plunger type. If the piston length is less than the length of the stroke, the pump is called a piston pump, otherwise it is called a plunger pump. Diaphragm pumps (or membrane pumps) are also considered as reciprocating pumps that are operated either mechanically or pneumatically. They can handle highly viscous fluids and fluids with a high percentage of solid contents (sludge and slurry). Corrosive and/or

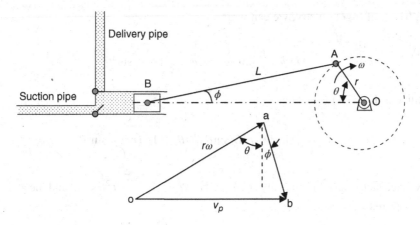

Figure 7.2   The piston pump mechanism and the velocity diagram

abrasive fluids (concrete, acids, and other chemicals) can be handled using especially designed diaphragm pumps.

The flow delivered by a reciprocating pump is always fluctuating in a periodic fashion, thus creating pressure pulsation. In order to deduce the relationship between the pump geometry and flow rate fluctuation, consider the single-cylinder, single acting piston pump shown in Figure 7.2 and let us use the following symbols:

r = crank radius
L = length of connecting rod
θ = crank angle
ϕ = angle between connecting rod and cylinder axis
ω = crank angular velocity
A_p = piston area
v_p = piston velocity
a_p = piston acceleration

In the above velocity diagram, the vector $\overrightarrow{oa}$ represents velocity of A relative to O and the vector $\overrightarrow{ob}$ represents the velocity of the piston (velocity of B relative to O). Using the sine rule, we can write

$$\frac{v_p}{r\omega} = \frac{\sin(\theta+\phi)}{\cos\phi} \quad \Rightarrow \quad v_p = r\omega\sin(\theta+\phi)/\cos\phi$$

Simplify the above equation to obtain

$$v_p = r\omega\left(\sin\theta + \cos\theta\frac{\sin\phi}{\cos\phi}\right) \tag{7.1}$$

Using basics of trigonometry, we can write

$$\sin\phi = \frac{r}{L}\sin\theta \rightarrow \cos\phi = \sqrt{1 - \frac{r^2}{L^2}\sin^2\theta}$$

Substitute in Eq. (7.1) to obtain

$$v_p = r\omega\left[\sin\theta + (r/L)\sin\theta\,\cos\theta / \sqrt{1 - (r/L)^2\sin^2\theta}\right]$$

In such mechanisms, (r/L) is usually small so that $\sqrt{1 - (r/L)^2\sin^2\theta} \approx 1.0$ and the above equation can be reduced to

$$v_p \approx r\omega\left[\sin\theta + \frac{r}{2L}\sin 2\theta\right] \tag{7.2}$$

Using Eq. (7.2), we can determine the maximum and minimum piston velocities as well as the crank angles, θ, at which they occur. This results in the following:

$$\begin{aligned} (v_p)_{min.} &= 0 \quad \text{and occurs at} \quad \theta = 0,\,\pi,\,2\pi,\,3\pi \\ (v_p)_{max.} &\approx r\omega \text{ and occurs near } \theta = \pi/2,\,3\pi/2 \end{aligned} \tag{7.3}$$

The piston acceleration, a_p, can be obtained from

$$a_p = \frac{dv_p}{dt} = \frac{dv_p}{d\theta}\frac{d\theta}{dt} = \omega\frac{dv_p}{d\theta}$$

Using Eq. (7.2) together with the above equation, we get

$$a_p \approx r\omega^2\left[\cos\theta + \frac{r}{L}\cos 2\theta\right] \tag{7.4}$$

From Eq. (7.4), we can easily see that the maximum acceleration occurs at $\theta = 0$ and the minimum acceleration (maximum retardation) occurs at $\theta = \pi$. Accordingly, we can write

$$(a_p)_{max.} \approx r\omega^2\left[1 + \frac{r}{L}\right] \quad \text{and} \quad (a_p)_{min.} \approx -r\omega^2\left[1 - \frac{r}{L}\right] \tag{7.5}$$

Expressions for the piston velocity and acceleration can be also deduced analytically as follows. Suppose that the crank is rotated an angle θ and the piston moved a corresponding distance x from the outer dead center (ODC), as shown in Figure 7.3, then the relation between x and θ can be expressed as:

$$x = L + r - (L\cos\phi + r\cos\theta) = r(1 - \cos\theta) + L(1 - \cos\phi)$$

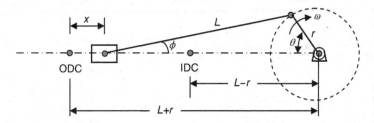

Figure 7.3   The pump mechanism showing the inner and outer dead centers (IDC and ODC)

$$\text{Accordingly, } v_p = \frac{dx}{dt} = r\sin\theta\frac{d\theta}{dt} + L\sin\phi\frac{d\phi}{dt} \tag{I}$$

Knowing that $r\sin\theta = L\sin\phi$, we can differentiate both sides with respect to time to obtain

$$r\cos\theta\frac{d\theta}{dt} = L\cos\phi\frac{d\phi}{dt}$$

Substitute in Eq. (I),

$$v_p = \frac{dx}{dt} = \omega r\sin\theta + L\sin\phi\left(\frac{\omega r}{L}\right)\frac{\cos\theta}{\cos\phi} = \omega r\frac{\sin\theta\cos\phi + \sin\phi\cos\theta}{\cos\phi}$$

$$\text{Simplify to obtain } \frac{v_p}{r\omega} = \frac{\sin(\theta+\phi)}{\cos\phi} \tag{II}$$

Finally, Eq. (II) is exactly the same as Eq. (7.1), deduced directly from the velocity diagram.

The flow rate in the suction and delivery pipes is controlled by the piston movement during the suction and delivery strokes. The variation of the flow rate with time during the delivery stroke can be expressed as

$$Q = A_p v_p \approx A_p r\omega\left[\sin\theta + \frac{r}{2L}\sin 2\theta\right] \tag{7.6}$$

If a single-cylinder, single-acting pump supplies a pipeline directly (without using any air vessel), the flow rate will change during every crank revolution from 0 to Q_{max} as shown in Figure 7.4. It is clear from the figure that the pipeline becomes inactive for 50% of the time, causing severe flow rate fluctuation.

7.3   Pressure Variation during Suction and Delivery Strokes

The pressure inside the pump cylinder varies according to the conditions at the suction and delivery ports, which may depend in turn on the flow rate fluctuations. To study the effect of flow unsteadiness on pressure variation, consider first the simple case of accelerating flow in a straight inclined pipe, as shown in Figure 7.5. Neglecting viscous forces (as a first

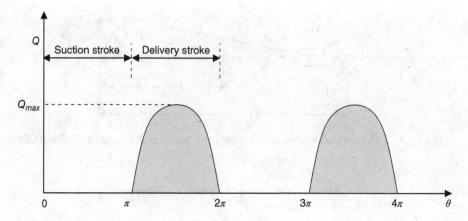

Figure 7.4   Variation of flow rate with crank angle θ

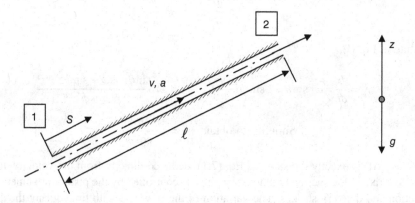

Figure 7.5   Unsteady flow in an inclined straight pipe

approximation), we can write the equation of motion along a streamline – which is known as Euler's equation – in the form

$$\frac{d}{ds}(p+\gamma z) = -\rho a_s$$

where a_s is the acceleration in the streamline direction. From the above equation, we can write

$$-dp = \rho a\,ds + \gamma\,dz$$

Integrate both sides between sections 1 and 2 to obtain

$$\int_{p_1}^{p_2} -dp = \int_{s_1}^{s_2} \rho a\,ds + \int_{z_1}^{z_2} \gamma\,dz$$

Therefore, $p_1 - p_2 = \rho a \ell + \gamma (z_2 - z_1)$

Or

$$\frac{p_1 - p_2}{\gamma} = \frac{a}{g}\ell + (z_2 - z_1) \tag{7.7}$$

If the effect of friction (due to viscosity) is also considered, the equation becomes

$$\frac{p_1 - p_2}{\gamma} = \frac{a}{g}\ell + (z_2 - z_1) + h_L \tag{7.8}$$

In Eq. (7.8), the first term on the right side $(a\ell/g)$ represents the pressure rise due to the inertia effect (or acceleration term) and the second term represents the effect of gravity, while the third term (h_L) represents the friction head loss (viscous effects) between sections 1 and 2. The kinetic energy term is missing since the c-s area is invariant and the flow is incompressible $(V_1 = V_2)$. In the special case of steady flow, Eq. (7.8) becomes exactly the same as the energy equation commonly used for incompressible steady flows.

7.3.1 Pressure Variation during the Suction Stroke and NPSHA

Consider the system shown in Figure 7.6 and let $A_s = \pi D_s^2 / 4$ be the c-s area of the suction pipe and $A_d = \pi D_d^2 / 4$ be the c-s area of the delivery pipe. Also let v_s and v_d represent the velocity in the suction and delivery pipes, respectively. Apply the continuity equation to obtain

$$v_s = v_p (A_p / A_s) \quad \text{therefore,} \quad \frac{dv_s}{dt} = \frac{dv_p}{dt}(A_p / A_s) \;\Rightarrow\; a_s = a_p (A_p / A_s)$$

Substitute from Eqs (7.2) and (7.4) in the above expressions to obtain

$$v_s \approx \frac{A_p}{A_s} r\omega \left[\sin\theta + \frac{r}{2L}\sin 2\theta \right] \tag{7.9}$$

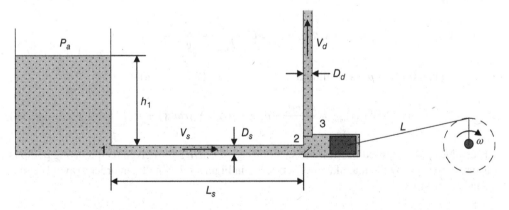

Figure 7.6   Schematic sketch of the pumping system during the suction stroke

$$a_s \approx \frac{A_p}{A_s} r\omega^2 \left[\cos\theta + \frac{r}{L} \cos 2\theta \right]$$ (7.10)

To obtain an expression for the pressure variation immediately before the suction valve (point 2), apply Eq. (7.8) between points 1 and 2 shown in Figure 7.6 to get

$$\frac{p_1 - p_2}{\gamma} = \frac{a_s}{g} \ell_s + (z_2 - z_1) + h_{Ls}$$

But $p_1 = p_a + \gamma h_1$, $z_2 - z_1 = 0$, and $h_{Ls} = \frac{f \ell_s v_s^2}{2g D_s}$, therefore

$$\frac{p_2}{\gamma} = \frac{p_a}{\gamma} + h_1 - \frac{a_s}{g} \ell_s - \frac{f \ell_s v_s^2}{2g D_s}$$ (7.11)

where p_a is the prevailing pressure at the fluid free surface in the suction reservoir. Substituting from Eqs (7.9) and (7.10) in the above equation, we obtain

$$\frac{p_2}{\gamma} = \frac{p_a}{\gamma} + h_1 - \frac{\ell_s}{g} \left[\frac{A_p}{A_s} r\omega^2 \left(\cos\theta + \frac{r}{L} \cos 2\theta \right) + \frac{f}{2D_s} \left\{ \frac{A_p}{A_s} r\omega \left(\sin\theta + \frac{r}{2L} \sin 2\theta \right) \right\}^2 \right]$$ (7.12)

Equation (7.12) describes the variation of pressure immediately upstream of the suction valve with angle θ. In order to obtain the minimum pressure at point 2, we write

$$dp_2/d\theta = 0 \quad \text{which gives} \quad \theta = 0 \text{ at which } v_s = 0$$

$$\text{Accordingly,} (p_2/\gamma)_{min.} = (p_a/\gamma) + h_1 - \frac{\ell_s}{g} \left[(A_p/A_s) r\omega^2 \{ 1 + (r/L) \} \right]$$ (7.13)

To avoid pump cavitation, we must have $NPSHA > NPSHR$ at all crank angles. According to the definition of $NPSHA$, we can write

$$(NPSHA)_{min.} = \left(\frac{p_2}{\gamma} \right)_{min.} + \frac{v_s^2}{2g} - \frac{p_v}{\gamma}$$

and since $v_s = 0$ when p_2 is minimum, we can use Eq. (7.13) to write

$$(NPSHA)_{min.} = \frac{p_a - p_v}{\gamma} + h_1 - \frac{\ell_s}{g} \left[(A_p/A_s) r\omega^2 \{ 1 + (r/L) \} \right]$$ (7.14)

The value of the required net positive suction head ($NPSHR$) for a given reciprocating pump is usually provided by the manufacturer, as shown in Figure 7.7. Finally, in order to avoid cavitation, we must have

$$(NPSHA)_{min.} > NPSHR$$ (7.15)

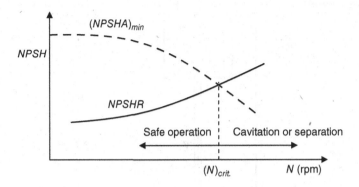

Figure 7.7   Variation of *NPSHA* and *NPSHR* with crank speed showing the critical speed $(N)_{crit}$

Based on Eqs (7.12) and (7.13), the lowest pressure in the pump suction side occurs at point 2 at the beginning of the suction stroke. Further pressure reduction occurs between point 2 and the piston surface (due to inertia and viscous effects). So cavitation starts at the surface of the piston (or plunger) when the minimum pressure falls below the fluid vapor pressure ($p_{min.} < p_v$). Formation of the vapor cavity creates separation between the pumped liquid and the piston surface due to the presence of a buffer layer of fluid vapor. For that reason, the term cavitation in piston or plunger pumps is sometimes referred to as flow separation. The *NPSHR* depends on the pump geometry, crank speed, and fluid properties. The variation of *NPSHR* with crank speed is obtained experimentally for a given pumped fluid.

At the end of the suction stroke, maximum retardation occurs in the suction pipe, causing maximum pressure at the pump suction valve. If we put $\theta = \pi$ in Eq. (7.12), we obtain

$$(p_2/\gamma)_{max.} = \frac{p_a}{\gamma} + h_1 + \frac{\ell_s}{g}\left[\frac{A_p}{A_s}r\omega^2\left(1 - \frac{r}{L}\right)\right] \tag{7.16}$$

7.3.2   Pressure Variation during the Delivery Stroke

The delivery stroke occupies the crank angle $\theta = \pi$ to 2π. The flow rate delivered by the pump is zero at the beginning and end of the delivery stroke and reaches a maximum in between, as indicated in Figure 7.4. Figure 7.8 shows a schematic of the delivery side of the pump during the delivery stroke where point 3 represents the condition immediately downstream of the delivery valve and p_b represents the pressure prevailing at the fluid free surface in the delivery reservoir. Using an approach similar to that used in the analysis of the suction stroke, we can deduce the following expression for the pressure variation at the delivery valve:

$$\frac{p_3}{\gamma} = \frac{p_b}{\gamma} + h_2 - \frac{\ell_d}{g}\left[\frac{A_p}{A_d}r\omega^2\left(\cos\theta + \frac{r}{L}\cos 2\theta\right) - \frac{f}{2D_d}\left\{\frac{A_p}{A_d}r\omega\left(\sin\theta + \frac{r}{2L}\sin 2\theta\right)\right\}^2\right] \tag{7.17}$$

where θ varies during the delivery stroke in the range between π and 2π.

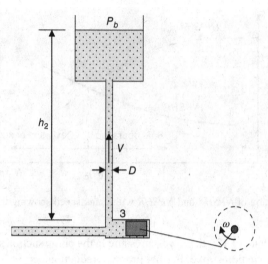

Figure 7.8 A schematic of the delivery side

By examining Eq. (7.17), it becomes clear that the maximum and minimum pressures during the delivery stroke occur at $\theta = \pi$ and $\theta = 2\pi$, respectively, and can be expressed as

$$(p_3/\gamma)_{max} = \frac{p_b}{\gamma} + h_2 + \frac{\ell_d}{g}\left[\frac{A_p}{A_d}r\omega^2\left(1-\frac{r}{L}\right)\right] \qquad (7.18)$$

$$(p_3/\gamma)_{min} = \frac{p_b}{\gamma} + h_2 - \frac{\ell_d}{g}\left[\frac{A_p}{A_d}r\omega^2\left(1+\frac{r}{L}\right)\right] \qquad (7.19)$$

The pressure variation inside the cylinder is as shown in the indicator diagram given in Figure 7.9 for both ideal {a-b-c-d} and actual {a'-b'-c'-d'} cases. In the ideal case, both inertial and viscous effects are neglected. The work done per stroke is represented by the area enclosed by the diagram a'b'c'd'.

The power required to drive the pump can be approximated by

$$B.P. = \frac{\gamma Q}{\eta_{mech}}\left[\frac{p_b-p_a}{\gamma} + (h_2-h_1) + \frac{2}{3}(h_{Ls}+h_{Ld})_{max}\right] \qquad (7.20)$$

where Q is the volume flow rate supplied by the pump, $(h_{Ls})_{max.}$ and $(h_{Ld})_{max.}$ are the maximum friction head loss during the suction and delivery strokes, respectively.

7.4   Use of Air Vessels in Reciprocating Pump Systems

In piston or plunger pump systems, the pressure pulsation due to the severe fluctuation in the flow rate during the suction and delivery strokes represent one of the main problems. These fluctuations create additional power loss due to fluid friction in the piping system in addition

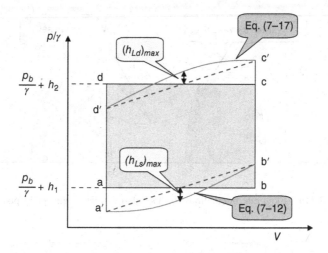

Figure 7.9   The pressure indicator diagram in the cylinder

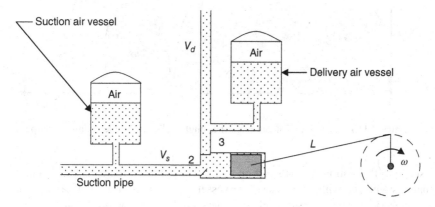

Figure 7.10   Air vessels fitted at the suction and delivery sides of a piston pump

to vibration and noise. One way to reduce these pressure fluctuations in the suction and delivery sides of the reciprocating pump is to use air vessels as shown in Figure 7.10. The air vessel is a small reservoir fitted to the suction (and/or delivery) side of the pump and contains pressurized air (or gas) above the pumped liquid. Liquid flows in and out of the air vessel depending on the pressure variation during the suction (or delivery) strokes.

Fitting an air vessel at the suction side helps to reduce the pressure drop due to the inertia effect (which tends to increase the minimum attainable pressure in the suction pipe) and thus allows the pump to be driven at a higher speed without cavitation (separation) problems. On the other hand, the air vessel at the delivery side helps to reduce the pressure increase due to the inertia effect, resulting in a reduction in the maximum attainable pressure in the delivery side. In addition, the installations of the suction and delivery air vessels tend to reduce the velocity

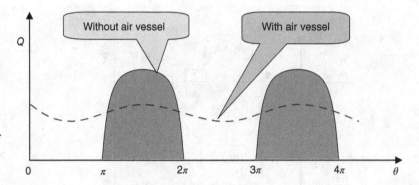

Figure 7.11a   Flow rate fluctuations for a single-cylinder single-acting piston pump.

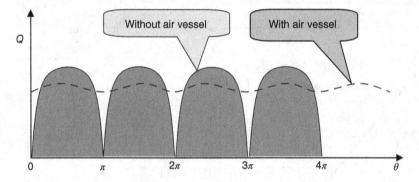

Figure 7.11b   Flow rate fluctuations for a single-cylinder double-acting piston pump

(or discharge) fluctuations in both suction and delivery pipes. Figure 7.11 shows the flow rate fluctuations with and without the use of air vessels for the two cases of single-cylinder single-acting pump (Figure 7.11a) and single-cylinder double-acting pump (Figure 7.11b).

The main advantages of using air vessels are:

1. reducing flow rate fluctuations in the suction and delivery sides of the pump
2. decreasing the friction head loss due to decreasing the maximum velocity in the pipe – this effect results in less work done against friction leading to an improvement in the system efficiency
3. reducing the pressure drop due to the inertia effect at the suction and delivery sides of the pump, which enables the pump to run at higher speeds with no cavitation problems.

7.5   Performance Characteristics of Reciprocating Pumps

Reciprocating pumps are usually used for high head and low flow rate requirements. If those two factors are combined with the low crank speed of such pumps, they lead to low specific speed ($N_s < 500$). In reciprocating pumps, the flow rate depends mainly on the speed of rotation and the pump shape and size, with little dependence on the pump head. The theoretical flow rate, $Q_{th.}$, is given by

$$Q_{th.} = mV \qquad (7.21a)$$

where m is the number of delivery strokes/second and V is the volume per cylinder per stroke. The above equation can also be written in the form

$$Q_{th.} = \frac{N}{60} \times n_{cyl.} \times k \left[\frac{\pi}{4} D_p^2 (2r) \right] \qquad (7.21b)$$

where N is the crank speed in rpm, $n_{cyl.}$ is the number of cylinders, k is a constant ($k = 1$ for single-acting and $k = 2$ for double-acting), D_p is the piston diameter, and $2r$ is the length of the stroke. The length of the stroke should be determined from the pump mechanism.

The actual flow rate differs from the theoretical one due to leakage. In piston and plunger pumps, leakage through the clearance between the piston and cylinder is normally very small but may be appreciable when there has been significant wear in the piston and/or the cylinder. Leakage also occurs through the suction and delivery valves, with quantities depending on the valve design and actuation. Some valves are spring-loaded automatic valves and they open due to the pressure difference across the valve. Leakage may occur through these valves due to backflow because of the insufficient time allowed for valve closure, especially at high pump speed. In other pumps, valves are operated by a timing mechanism (cam operated). In these pumps, leakage occurs because of the overlapping between the opening and closure times (crank angles) of the suction and delivery valves. During this overlapping time one valve is closing and the other one is opening allowing backflow from the high pressure side (delivery) to the low pressure side (suction). This backflow depends on the valve design, overlapping period, and the pressure difference between the delivery and suction sides (or pump total head).

The difference between the theoretical and actual flow rates is usually expressed either in terms of the slip, S, or in terms of the volumetric efficiency, $\eta_{vol.}$, defined as

$$S = \frac{Q_{th} - Q_{act}}{Q_{th}} \times 100 \qquad (7.22)$$

$$\eta_{vol.} = \frac{Q_{act}}{Q_{th}} \times 100 \qquad (7.23)$$

At constant speed N, the value of the slip S depends on the amount of leakage which in turn depends on the delivery pressure. As the pump head H increases, the slip S increases, as shown in Figure 7.12. The head developed by the reciprocating pump depends mainly on the system characteristics. To protect such pumps against excessively high pressure (which could occur due to failure of various of the system components), they are usually provided with a bypass, as shown in Figure 7.13.

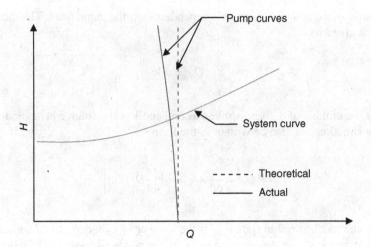

Figure 7.12 Pump and system $H–Q$ curves

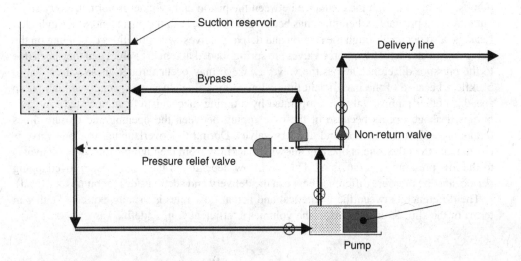

Figure 7.13 The use of a bypass for pump protection and also for flow rate control

7.6 Flow Rate Control

When reciprocating pumps are used in a pumping system, the flow rate is controlled either by controlling the driver speed or by changing the length of stroke, in some designs. In principle, the bypass can also be used for flow rate control, but it results in a considerable power loss as well as increase of the fluid temperature.

The reciprocating pumps can be safely operated at capacities much lower than their rated capacity without any problem (unlike centrifugal pumps) by using the first two types of flow rate control (driver speed or stroke adjustment). The efficiency at partial flow rates is usually much higher than that obtained when using centrifugal pumps (Figure 7.14).

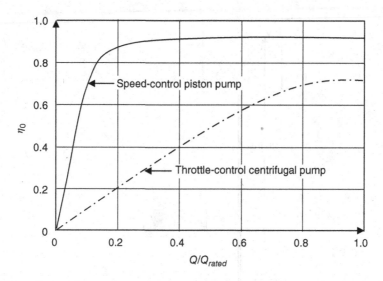

Figure 7.14 The variation of overall efficiency with flow rate for a piston pump

Example 7.1

A single-cylinder single-acting plunger pump is used to deliver water from a sump to a delivery reservoir as shown in Figure 7.15. The following data is given for the pump and the piping system:

Plunger diameter = 12 cm	Suction pipe: D_s = 75 mm and L_s = 4 m
Length of stroke = 15 cm	Delivery pipe: D_d = 75 mm and L_d = 22 m
Length of connecting rod = 22.5 cm	Friction coefficient for all pipes = 0.012
Crank speed = 60 rpm	Vapor pressure = 5.6 kPa abs
NPSHR at 60 rpm = 0.75 m	

a. Determine the minimum pressure in the cylinder and state the corresponding crank angle.
b. Determine the maximum pressure in the cylinder and state the corresponding crank angle.
c. What is the maximum permissible depth of water in the sump if cavitation is to be avoided?
d. What is the power required to drive the pump assuming η_{mech} = 94% and a slip of 2%?

Solution

$$\omega = 2\pi N/60 = 2\pi \, \text{rad/s}, \ r = 0.075 \text{ m}, \ L = 0.225 \text{ m}$$

$$r\omega^2 = 0.075(2\pi)^2 = 2.96 \, \text{m/s}^2,$$

$$\frac{r}{L} = \frac{7.5}{22.5} = \frac{1}{3}, \ \frac{A_p}{A_s} = \frac{A_p}{A_d} = \left(\frac{12}{7.5}\right)^2 = 2.56$$

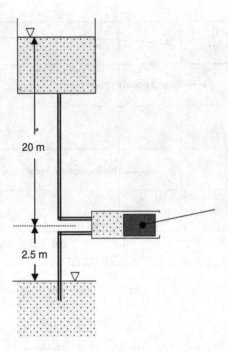

Figure 7.15 Diagram for Example 7.1

a. Using Eq. (7.13), we can write

$$\left(\frac{p_2}{\gamma}\right)_{min} = \frac{p_a}{\gamma} + h_1 - \frac{\ell_s}{g}\left[\frac{A_p}{A_s}r\omega^2\left(1+\frac{r}{L}\right)\right]$$

$$= \frac{101}{9.81} - 2.5 - \frac{4}{9.81}\times 2.56\times 2.96\left(1+\frac{1}{3}\right) = 3.68\,\text{m}$$

Therefore $(p_2)_{min.} = 3.68\times 9.81 = \underline{36.1\,\text{kPa}\text{ and occurs at}\,\theta = 0}$

b. Recall Eq. (7.18),

$$\left(\frac{p_3}{\gamma}\right)_{max.} = \frac{p_b}{\gamma} + h_2 + \frac{\ell_d}{g}\left[\frac{A_p}{A_d}r\omega^2\left(1-\frac{r}{L}\right)\right] = \frac{101}{9.81} + 20 + \frac{22}{9.81}\times 2.56\times 2.96\left(1-\frac{1}{3}\right) = 41.6\,\text{m}$$

Therefore $(p_3)_{max.} = 41.6\times 9.81 = \underline{408\,\text{kPa}\text{ and occurs at}\,\theta = \pi}$

c. To determine the maximum allowable depth (h_1), we write $(NPSHA)_{min} = NPSHR$, and using Eq. (7.14) we can write

$$\frac{p_a-p_v}{\gamma}+h_1-\frac{\ell_s}{g}\left[\frac{A_p}{A_s}r\omega^2\left(1+\frac{r}{L}\right)\right]=0.75$$

Substitute from the above data, $\dfrac{101-5.6}{9.81}+h_1-4.12=0.75 \;\rightarrow\; h_1=-4.86$ m

Therefore, the maximum allowable depth is 4.86 m

d. Using Eq. (7.21b), we can write

$$Q_{th.}=\frac{N}{60}\times n_{cyl.}\times k\left[\frac{\pi}{4}D_p^2(2r)\right]=\frac{60}{60}\times 1\times 1\left[\frac{\pi}{4}(0.12)^2(0.15)\right]=1.694\times 10^{-3}\,\mathrm{m^3/s}$$

$$Q_{Act.}=Q_{th.}(1-S)=1.694\times 10^{-3}\times 0.98=1.66\times 10^{-3}\,\mathrm{m^3/s} \tag{Q}$$

$$(v_s)_{max}=(v_d)_{max}=(v_p)_{max}(A_p/A_s)=2\pi\times 0.075\times 2.56=1.2\,\mathrm{m/s}$$

$$(h_{Ls})_{max}+(h_{Ld})_{max}=f\frac{(L_s+L_d)}{D}\frac{(V_s)_{max}^2}{2g}\ \text{because}\ D_s=D_d\ \text{and}\ V_s=V_d$$

Therefore $(h_{Ls})_{max}+(h_{Ld})_{max}=0.012\dfrac{26}{0.075}\dfrac{(1.2)^2}{2\times 9.81}=0.3\,\mathrm{m}$

Using Eq. (7.20), we can write

$$B.P.=\frac{\gamma Q}{\eta_{mech}}\left[\frac{p_b-p_a}{\gamma}+(h_2-h_1)+\frac{2}{3}(h_{Ls}+h_{Ld})_{max}\right]$$

$$=\frac{9.81\times 1.66\times 10^{-3}}{0.94}\left[20+2.5+\frac{2}{3}(0.3)\right]=\underline{0.393\,\mathrm{kW}}$$

Example 7.2

A single-cylinder, single-acting piston pump is used to deliver hot oil (sp. gr. = 0.85, $\mu=0.01$ N·s/m², and $p_v=90$ kPa abs.) from an open tank A to a pressurized tank B as shown in Figure 7.16. The pump is driven at a crank speed of 45 rpm. The following data are given for the pump and piping system:

Piston diameter = 10 cm,	Suction pipe: $D_s=80$ mm and $L_s=5$ m,
Length of stroke = 15 cm,	Delivery pipe: $D_d=80$ mm and $L_d=7$ m,
Length of connecting rod = 30 cm,	Friction coefficient for all pipes = 0.05.
NPSHR at 45 rpm = 0.6 m,	

a. Determine the minimum pressure in the cylinder and state the corresponding crank angle.
b. What is the minimum value of the static suction head if cavitation is to be avoided?
c. What is the power required to drive the pump assuming $\eta_{mech}=92\%$ and $S=3\%$?

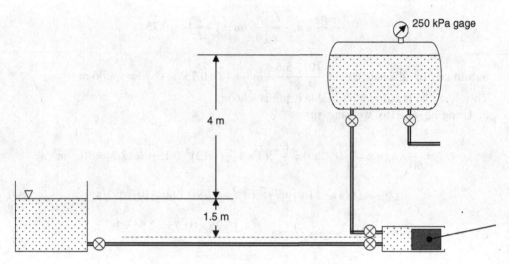

Figure 7.16 Diagram for Example 7.2

Solution

$$\omega = 2\pi N/60 = 1.5\pi \, \text{rad/s}, \, r = 0.075 \, \text{m}, \, L = 0.3 \, \text{m}, \, \gamma_{fluid} = 0.85 \times 9.81 = 8.34 \, \text{kN/m}^3$$

$$r\omega^2 = 0.075(1.5\pi)^2 = 1.665 \, \text{m/s}^2, \, \frac{r}{L} = \frac{7.5}{30} = 0.25, \, \frac{A_p}{A_s} = \frac{A_p}{A_d} = (10/8)^2 = 1.562$$

a. Using Eq. (7.13), we can write

$$(p_2/\gamma)_{min} = \frac{p_a}{\gamma} + h_1 - \frac{\ell_s}{g} \left[\frac{A_p}{A_s} r\omega^2 \{1 + (r/L)\} \right]$$

$$= \frac{101}{8.34} + 1.5 - \frac{5}{9.81}[1.562 \times 1.665\,(1 + 0.25)] = 11.96 \, \text{m}$$

Therefore $(p_2)_{min} = 8.34 \times 11.96 = \underline{99.7 \, \text{kPa and occurs at}}\ \theta = 0$

b. The minimum value of the static suction head satisfies the condition $(NPSHA)_{min} = NPSHR$ and by using Eq. (7.14), we can write

$$\frac{p_a - p_v}{\gamma} + h_1 - \frac{\ell_s}{g} \left[\frac{A_p}{A_s} r\omega^2 \left(1 + \frac{r}{L} \right) \right] = 0.6$$

Or $\dfrac{101 - 90}{8.34} + h_1 - \dfrac{5}{9.81}[1.562 \times 1.665(1 + 0.25)] = 0.6 \Rightarrow \underline{h_1 = 0.93 \, \text{m}}$

Therefore, the minimum allowable height of the fluid in the suction tank (above the cylinder axis) is 0.93 m.

c.

$$Q_{Act.} = Q_{th.}\,(1-S) = \frac{N}{60} \times n_{cyl.} \times k\left[(\pi/4)D_p^2\,(2r)\right](1-S)$$

$$= \frac{45}{60} \times 1 \times 1\left[\frac{\pi}{4}(0.1)^2(0.15)\right](0.97) = 8.57 \times 10^{-4}\,\text{m}^3/\text{s}$$

$$(v_s)_{max.} = (v_d)_{max.} = (v_p)_{max.}\frac{A_p}{A_s} = 1.5\pi \times 0.075 \times 1.562 = 0.552\,\text{m/s}$$

$$(h_{Ls})_{max.} + (h_{Ld})_{max.} = f\frac{(L_s + L_d)\,(V_s)^2_{max.}}{D}\frac{1}{2g}\ \text{since}\ D_s = D_d$$

Therefore, $(h_{Ls})_{max.} + (h_{Ld})_{max.} = 0.05\frac{12}{0.08}\frac{(0.552)^2}{2 \times 9.81} = 0.12\text{m}$

Using Eq. (7.20), we can write

$$B.P. = \frac{\gamma Q}{\eta_{mech.}}\left[\frac{p_b - p_a}{\gamma} + (h_2 - h_1) + \frac{2}{3}(h_{Ls} + h_{Ld})_{max.}\right]$$

$$= \frac{8.34 \times 8.57 \times 10^{-4}}{0.92}\left[\frac{250}{8.34} + 4 + \frac{2}{3}(0.12)\right] = \underline{0.265\,\text{kW}}$$

Example 7.3

A single-cylinder double-acting piston pump is used to transport a hydraulic fluid ($\rho = 875\,\text{kg/m}^3$, $\mu = 0.015$ Pa.s, $p_v = 16$ kPa abs.) from an open reservoir to a pressurized one. The pump location is 2.8 m below the fluid free surface in the suction reservoir and the static delivery head is 50 m.

a. Considering the *NPSHR* curve given in Figure 7.17, and assuming no air vessels are used in the system, determine the maximum crank speed for this pump if cavitation is to be avoided.
b. In order to increase the flow rate without cavitation, it is suggested to install an air vessel in the pump suction side. In this case, will it be possible to operate the pump at double the speed obtained in part (a) without cavitation?

Plunger diameter, $D_p = 80$ mm,	Suction pipe: $D_s = 50$ mm and $L_s = 1.8$ m,
Crank radius, $r = 50$ mm,	Delivery pipe: $D_d = 50$ mm and $L_d = 150$ m,
Length of connecting rod, $L = 20$ cm,	Friction coefficient for all pipes = 0.04
Pump volumetric efficiency = 98%.	

Hint: The flow in part (b) may be assumed steady.

Solution

a. To determine $(NPSHA)_{min}$, we can use Eq. (7.14) to write

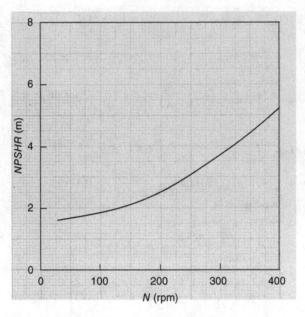

Figure 7.17 Diagram for Example 7.3

$$(NPSHA)_{min} = \frac{p_a - p_v}{\gamma} + h_{ss} - \frac{L_s}{g}\left[\frac{A_p}{A_s}r\omega^2\left(1 + \frac{r}{L}\right)\right]$$

$$\Rightarrow (NPSHA)_{min} = \frac{101 - 16}{8.58} + 2.8 - \frac{1.8}{9.81}\left[\left(\frac{8}{5}\right)^2 0.05\,\omega^2\left(1 + \frac{0.05}{0.20}\right)\right]$$

Simplify, $(NPSHA)_{min} = 12.7 - 0.0293\,\omega^2$

Replace ω with $N \Rightarrow (NPSHA)_{min.} = 12.7 - 0.0293\left(\frac{2\pi N}{60}\right)^2 = 12.7 - 0.0293\left(\frac{2\pi N}{60}\right)^2$

Simplify $\Rightarrow (NPSHA)_{min.} = 12.7 - 3.22 \times 10^{-4} N^2$

N (rpm)	100	125	150	175	200
$(NPSHA)_{min.}$ (m)	9.48	7.67	5.45	2.84	−0.18

The point of intersection of the *NPSHA* and *NPSHR* vs *N* curves shown in Figure 7.18 gives the critical speed below which no cavitation will occur. Accordingly, $N_{crit.} = 180\,\text{rpm}$

b. Assuming that the flow will be approximately steady after using the air vessel, we can obtain a new expression for *NPSHA* as follows:

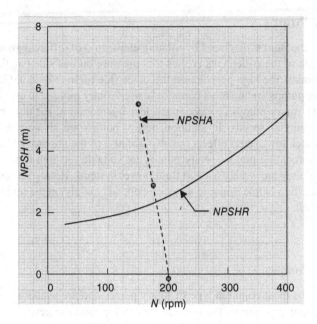

Figure 7.18 *NPSHA* and *NPSHR* vs *N* curves for Example 7.3

$$NPSHA = \frac{p_a - p_v}{\gamma} + h_{ss} - \sum h_{Ls} = \frac{p_a - p_v}{\gamma} + h_{ss} - f\frac{L_s V_s^2}{2gD_s}$$

$$NPSHA = \frac{101 - 16}{8.58} + 2.8 - 0.02\frac{1.8Q^2}{2gA_s^2 D_s} = 12.7 - 0.04\frac{1.8Q^2}{2g(\pi/4)^2 D_s^5}$$

$$= 12.7 - 0.04\frac{1.8Q^2}{2 \times 9.81(\pi/4)^2(0.05)^5} = 12.7 - 19037Q^2$$

But Q can be expressed in terms of N as follows:

$$Q = \frac{N}{60}nkV_{stroke} = \frac{N}{60} \times 1 \times 2\left[\frac{\pi}{4}D_p^2(2r)\right]$$

The new speed is double the speed obtained in part (a), therefore, N = 360 rpm. Substitute in the above equation to obtain

$$Q = \frac{360\pi}{60}(0.08)^2(0.05) = 6.05 \times 10^{-3}\,\text{m}^3/\text{s}$$

Therefore, *NPSHA* = 12.7 – 19037 Q^2 = 12.7 – 19037 $(6.05 \times 10^{-3})^2$ = 12 m

Finally, at the new speed (N = 360 rpm), the *NPSHR* obtained from the figure is 4.5 m while *NPSHA* = 12 m. Therefore, *NPSHA* > *NPSHR* and there will be no cavitation.

7.7 Rotary Pumps

In rotary pumps, the fluid is displaced by rotating element(s) of different geometries inside the pump. The motion of these elements creates low pressure on one side (suction side) and supplies the fluid against the high pressure on the other side (delivery side). Gear pumps, screw pumps, and vane pumps are the most commonly used rotary pumps. In comparison with reciprocating pumps, rotary pumps produce much less fluctuation in the flow rate, and they can operate at much higher speeds. In addition, rotary pumps are simpler in design, not only because of the much low inertia effects but also because of the absence of inlet and discharge valves. Like reciprocating pumps, rotary pumps are suitable for high head and low flow rate requirements and they are widely used in fluid power systems. Rotary pumps are also used as fluid meters since the volume of liquid supplied depends mainly on the number of revolutions of the driving shaft.

7.7.1 Gear Pumps

Gear pumps are the most common type of rotary pump, and are available in three basic configurations: the spur gear pump, the internal gear pump, and the gerotor pump. Figure 7.19 shows the basic configuration for a typical spur gear pump (external mesh) in which the driving and driven gears are enclosed in a casing with minimum clearance at the gear tips. The pumped fluid moves from the low pressure (suction) side to the high pressure (delivery) side through the space between the gear teeth and the casing.

The volume of fluid displaced per revolution of the driving shaft is theoretically constant and is determined by the geometry and size of the two gears. Leakage losses in gear pumps are caused by the fluid leaking between the mating surfaces and by jet losses at the line of contact between the teeth. The volumetric efficiency of these pumps ranges from 80 to 96% or more.

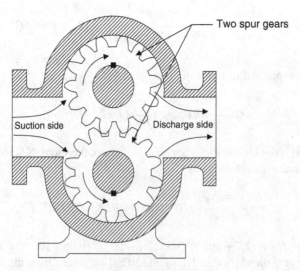

Figure 7.19 Configuration of an external gear pump

Due to the clearance space between the gears and side plates and also between the gear tips and the housing, more fluid slippage occurs as the delivery pressure increases.

If the area enclosed by two adjacent teeth and the casing is a and the axial length of the gear is ℓ, then the volume of fluid enclosed between adjacent teeth is $a\ell$ and the total volume moved by one gear in one revolution is $a\ell n$, where n is the number of teeth. If the pump is driven at a speed N, then the pump actual flow rate is given by

$$Q = 2\eta_{vol.}\, a\ell n\, N \tag{7.24}$$

where $\eta_{vol.}$ is the volumetric efficiency.

Gear pumps are used for flow rates up to about 400 m³/hr and for delivery pressures up to 17 MPa. The volumetric efficiency of gear pumps is about 96% for moderate pressures (up to 4 MPa) but decreases as the pressure rises.

The overall efficiency takes into account the mechanical and hydraulic losses and can be expresses as

$$\eta_o = \frac{\gamma Q H}{B.P.} = \eta_{vol.}\,\eta_{mech.}\,\eta_{hyd.} \tag{7.25}$$

where $\eta_{mech.}$ is the mechanical efficiency and $\eta_{hyd.}$ is the hydraulic efficiency. The hydraulic losses represent the energy loss due to fluid friction as it moves through the pump. The flow rate in this pump can be easily controlled by the speed of the prime mover. Figure 7.20 shows typical performance curves for a gear pump at different speeds. It is clear from the figure that the flow rate decreases as the delivery pressure increases and that is mainly because of the increase of leakage.

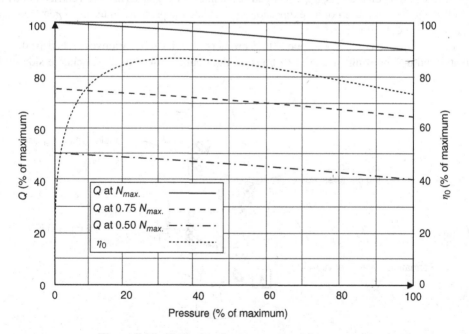

Figure 7.20   Typical performance curves of a gear pump

7.7.2 Some Design Considerations of Gear Pumps

The use of gear pumps to deliver fluids at high pressure necessitates some special design considerations. Because of the high pressure difference between the suction and discharge sides, which occupy opposite locations (see Figure 7.19), the spur gears are loaded by a large unbalanced pressure force. This force tends to push the gears toward the suction side of the casing and that requires the use of heavy duty shafts and bearings. Some gear pumps utilize journal bearings that are lubricated by the pumped fluid, but the majority use heavy-duty roller bearings. Because of this large unbalanced force, it is important to avoid operating a gear pump continuously at higher than the design pressure since this may cause surface-to-surface contact between the gears and the casing on the suction side. Operation at high pressures normally results in increase in fluid leakage and a considerable reduction in the volumetric efficiency. Hong *et al.* [1] proposed a model for internal leakage in gear pumps.

An additional pressure force on each gear is created by the high-pressure fluid trapped (squeezed) between the meshing teeth as shown in Figure 7.21. With gear rotation, the trapped fluid is first compressed (region 2) until reaching maximum pressure (when the trapped volume reaches its minimum (region 3)) and then the pressure decreases as the trapped volume increases (region 4). The pressure in region 4 may reach the vapor pressure and a vapor cavity will be formed. With further rotation, the trapped fluid is released (due to the opening of the trapped volume) and the vapor cavity collapses creating a strong pressure wave. The additional dynamic load on the pump shaft and bearings due to the high pressure of the trapped fluid as well as the strong pressure waves may be the cause of pump failure. Also, the return of the high-pressure trapped fluid from the delivery side of the pump to the suction side is one of the main sources of fluid leakage, and represents part of the energy loss. A full investigation of the effect of the trapped oil on the pressure generated in the inter-tooth spaces has been carried out by Ali [2] through the derivation of an expression to calculate the pressure in terms of gear geometry and operating conditions.

In order to overcome the trapped fluid problem, special relief grooves milled in the side plates and end housings are used to transmit the trapped liquid to the discharge side of the

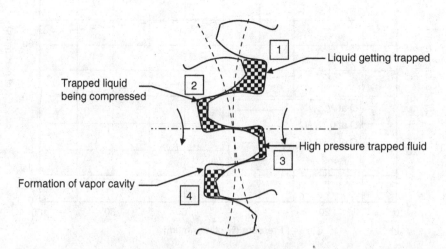

Figure 7.21   Compression and expansion of the fluid trapped between the teeth

pump. This tends to minimize the pressure increase due to liquid compression and eliminates cavitation pressure waves.

The liquid movement from the suction side is in the form of pockets of liquid trapped between the gear teeth and the casing. The intermittent arrival of these pockets to the delivery side causes pressure pulsations due to the flow rate fluctuations. This flow rate fluctuation gets less with the increase of the number of teeth and also with the increase of the speed of rotation [3]. The use of helical gears is found to solve the problem of the high pressure trapped fluid and to reduce the pressure pulsation due to flow rate fluctuations [4]. The level of noise in using helical gears is also much less.

7.7.3   Internal Gear Pumps

Two different designs of internal gear pumps are shown in Figures 7.22 and 7.23. In these designs, the pump has one gear with internally cut teeth that meshes with a smaller, externally cut gear. Figure 7.22 shows standard involute gears with a crescent-shaped divider that separates the inlet and outlet sides. However, in the gerotor type (Figure 7.23) a divider is not required. The gear contour is such that the line contact is always maintained between each external tooth and the internal gear. As they pass the suction port, the space between the gears increases, pulling liquid into the pump. This space is reduced at the discharge, forcing liquid out of the pump. An interesting feature of this pump is that it has low sliding velocity at the gear contact lines.

Gear pumps can operate at speeds ranging from a few hundred to several thousand rpm. Practical speed limitations are due to noise level, bearing life, and the hydraulic losses resulting from the fluid high velocities. Typical applications of gear pumps include process fluid transfer in chemical industries, fuel pumps for oil-fired furnaces, and supplying hydraulic power for industrial and mobile machinery.

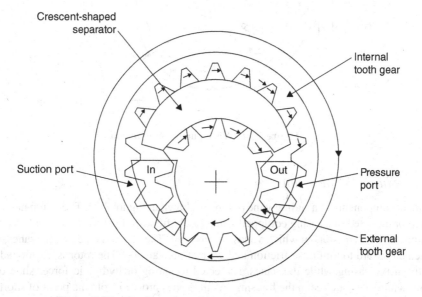

Figure 7.22   Configuration of internal gear pump (involute type)

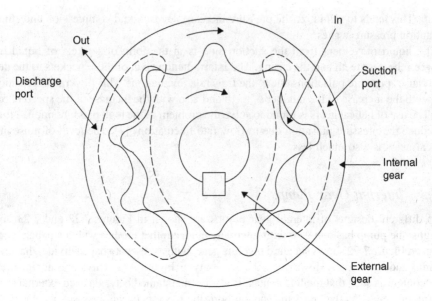

Figure 7.23   Configuration of internal gear pump (gerotor type)

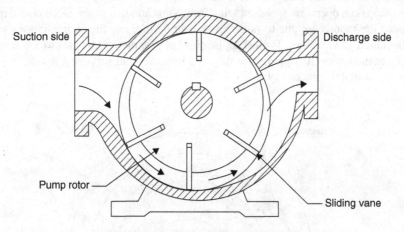

Figure 7.24   Vane pump configuration

7.7.4   Sliding Vane Pumps

The main components of a sliding vane pump are shown in Figure 7.24. The pump has a cylindrical rotor mounted in a casing equipped with closely fitted end plates. The rotor has a number of radial (or slanted) slots in which vanes are inserted. The slots, and hence the vanes, may be slanted backward to minimize friction and wear at the vane tip. The rotor is positioned eccentrically in the casing while the vanes, subjected to spring or hydraulic force, slide outward against a hardened surface in the housing. As the vanes proceed from the point of shortest distance between the rotor and housing (the topmost point), the space between the two increases and gets filled with liquid from the suction port. As the cycle continues, the spaces between the vanes become smaller and the liquid is discharged through the opposite discharge port. The number of vanes may vary from one pump to another and may be as high as 12.

Vane pumps may have either fixed or variable displacement, resulting in constant or variable capacity. The variable displacement design has a mechanism for changing the eccentricity between the cam ring that contacts the vane tips and rotor. Decreasing the eccentricity reduces the amount of fluid displaced during each revolution of the rotor. The cam ring may be moved manually or in reaction to hydraulic pressure. Variable displacement pumps are used where flow rate requirements vary considerably. It is often more practical to change the pump displacement rather than alter the speed of the prime mover, which is usually an electric motor or governed engine.

The theoretical flow rate delivered by the pump is obtained from

$$Q_{th} = 2LenDN \sin(\pi/n) \tag{7.26}$$

where L is the vane width, e is the eccentricity, n is the number of vanes, D is the casing inner diameter, and N is the rotor speed. The actual flow rate is less than Q_{th} because of leakage, and can be obtained from

$$Q = \eta_{vol} Q_{th} \tag{7.27}$$

To minimize leakage, the vanes must be kept in contact with the cam ring. Centrifugal force may not be enough, and various arrangements involving spring and pressure loading are employed. Such spring force should be optimized because the vanes must be supported on a liquid film that should be thick enough to prevent metal-to-metal contact and thin enough to minimize leakage. The volumetric efficiency of vane pumps is generally above 90%. Typical performance curves of a vane pump are shown in Figure 7.25.

In a simple sliding vane pump, the inlet and outlet ports are placed on opposite sides, thus causing the same type of hydraulic unbalance as is present in gear pumps (Section 7.7.2). The same limitations for operating the pump at pressures higher than the design pressure apply. Vane pumps are used over roughly the same pressure range as gear pumps, but are found most frequently in high-pressure applications (10–17 MPa). Sliding vane pumps are commonly used as the fluid power source in industrial hydraulic machinery.

7.7.5   Other Rotary Pumps

Other rotary pumps include screw pumps, lobe pumps, and rotary piston pumps. The screw pumps shown in Figure 7.26, have either two or three rotors with helical threads. The rotors mesh so that their contact points progress axially down the screws, creating the pumping action. These pumps produce continuous flow with relatively little pulsation or fluid agitation. Screw pumps have high efficiency and low noise level. They are available in speed ranges up to 3500 rpm, pressures up to 17 MPa, and flow rates that may exceed 6 L/s. Applications range from process transfer pumps to hydraulic power pumps for submarine systems.

The pressure variation around the pump rotors produces radial and axial hydraulic unbalanced forces. The mechanical and hydraulic methods used for balancing these forces

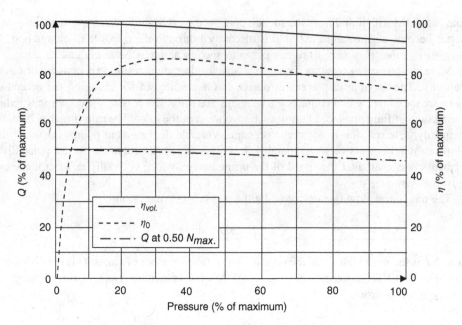

Figure 7.25   Typical performance curves of a sliding vane pump

differ from one design to another. One of the fundamental design differences lies in the method of driving the rotors while maintaining the running clearances between them. The two commonly used basic designs are

1. Timing the motion of the rotors by external means for synchronizing the motion of the threads, and also for providing the driving force on each rotor. In this design, the threads are not contacting each other or contacting the housing bores. Figure 7.26a shows a typical screw pump with two **timed rotors**.
2. The *untimed rotors* design relies on the precision of the screws for proper meshing and for the transmission of rotary motion. In this design, the housing bores are used as journal bearings for supporting the rotors. Figure 7.26b shows a typical screw pump with three **untimed rotors**.

Lobe pumps have the same operation principle as that of gear pumps. Each rotor may have two to four lobes (Figure 7.27 shows a three-lobe design). However, in these pumps, timing gears are employed to hold the lobes in place, and the lobes are positioned so that there is a small clearance between themselves and between each one and the housing at any mesh position. Sometimes helical lobes are used in lower pressure applications. Lobe pumps find important applications in the food processing industries. These pumps are usually low-speed types constructed of stainless steel. They are designed so that relatively large pockets are formed by the rotation of the rotors. In this way, the suspended solids often present in food products may be pumped without being sheared or crushed.

Other rotary displacement pumps include rotary piston pumps that are available in two main configurations; the first has radial piston stroke and the second has an axial-piston stroke as

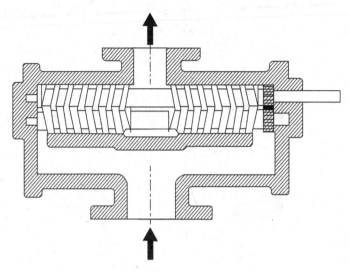

Figure 7.26a   Schematic of a screw pump with two timed rotors (double-end arrangement).

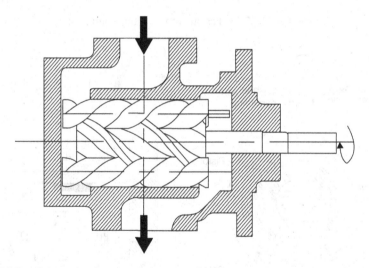

Figure 7.26b   Schematic of a screw pump with three untimed rotors (single-end arrangement)

shown in Figures 7.28a and b. They have the highest volumetric efficiencies of all displacement pumps and are mostly used in high-pressure (10–20 MPa) applications in hydraulic power systems. Some axial-piston pumps develop pressures up to 50 MPa.

In the radial type, spring-loaded pistons are housed in a cylinder block that is eccentric to the pump housing (Figure 7.28a). As the block rotates, the pistons move radially in and out of their respective cylinders, producing the pumping action. In another design, pistons are contained in a stationary housing and spring loaded against a rotating cam. The stroke of each piston is determined by the eccentricity, e, which is the distance between the center of the cylinder block and the center of the casing. Since the stroke is equal to $2e$, the theoretical flow rate is given by

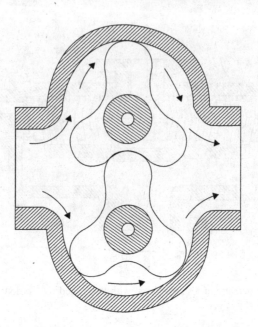

Figure 7.27   Schematic of lobe pump configuration

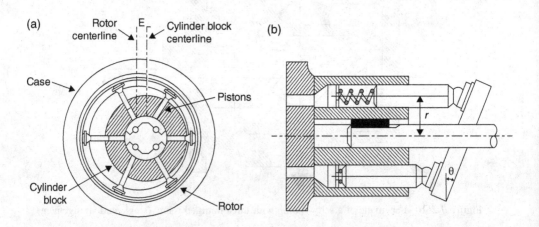

Figure 7.28   Schematic of rotary piston pump; (a) radial-stroke type and (b) axial-stroke type

$$Q_{th} = \frac{1}{2} e \pi d^2 n N \tag{7.28}$$

where e is the eccentricity, d is the cylinder diameter, n is the number of cylinders, and N is the speed of rotation. The actual flow rate is less than Q_{th} because of leakage. Since the flow rate depends on the eccentricity, it is possible to design these pumps in order to have variable eccentricity and thus variable flow rate when operating at the same speed.

In the axial-piston arrangement (Figure 7.28b), the cylinders are placed parallel to the pump driving shaft. They are pressed by springs against a plate that is inclined at an angle, θ, to the shaft axis. When operating this pump, either the driving shaft rotates the cylinder block against a non-rotating angle plate, or it rotates an angle plate against a non-rotating cylinder block. The result causes a reciprocating axial motion for each piston. For these pumps, the theoretical flow rate is given by

$$Q_{th} = \frac{1}{2} \pi d^2 r n N \tan \theta \qquad (7.29)$$

where d is the piston diameter, r is the cylinder centerline radius, n is the number of cylinders, and N is the speed of rotation. By changing the angle, θ, the pump capacity can be controlled. The flow rate can also be controlled by the speed of the prime mover in both radial- and axial-rotary piston pumps. To achieve high volumetric efficiencies (over 95%) these pumps require small clearances and very accurate machining. Piston clearances are usually held to less than 25 μm. The rate of leakage is indicative of the piston wear and can be used for condition monitoring of the pump performance [5].

References

1. Hong X, Lijuan Y, Baode J, Zhiliang T. (2009) Research of internal leakage theory model in the exterior meshing gear pump. First International Conference on Intelligent Human-Machine Systems and Cybernetics, 26–27 Aug 2009.
2. Ali, K.H. (1989) The design and performance of gear pumps with particular reference to marginal suction condition, PhD thesis. Cranfield Institute of Technology.
3. Manring, N.D. and Kasaragadda, S.B. (2003) The theoretical flow ripple of an external gear pump. *ASME J Dyn Sys Meas Con*, **125**, 397–404.
4. Mitome, K. and Seki, K. (1983) A new continuous contact low-noise gear pump. *ASME J Mech Trans Autom*, **105** (IV), 736–741.
5. Li, Z. (2005) Condition monitoring of axial piston pump. MSc thesis. ME Department, University of Saskatchewan, Canada.

Problems

7.1 A five-cylinder double-acting piston pump is driven at a speed of 120 rpm and delivers a flow rate of 0.36 L/s. Determine the slip, S.

Data:	Piston diameter = 25 mm
	Crank radius = 20 mm
	Length of connecting rod = 200 mm
	Diameter of all pipes = 75 mm

7.2 A horizontal single-cylinder reciprocating pump has a plunger of diameter 150 mm and a stroke of 300 mm. The connecting rod length is 0.6 m. The suction pipe is 100 mm in diameter and 7 m long. The fluid free surface in the suction tank is 3 m above the cylinder

centerline and the tank is open to the atmosphere. The fluid pumped has a density of
870 kg/m^3, kinematic viscosity of 3.2×10^{-5} m^2/s, and a vapor pressure of 4.8 kPa abs.
If the pump is operating at 30 rpm, find the pressure at the pump inlet at the beginning,
middle, and end of the suction stroke.
If the *NPSHR* at 30 rpm is 1.2 m, investigate the system with respect to cavitation. Also
determine the minimum permissible static suction head if cavitation is to be avoided.
Assume that the length of the suction pipe is unchanged.
Note: Consider the friction coefficient, $f = 0.012$.

7.3 The pump described in the above problem is supplying the pumped fluid to a pressurized
tank in which the pressure is 350 kPa gage and the fluid free surface is 8 m above the cylinder centerline. The delivery pipe is 100 mm in diameter and 24 m long. Knowing that the
pump has a slip factor of 0.03 and a mechanical efficiency of 92%, determine:

a. average head developed by the pump
b. power required to drive the pump.

7.4 A single-cylinder piston pump is used to pump a certain fluid (sp. gr. = 0.85 and vapor pressure = 32 kPa abs). The pump location is 4 m below the fluid free surface in the suction tank
and 32 m below the fluid free surface in the delivery tank. Both tanks are open to the atmosphere. The pump is driven at a crank speed of 90 rpm. Other details of the pump and piping
system are:

piston diameter = 10 cm
length of stroke = 8 cm
length of connecting rod = 20 cm
the *NPSHR* at 90 rpm = 0.9 m
suction pipe : $D_s = 5$ cm and $L_s = 5$ m
delivery pipe: $D_d = 5$ cm and $L_d = 36$ m
friction coefficient for all pipes = 0.02.

Assuming no air vessels are used in the system, determine:

a. maximum pressure in the cylinder and the corresponding crank angle, θ
b. minimum permissible height of fluid in the suction tank (hss) for cavitation to be
   avoided
c. power required to drive the pump, assuming a mechanical efficiency of 90% and a
   slip of 3%.

7.5 A single-cylinder piston pump is used to deliver hot oil (sp. gr. = 0.85, $\mu = 0.01$ N.s/m^2, and
vapor pressure $p_v = 60$ kPa abs.) from an open tank A to a pressurized tank B as shown in
Figure 7.29. The pump is driven at a speed of 60 rpm. The following data are given for the
pump and piping system:

Piston diameter = 12 cm	Delivery pipe: $D_d = 8$ cm and $L_d = 25$ m
Length of stroke = 8 cm	Friction coefficient for all pipes = 0.04
Length of connecting rod = 22 cm	Mechanical efficiency = 0.92
Suction pipe: $D_s = 8$ cm and $L_s = 4$ m	Slip = 0.03

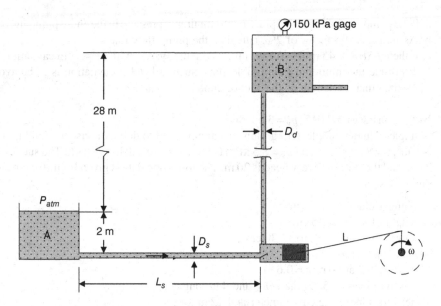

Figure 7.29   Schematic sketch of the pumping system for Problem 7.5

a. Determine the average head developed by the pump.
b. What is the power required to drive the pump?
c. Considering the *NPSHR–N* curve shown in Figure 7.30, determine the maximum operating speed if cavitation is to be avoided.

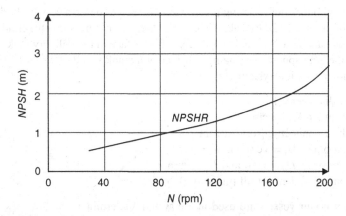

Figure 7.30   Variation of *NPSHR* with the pump speed for Problem 7.5

7.6 A single-cylinder double-acting piston pump has a piston of diameter of 150 mm and a stroke of 200 mm. The connecting rod length is 0.5 m. The suction pipe is 100 mm in diameter and 7 m long. The static suction head is 3 m and the suction tank is open to the atmosphere.

a. If the pump is working at 45 rpm, find the minimum pressure in the entire pumping system.
b. Assuming a slip factor of 2%, calculate the pump flow rate.
c. If the *NPSHR* at 45 rpm is 0.8 m, investigate the system with respect to cavitation. Also determine the minimum permissible static suction head if cavitation is to be avoided. Assume that the length of the suction pipe is unchanged.

Note: Consider $f = 0.015$, $p_v = 3.5$ kPa.

7.7 A single-cylinder, single-acting plunger pump is used to deliver a viscous fluid ($\mu = 0.02$ N·s/m², $\gamma = 8.5$ kN/m³, and $p_v = 8.5$ kPa) from a sump to a delivery tank. The suction lift is 2.5 m and the static delivery head is 20 m. The following data is given for the pump and the piping system:

plunger diameter = 12 cm
length of stroke = 15 cm
length of connecting rod = 30 cm
crank speed = 60 rpm
the *NPSHR* at 60 rpm = 0.6 m
suction pipe is 7.5 cm diameter and 4 m long
delivery pipe is 7.5 cm diameter and 22 m long
the pipes are all commercial steel pipes

Assuming a mechanical efficiency of 94% and a slip of 2%, determine:

a. minimum and maximum pressures in the cylinder and the corresponding crank position for each,
b. maximum permissible depth of fluid in the sump if cavitation is to be avoided
c. the power required to drive the pump.

7.8 A single-cylinder double-acting piston pump is used to pump a certain fluid (sp. gr. = 0.85 and vapor pressure = 24 kPa abs). The pump location is 5 m below the fluid free surface in the suction tank and 28 m below the fluid free surface in the delivery tank. Both tanks are open to the atmosphere. The pump is driven at a crank speed of 120 rpm. Other details of the pump and piping system are:

piston diameter = 10 cm
length of stroke = 8 cm
length of connecting rod = 20 cm
suction pipe : $D_s = 5$ cm and $L_s = 5$ m
delivery pipe: $D_d = 5$ cm and $L_d = 36$ m
friction coefficient for all pipes = 0.015

Assuming no air vessels are used in the system, determine:

a. power required to drive the pump, assuming $\eta_{mech} = 90\%$ and slip, $S = 3\%$
b. maximum pressure in the cylinder
c. minimum value of *NPSHA*.
d. If a large air vessel is fitted in the suction side close to the pump such that the flow in the suction pipe can be assumed steady, what will be the new value of *NPSHA*?

8

Introduction to Fans and Compressors

8.1 Introduction

Gas movers are very widely used in many industries such as petroleum and petrochemical industries, gas transport pipelines, jet engines, gas turbines, ventilation and air conditioning systems, and many others. Some compressors consume huge amounts of power that may reach 15 MW or more, and with a discharge gas pressure as high as 180 bars or more. Recently, multiple compressors have been designed to handle a number of gases at different pressures and flow rates. Such machines are driven by a single prime mover through a gear box with many output driving shafts rotating at different speeds. Several compressors are driven by these shafts to handle different gases.

Although the main components of fans and compressors are very much the same as those of pumps (shaft, impeller, diffuser, casing, seals, etc.) the shape of these components are different. This is mainly because flow in pumps is always incompressible while flow of gases in gas movers may be compressible or incompressible. Low flow velocities exist in pumps and fans while flow velocities in compressors are comparable to the speed of sound. The materials used for manufacturing compressors are different from those used in pumps. While pumps can be used to handle liquids that may contain some solid particles (such as river water, sea water, or crude oil), compressors are very sensitive to the presence of solid particles in the gas. Due to the high speeds inside the compressor, solid particles impacting the impeller vanes create severe erosion. This is one of the main problems occurring in the compressor blades of jet engines and gas turbines operating in desert environments. Similar to pumps, fans and compressors are designed in radial-type and axial-type configurations. Although axial-flow impellers are hardly found in multistage pumps, they are very common in multistage compressors, as for example, jet engine compressors. Also, some multistage compressors are designed to have a combination of radial and axial flow impellers. The following sections include detailed analysis of centrifugal fans and compressors.

Pumping Machinery Theory and Practice, First Edition. Hassan M. Badr and Wael H. Ahmed.
© 2015 John Wiley & Sons, Ltd. Published 2015 by John Wiley & Sons, Ltd.

8.2   Centrifugal Fans

Fans are gas movers that develop a very small pressure increase and are widely used in ventilation, air-conditioning systems, air-cooled heat exchangers, cooling towers, and many others. Compressors are also gas movers but are characterized by higher pressure increase. It is widely accepted that the gas mover is called a fan if the density change through the machine is less than 5%, otherwise it is called a compressor. However, according to the ASME code, the point of division is a 7% density increase, which corresponds to an isentropic compression ratio of about 1.1. The term 'blower' is sometimes used commercially for gas movers when the pressure rise is between 2 and 10 psi considering atmospheric conditions at the suction nozzle. Fans are normally designed as a single-stage machine while blowers and compressors may have more than one stage.

The analysis of fluid flow and impeller design considerations are the same as in centrifugal pumps since the density change is very small and the flow can be considered incompressible. The performance curves are qualitatively the same, although the total (stagnation) pressure rise Δp_o is usually used instead of the total head H in the case of pumps. Of course, there are no cavitation problems in centrifugal fans or compressors.

The analysis of energy transfer from the fan rotor to the fluid is very much the same as previously discussed in centrifugal pumps. The reduction of the whirl component of flow velocity at the vane exit due to the circulatory flow effect is normally referred to as the '*slip*.' The slip causes less head and also less input power. The slip decreases with the increase of the number of vanes and with the decrease in the vane angle β_2. It also decreases with the decrease of impeller width. The slip factor correlations for centrifugal impellers obtained by different researchers were discussed by Dixon [1]. Several models based on analytical and experimental studies for the prediction of the slip factor were proposed by many researchers as for example the work by Sharma *et al.* [2] and Guo and Kim [3]. A comprehensive review of the earlier work done prior to 1967 was reported by Wiesner [4].

Two sectional views of a centrifugal fan impeller are shown in Figure 8.1, and the inlet and exit-velocity diagrams for a typical impeller are shown in Figure 8.2 for the case of radial inflow (no whirl at inlet). In most cases, fans have only a volute casing but with no diffusing vanes. The rest of the components are exactly the same for a centrifugal pump, but the diameter ratio D_2/D_1 is usually

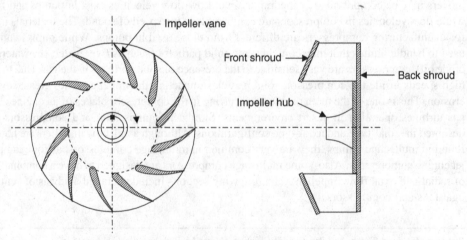

Figure 8.1   Centrifugal fan impeller with backward-curved vanes

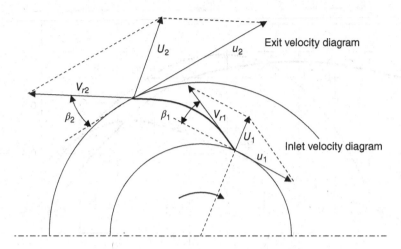

Figure 8.2 Inlet and exit velocity diagrams for a centrifugal fan with radial inflow

much smaller than that for a pump because of the shorter vanes. Since the pressure rise in these machines is small, the work done by the vanes is less than that in pumps, and this explains the need for shorter vanes. Fans are normally driven at speeds comparable to pump speeds (a few hundred rpm upto 3000 rpm). Large fans (such as cooling tower fans) are normally driven at low speeds while small fans used for cooling electronic equipment are driven at higher speeds. The volume flow rate supplied by a centrifugal fan is highly dependent on the fan speed, and the same applies to the pressure increase through the machine. Typical performance curves for a fan with backward-curved vanes (when operating at constant speed) are shown in Figure 8.3.

Centrifugal fans may be equipped with guide vanes at inlet (ahead of the impeller vanes). Such vanes are sometimes used to increase the inlet whirl which leads to reduction in both the head developed and the input power. The use of movable guide vanes at inlet makes it possible to operate the fan at slightly reduced capacity (may reach 75% of normal capacity) while maintaining almost the same efficiency. These guide vanes are used with forward, radial, or backward-curved vane impellers. The velocity diagrams at the vane inlet with and without guide vanes are shown in Figure 8.4.

Example 8.1
It is required to design a centrifugal fan to be used for air circulation in a water cooling tower. The fan is designed to operate at 1200 rpm and delivers 16 m³/s against a stagnation pressure difference (Δp_o) of 2.8 kPa. This operating condition represents the design point at which there are no shock losses and no whirl at inlet. The fan draws air from the atmosphere ($\rho = 1.2$ kg/m³) and the flow enters the impeller radially (with no prewhirl) at the design point. The impeller diameters at vane inlet and vane exit are considered to be 0.9 m and 1.35 m, respectively. The radial velocity component through the impeller is kept constant at 20 m/s. Considering $\eta_{vol.} = 96\%$, $\eta_{vane} = 88\%$, $\eta_{hyd.} = 85\%$, $\eta_{mech.} = 95\%$, determine:

a. vane angles at inlet and at exit (β_1 and β_2)
b. impeller width at inlet and at exit (b_1 and b_2)
c. fan power consumption

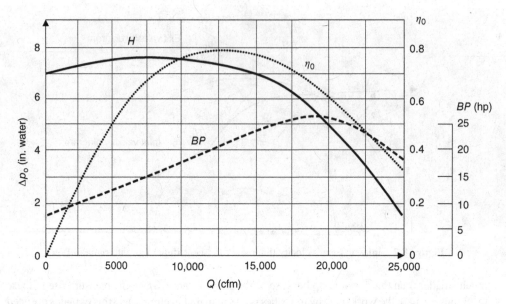

Figure 8.3 Typical performance curves for a centrifugal fan with backward-curved vanes operating at a constant speed

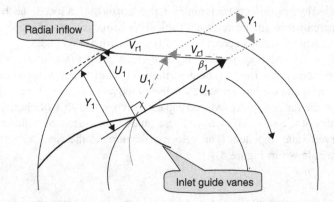

Figure 8.4 Inlet velocity diagrams for a centrifugal fan showing the use of inlet guide vanes for flow rate control

Solution

Data and preliminary calculations

$\rho = 1.2\,\text{kg/m}^3$, $\gamma = 11.8\,\text{N/m}^3$, $Q = 16\,\text{m}^3/\text{s}$, $N = 1200\,\text{rpm}$, $\Delta p_o = 2.8\,\text{kPa}$, $r_1 = 0.45\,\text{m}$,

$r_2 = 0.675\,\text{m}$, $Y_1 = Y_2 = 20\,\text{m/s}$, $\eta_{vol.} = 96\%$, $\eta_{vane} = 88\%$, $\eta_{hyd.} = 85\%$, $\eta_{mech.} = 95\%$,

$\omega = 125.7\,\text{rad/s}$, $u_1 = \omega r_1 = 56.5\,\text{m/s}$, $u_2 = \omega r_2 = 84.8\,\text{m/s}$, $H = \dfrac{\Delta p_o}{\gamma} = \dfrac{2800}{11.8} = 238\,\text{m}$

a. To determine β_1 and β_2, we proceed as follows:

$$\tan \beta_1 = \frac{Y_1}{u_1} = \frac{20}{56.5} \quad \Rightarrow \quad \underline{\beta_1 = 19.5^\circ}$$

$$H_e = \frac{u_2 V_2 - u_1 V_1}{g} = \frac{H}{\eta_{vane}\eta_{hyd.}} \quad \Rightarrow \quad \frac{84.8 V_2 - 0_1}{9.81} = \frac{238}{0.88 \times 0.85} \quad \Rightarrow \quad \underline{V_2 = 36.8\,\text{m/s}}$$

$$\tan \beta_2 = \frac{Y_2}{u_2 - V_2} = \frac{20}{84.8 - 36.8} \quad \Rightarrow \quad \underline{\beta_2 = 22.6^\circ}$$

b. To determine b_1 and b_2, we proceed as follows:

$$Q_{th} = \frac{Q}{\eta_{vol.}} = \pi D_1 b_1 Y_1 \quad \Rightarrow \quad \frac{16}{0.96} = \pi \times 0.9 \times b_1 \times 20 \quad \Rightarrow \quad \underline{b_1 = 0.295\,\text{m}}$$

$$Q_{th} = \frac{Q}{\eta_{vol.}} = \pi D_2 b_2 Y_2 \quad \Rightarrow \quad \frac{16}{0.96} = \pi \times 1.35 \times b_2 \times 20 \quad \Rightarrow \quad \underline{b_2 = 0.197\,\text{m}}$$

c. The brake power can be obtained from

$$BP = \frac{\gamma QH}{\eta_o} = \frac{Q \Delta p_o}{\eta_{vol.}\eta_{hyd.}\eta_{mech.}} = \frac{16 \times 2.8}{0.96 \times 0.85 \times 0.95} \quad \Rightarrow \quad \underline{BP = 57.8\,\text{kW}}$$

Example 8.2

A centrifugal fan is equipped with inlet guide vanes for flow rate control. The fan has a rated speed of 750 rpm and is designed to deliver air $(\rho = 1.2\,\text{kg/m}^3)$ at a rate of 4.25 m³/s when the flow has no prerotation at inlet. The fan impeller has the following dimensions:

Diameter at vane inlet, $D_1 = 525\,\text{mm}$	Outer diameter, $D_2 = 750\,\text{mm}$
Vane width at inlet, $b_1 = 172\,\text{mm}$	Vane width at exit, $b_2 = 100\,\text{mm}$
	Vane angle at exit, $\beta_2 = 70^\circ$

The pressure recovery in the volute casing amounts to 40% of the actual velocity head at the impeller exit $\left[(\Delta p)_{volute} = 0.4\left(\rho U_2^2/2\right)\right]$ and the leakage is negligibly small.

a. Determine the actual velocity and pressure at the fan discharge section.
b. Determine the fan brake power, BP.

Note: If needed, consider $\eta_{vane} = 88\%$, $\eta_{hyd.} = 85\%$, $\eta_{mech.} = 96\%$.

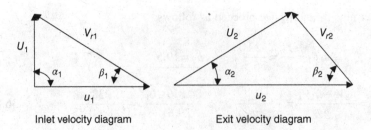

Inlet velocity diagram   Exit velocity diagram

Figure 8.5   Velocity diagrams for Example 8.2

Solution

Data:

$\omega = 78.5$ rad/s, $r_1 = 0.263$ m, $r_2 = 0.375$ m, $\beta_2 = 70°$, $\eta_{vane} = 88\%$, $\eta_{hyd.} = 85\%$, $\eta_{mech.} = 96\%$

a. $u_1 = \omega\, r_1 = 20.6$ m/s, $u_2 = \omega\, r_2 = 29.4$ m/s, $\gamma = \rho g = 1.2 \times 9.81 = 11.8$ N/m^3

$$Q_{th} = \frac{Q}{\eta_{vol.}} = \pi D_2 b_2 Y_2 \ \Rightarrow\ \frac{4.25}{1.0} = \pi \times 0.375 \times 0.1 \times Y_2 \ \Rightarrow\ \mathbf{Y_2 = 18.04\,m/s}$$

From Figure 8.5, we can write

$$V_2 = u_2 - Y_2 \cot \beta_2 = 29.4 - 18.04 \cot 70° = 22.9\,\text{m/s}$$

$$H_e = \frac{u_2 V_2 - u_1 V_1}{g} = \frac{29.4 \times 22.9 - 0}{9.81} = 68.6\,\text{m}$$

$$H_i = \eta_{vane} H_e = 0.88 \times 68.6 = \underline{60.4\,\text{m}}, \quad H = \eta_{hyd.} H_i = 0.85 \times 60.4 = \underline{51.3\,\text{m}}$$

$$H_i = \frac{u_2 V_2'}{g} \ \Rightarrow\ 60.4 = \frac{29.4 \times V_2'}{9.81} \ \Rightarrow\ \underline{V_2' = 20.2\,\text{m/s}}$$

$$U_2' = \sqrt{V_2'^2 + Y_2^2} \ \Rightarrow\ U_2' = \sqrt{20.2^2 + 18.04^2} \ \Rightarrow\ \underline{U_2' = 27.1\,\text{m/s}}$$

Let U_3 be the fluid velocity at the fan exit section, therefore

$$\frac{U_3^2}{2g} = 0.6 \frac{U_2'^2}{2g} \ \Rightarrow\ U_3 = \sqrt{0.6}\,U_2' = \underline{21\,\text{m/s}}$$

The pressure at the fan exit section (p_3) can be obtained as follows:

$$H = \frac{p_3 - p_{atm}}{\gamma} + \frac{U_3^2 - 0}{2g} \ \Rightarrow\ 51.3 = \frac{p_3 - 0}{11.8} + \frac{21^2}{2 \times 9.81}$$

$$\Rightarrow\ p_3 = 340\,\text{Pa gage} = \underline{0.34\,\text{kPa gage}}$$

b. The fan brake power can be obtained from

$$BP = \frac{\gamma QH}{\eta_o} = \frac{\gamma QH}{\eta_{vol.} \eta_{hyd.} \eta_{mech.}} = \frac{11.8 \times 10^{-3} \times 4.25 \times 51.3}{1.0 \times 0.85 \times 0.96}$$

$$\Rightarrow \underline{BP = 3.15 \, kW}$$

Example 8.3

Assume that the fan given in Example 8.2 is now equipped with inlet guide vanes for flow rate control (Figure 8.6). The flow rate is required to be reduced to 3.2 m³/s by using these guide vanes to impose prewhirl. What should be the flow angle at inlet (α_1)? In this case, the hydraulic efficiency is reduced to 80% due to the increase in shock losses. What will be the new brake power?

Solution

Let us first determine the inlet vane angle based on the data given in Example 8.2.

$$(Q_{th})_{old} = \frac{Q_{old}}{\eta_{vol.}} = \pi D_1 b_1 Y_1 \quad \Rightarrow \quad \frac{4.25}{1.0} = \pi \times 0.525 \times 0.172 \times (Y_1)_{old}$$

$$\Rightarrow \quad \underline{(Y_1)_{old} = 15 \, m/s}$$

$$\beta_1 = \tan^{-1} \frac{(Y_1)_{old}}{u_1} = \tan^{-1} \frac{15}{20.6} = 36°$$

After using the inlet guide vanes, the flow rate reduced to 3.2 m³/s, therefore

$$(Q_{th})_{new} = \frac{Q_{new}}{\eta_{vol.}} = \pi D_1 b_1 (Y_1)_{new} \quad \Rightarrow \quad \frac{3.2}{1.0} = \pi \times 0.525 \times 0.172 \times (Y_1)_{new} \quad \Rightarrow \quad \underline{(Y_1)_{new} = 11.3 \, m/s}$$

$$V_1 = u_1 - Y_1 \cot \beta_1 = 20.6 - 11.3 \cot 36° = 5.05 \, m/s$$

$$\alpha_1 = \tan^{-1} \frac{(Y_1)_{new}}{V_1} = \tan^{-1} \frac{11.3}{5.05} \quad \Rightarrow \quad \underline{\alpha_1 = 66°}$$

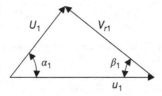

Figure 8.6 Inlet velocity diagram after using inlet guide vanes for flow rate control

To determine the new brake power, we first determine the new head as follows:

$$Y_2 = (Y_2)_{old} \frac{Q_{new}}{Q_{old}} = 18.04 \frac{3.2}{4.25} = 13.6 \, \text{m/s} \quad \text{and}$$

$$V_2 = u_2 - Y_2 \cot \beta_2 = 29.4 - 13.6 \cot 70° = 24.5 \, \text{m/s}$$

$$H_e = \frac{u_2 V_2 - u_1 V_1}{g} = \frac{29.4 \times 24.5 - 20.6 \times 5.05}{9.81} = 62.8 \, \text{m}$$

Therefore, $H_i = \eta_{vane} H_e = 0.88 \times 62.8 = 55.3 \, \text{m}$

$$(BP)_{new} = \frac{\gamma Q_{new} H_{new}}{\eta_o} = \frac{\gamma Q_{new} H_{i \, new}}{\eta_{vol.} \eta_{mech.}} = \frac{11.8 \times 10^{-3} \times 3.2 \times 55.3}{1.0 \times 0.96}$$

$$\Rightarrow \quad BP = 2.18 \, \text{kW}$$

8.3   Some Basic Concepts of High Speed Flow

8.3.1   The Speed of Sound, C, in an Ideal Gas

The speed of sound is defined as the speed of propagation of a small pressure wave. In general, the speed of propagation of a pressure wave is given by

$$C' = \sqrt{E_v/\rho}$$

where E_v is the fluid modulus of elasticity and ρ is the fluid density. For a sound wave, the pressure difference across the wave is extremely small and the process of wave propagation is considered isentropic. Accordingly, the speed of sound C becomes

$$C = \sqrt{[E_v]_s/\rho} = \sqrt{kp/\rho}$$

This is mainly because $E_v = kp$ for an ideal gas undergoing an isentropic process, where k is the specific heat ratio.

Now, by using the equation of state for an ideal gas, we can write

$$C = \sqrt{\frac{k(\rho RT)}{\rho}} = \sqrt{kRT} \qquad\qquad (8.1)$$

8.3.2 The Stagnation State

The stagnation state at a given section in a flow passage is a reference state that defines the reservoir condition from which the fluid expanded adiabatically and reversibly to reach the flow condition at the given section. It can also be defined as the state that will be reached if the flowing fluid is brought to rest through an isentropic process. Figure 8.7 shows a schematic of a hypothetical process through which part of the flowing fluid is brought to rest in an imaginary reservoir. The fluid properties in that reservoir define the stagnation state.

The stagnation state may also exist at a stagnation point in a real flow process as shown in Figure 8.8.

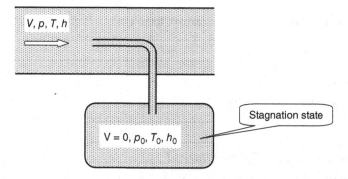

Figure 8.7 A hypothetical process from flow state to stagnation state

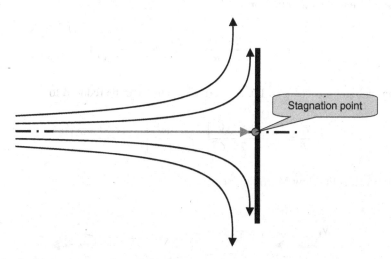

Figure 8.8 Streamlines for flow over a flat plate placed normal to the oncoming stream showing the stagnation point

8.3.3   The Stagnation Enthalpy, h_o

Consider the case of steady gas flow in the flow passage shown in Figure 8.9. If we write the energy equation between any two points 1 and 2 along the same streamline, then

$$q + h_1 + \frac{V_1^2}{2} + gz_1 = h_2 + \frac{V_2^2}{2} + gz_2 + w \tag{8.2}$$

For isentropic flow in a flow passage (no heat transfer, no friction, and no work done), the above equation can be reduced to

$$h_1 + \frac{V_1^2}{2} \simeq h_2 + \frac{V_2^2}{2} \tag{8.3}$$

If state 2 is replaced by the stagnation state 0, then

$$h + \frac{V^2}{2} = h_0 \tag{8.4}$$

where h_o is the stagnation enthalpy.

8.3.4   The Stagnation Temperature, T_o

Using Eq. (8.4) and assuming ideal gas behavior, then

$$h_0 = h + \frac{V^2}{2} \Rightarrow C_p T_0 = C_p T + \frac{V^2}{2} \text{ and } C_p = \frac{kR}{k-1}$$

where C_p is the specific heat at constant pressure and R is the gas constant. Therefore, we can write the above equation in the form

$$T_0 = T + \frac{(k-1)V^2}{2kR} = T \left[1 + \frac{(k-1)}{2} \frac{V^2}{kRT} \right]$$

Knowing that $kRT = C^2$ (Eq. (8.1)), the above equation can be reduced to

$$\frac{T_0}{T} = \left[1 + \frac{(k-1)}{2} \frac{V^2}{C^2} \right] \text{ or } \frac{T_0}{T} = \left[1 + \frac{(k-1)}{2} M^2 \right] \tag{8.5}$$

where $M = V/C$ is the flow Mach number.

Figure 8.9   Flow passage of irregular shape

8.3.5   The Stagnation Pressure, p_o

Since the process from the flow state to the stagnation state is isentropic (by definition), therefore $p_o/p = (T_o/T)^{k/(k-1)}$, and by using Eq. (8.5), we can write

$$\frac{p_0}{p} = \left[1 + \frac{k-1}{2}M^2\right]^{\frac{k}{k-1}} \tag{8.6}$$

Having defined two independent properties (p_o and T_o) or (p_o and h_o), the stagnation state becomes completely defined.

8.3.6   Important Features of Adiabatic Flow Processes

8.3.6.1   Isentropic Flow Process

Consider the case of an adiabatic expansion or compression flow process in a flow passage (1–2) as shown in Figure 8.10, and assume frictionless flow. In this case, the process is reversible and characterized by constant entropy (isentropic process). The application of Eq. (8.2) gives

$$h_1 + \frac{V_1^2}{2} = h_2 + \frac{V_2^2}{2} \quad \text{or} \quad h_{o1} = h_{o2}$$

and since $s_1 = s_2 = s_{o1} = s_{o2}$, the two stagnation states 01 and 02 are exactly the same. Hene we can show that all stagnation properties do not change with the direction of flow so long as the flow process is isentropic. It follows that $p_{o1} = p_{o2}$ and $\rho_{o1} = \rho_{o2}$.

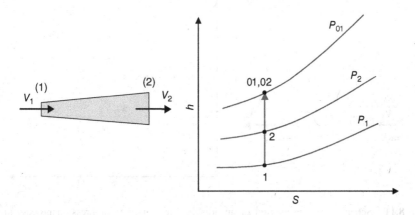

Figure 8.10   Schematic representation of an isentropic compression process on the h–s diagram

8.3.6.2 Process of Adiabatic Flow with Friction

Consider the case of adiabatic flow in a flow passage (1–2) associated with friction as shown in Figure 8.11. In this case, the process is irreversible and characterized by entropy increase. The application of Eq. (8.2) gives

$$h_1 + \frac{V_1^2}{2} = h_2 + \frac{V_2^2}{2} \quad \text{or} \quad h_{o1} = h_{o2}$$

Now, recalling the definition of entropy, $ds \geq \delta Q/T$, it follows that $ds > 0$ for adiabatic irreversible processes. Therefore $s_2 > s_1$ and consequently $s_{o2} > s_{o1}$. Accordingly, we can conclude that the two stagnation states 01 and 02 are different. This leads to $p_{o2} < p_{o1}$ and $\rho_{o2} < \rho_{o1}$.

Example 8.4

Air at a temperature of 47 °C and a pressure of 150 kPa abs. is flowing with a velocity of 250 m/s. Determine:

a. sonic velocity and the flow Mach number
b. stagnation temperature and stagnation pressure.

Solution

a. $C = \sqrt{kRT} = \sqrt{1.4 \times 287 \times (273 + 47)} = 359\,\text{m/s}$
   $M = V/C = 250/359 = 0.697$

b. Equation (8.6) $\Rightarrow \dfrac{p_0}{p} = \left[1 + \dfrac{k-1}{2}M^2\right]^{\frac{k}{k-1}} = \left[1 + \dfrac{1.4-1}{2}(0.697)^2\right]^{\frac{1.4}{1.4-1}} = 1.384$

   Therefore, $p_o = 207.5$ kPa

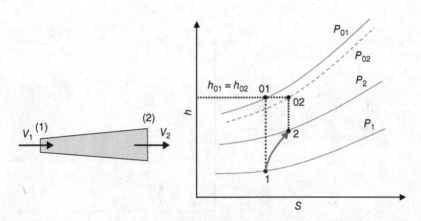

Figure 8.11 Schematic representation of an adiabatic irreversible compression process on the *h–s* diagram

Also, by using Eq. (8.5), we can write $\dfrac{T_0}{T} = 1 + \dfrac{k-1}{2}M^2 = 1 + \dfrac{1.4-1}{2}(0.697)^2 = 1.097$

Therefore, $T_o = 351.1 \text{ K} = 78.1°C$.

Example 8.5

Prove that, for small Mach numbers, the stagnation pressure can be obtained using incompressible flow relations $\left(p_0 = p + \tfrac{1}{2}\rho V^2\right)$ without losing much accuracy.

Solution

Equation (8.6) $\Rightarrow \dfrac{p_0}{p} = \left[1 + \dfrac{k-1}{2}M^2\right]^{\frac{k}{k-1}}$

Using power series expansion, the right-hand side of the above equation can be expressed as

$$\frac{p_0}{p} = 1 + \frac{k}{k-1}\frac{k-1}{2}M^2 + \frac{1}{2!}\frac{k}{k-1}\left(\frac{k}{k-1}-1\right)\left(\frac{k-1}{2}\right)^2 M^4 + \cdots$$

Simplify, $\dfrac{p_0}{p} = 1 + \dfrac{k}{2}M^2 + \dfrac{k}{8}M^4 + \cdots$

Now, if M is small (say up to $M = 0.25$), the first two terms on the right-hand side will have significant values but the following terms become very small and can be neglected. Accordingly, we can write

$$\frac{p_o}{p} \approx 1 + \frac{k}{2}M^2 = 1 + \frac{kV^2}{2C^2} \quad \Rightarrow \quad \text{and } C = \sqrt{kRT}$$

Therefore, $p_o \approx p + p\dfrac{k}{2}\dfrac{V^2}{kRT} = p + \dfrac{1}{2}\dfrac{p}{RT}V^2$ or $p_o \approx p + \dfrac{1}{2}\rho V^2$

The last expression is the one normally used in incompressible flows. In general, for small values of Mach number the assumption of incompressible flow will only result in a small error.

Example 8.6

Prove that the maximum change of density in a low Mach number flow (say up to $M = 0.25$) is very small.

Solution

The maximum density in any flow field occurs at the stagnation point, and so let us assume isentropic flow. Using ideal gas relations together with Eqs (8.5) and (8.6), we can write

$$\frac{p_0}{\rho} = \frac{p_o/RT_o}{p/RT} = (p_o/p)\,(T/T_o) = \left[1 + \frac{(k-1)}{2}M^2\right]^{\frac{k}{k-1}}\left[1 + \frac{(k-1)}{2}M^2\right]^{-1}$$

Simplify to obtain $\dfrac{\rho_0}{\rho} = \left[1 + \dfrac{(k-1)}{2}M^2\right]^{\frac{1}{k-1}}$

Considering air flow at Mach number $M = 0.25$,

$$\frac{\rho_0}{\rho} = \left[1 + \frac{(1.4-1)}{2}0.25^2\right]^{\frac{1}{1.4-1}} = 1.0315$$

Accordingly, the maximum change of density within the flow field will be

$$\frac{(\Delta\rho)_{max.}}{\rho} = \frac{\rho_0 - \rho}{\rho} = 3.15\%$$

This example explains why the flow can be assumed incompressible up to Mach number $M = 0.25$ or slightly higher.

Example 8.7

Assuming that the air flow process through a centrifugal fan is isentropic and the stagnation density ratio between exit and inlet is 1.07, calculate the stagnation pressure ratio between the fan exit and inlet.

Solution

Since the compression process is isentropic and using subscript 1 for inlet and 2 for exit, we can write

$$p_{o2}/p_{o1} = (T_{o2}/T_{o1})^{\frac{k}{k-1}} \quad \text{and} \quad \frac{\rho_{o2}}{\rho_{o1}} = \frac{(p_{o2}/RT_{o2})}{(p_{o1}/RT_{o1})} = \frac{p_{o2}}{p_{o1}}\frac{T_{o1}}{T_{o2}}$$

Therefore, $\dfrac{\rho_{o2}}{\rho_{o1}} = \dfrac{p_{o2}}{p_{o1}}\left(\dfrac{p_{o1}}{p_{o2}}\right)^{\frac{k-1}{k}} \Rightarrow \dfrac{p_{o2}}{p_{o1}} = \left(\dfrac{\rho_{o2}}{\rho_{o1}}\right)^{k} = (1.07)^{1.4} \simeq 1.1$

Alternative solution

Since $pv^k = Const.$ for an isentropic process, it follows that $p/\rho^k = Const.$

Now, the process from flow state to stagnation state is isentropic by definition. So we can write

$$p_o/\rho_o^k = Const. \quad \Rightarrow \quad \frac{p_{o2}}{p_{o1}} = \left(\frac{\rho_{o2}}{\rho_{o1}}\right)^{k} = (1.07)^{1.4} \simeq 1.1$$

8.3.7 Shockwaves

Shockwaves can occur in high speed centrifugal or axial flow compressors. Shockwaves are undesirable in compressors because they result in a loss of stagnation pressure. In

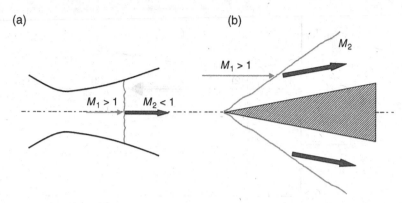

Figure 8.12 Schematic of (a) normal shock in a convergent-divergent nozzle and (b) oblique shocks in flow over a 2-D wedge

general, shockwaves can exist in totally enclosed flows (as in nozzles) and also in external flows (as in flow over a supersonic aircraft). The shock plane is very thin and may be perpendicular to the direction of flow, as in normal shocks, or may be inclined to the direction of flow, as in oblique shocks. Figure 8.12 shows a normal shock in a convergent–divergent (C-D) nozzle and oblique shocks in supersonic flow over a two-dimensional (2-D) wedge.

The main features of flow through shockwaves can be summarized as follows:

The shockwave thickness is very small (a small fraction of a millimeter).
The flow before any shockwave must be supersonic ($M_1 > 1.0$).
The flow after a normal shock is always subsonic ($M_2 < 1.0$), but flow after an oblique shock may be subsonic or supersonic.
The higher the Mach number before the shock the lower the Mach number after it.
Flow through the shock is characterized by severe reduction in flow velocity (the stronger the shock the higher the reduction in flow velocity).
Flow through the shock is accompanied by an instantaneous increase in fluid pressure and temperature.
Conversely, the stagnation pressure always decreases through any shockwave ($p_{o2} < p_{o1}$).
The flow process through the shock is irreversible.
The degree of irreversibility depends on the Mach number before the shock as well as the shape of the shock.
Flow through a shock may be considered adiabatic, and accordingly there is no change of stagnation temperature ($T_{o2} = T_{o1}$).
The shockwave strength, ξ, is defined as $\xi = \frac{p_2 - p_1}{p_1}$ ($\xi \ll 1$ for weak shocks).
The flow direction is the same before and after a normal shock, but the flow direction changes through an oblique shock.
Although shockwaves are undesirable, they are sometimes unavoidable.

Consider the simple case of flow through a convergent–divergent nozzle which is supplied with compressed gas from a large reservoir in which the pressure is p_{o1} and the temperature is T_{o1}, as shown in Figure 8.13. Flow starts in the nozzle when the back pressure, p_b, is less than p_{o1}.

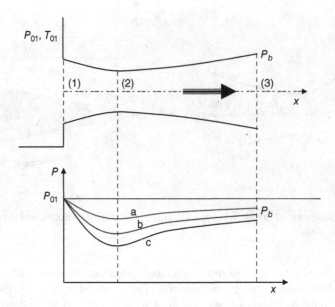

Figure 8.13 Pressure variation along the axis of a C-D nozzle when operating in the Venturi mode for different values of the back pressure (p_b)

As the back pressure gets lower, the velocity through the nozzle (and the mass flow rate) gets higher. So long as the flow in the entire nozzle is subsonic, the maximum velocity occurs at the throat (as in a Venturi tube) at which the pressure and density are minimum. The pressure distribution depends on the value of the back pressure, p_b. Typical pressure distributions are shown in the figure (curves a, b, and c). This mode of nozzle operation is called the Venturi mode.

The maximum value of the velocity at the throat is reached when the Mach number there reaches unity (i.e. $(M_2)_{max} = 1$). This occurs when the back pressure reaches p_{b1} and the corresponding pressure distribution is shown in Figure 8.14 (curve d). Any further decrease in the back pressure does not affect, in any way, the flow in the convergent part of the nozzle since the pressure wave cannot travel upstream of the throat anymore. When this condition is reached ($M_2 = 1$), the nozzle is said to be choked and this means that the flow rate through the nozzle cannot be increased anymore. The decrease in the back pressure below p_{b1} affects only the velocities and pressures in the diverging part of the nozzle (between sections 2 and 3). There is a unique value of the back pressure (say p_{b2}) which supports a full expansion in the diverging part (with no shock or expansion waves anywhere). The value of p_{b2} depends on the reservoir pressure and the nozzle geometry. The corresponding pressure distribution is shown in Figure 8.14 (curve e). Any back pressure in between p_{b1} and p_{b2} will cause a shockwave to occur either inside the nozzle or at its exit section. The location and strength of the shock depends on the value of the back pressure, p_b. In this case, shockwaves cannot exist in the converging part of the nozzle.

In the range of back pressures less than p_{b1} ($p_b \leq p_{b1}$) the nozzle becomes choked and the mass flow rate reaches a maximum that cannot be exceeded provided that the reservoir conditions (p_{o1} and T_{o1}) are unchanged. In this case, the variation of the mass flow rate with the back pressure (represented by the ratio p_b/p_{o1}) becomes as shown in Figure 8.15.

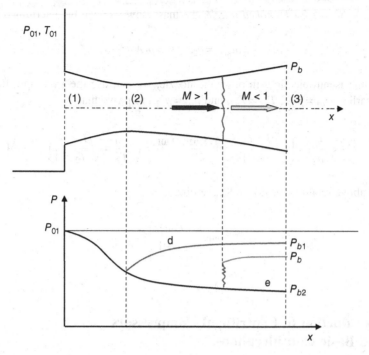

Figure 8.14 Pressure variation along the nozzle axis for different back pressures (with and without shockwaves)

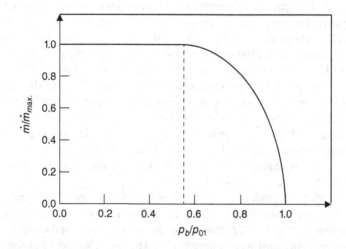

Figure 8.15 The variation of the mass flow rate through the nozzle with the change in the back pressure

The degree of irreversibility in the case of flow through a shock can be measured by the entropy increase, Δs, which can be calculated from

$$\frac{p_{o2}}{p_{o1}} = e^{-\frac{\Delta s}{R}} \tag{8.7}$$

Since $M_2 = V_2/C_2 = 1$ for a *choked nozzle*, the mass flow rate can be determined from

$$\dot{m}_{\max} = \rho_2 A_2 V_2 = \rho_2 A_2 C_2 \qquad (8.8)$$

If we assume isentropic flow through the nozzle, we can use the isentropic flow relations obtained earlier ($p_{o1} = p_{o2}$, $T_{o1} = T_{o2}$, and $\rho_{o1} = \rho_{o2}$) to prove that

$$\frac{\rho_{02}}{\rho_2} = \frac{\rho_{01}}{\rho_2} = \left[1 + \frac{k-1}{2}\right]^{\frac{1}{k-1}} \text{ and } \frac{C_{02}}{C_2} = \frac{C_{01}}{C_2} = \sqrt{\frac{T_{o1}}{T_2}} = \sqrt{\frac{T_{o2}}{T_2}} = \left[1 + \frac{k-1}{2}\right]^{\frac{1}{2}}$$

Using the above relations in Eq. (8.8), we obtain

$$\dot{m}_{\max} = \frac{\rho_{o1} A_2 C_{o1}}{\left(\dfrac{k+1}{2}\right)^{\frac{k+1}{2(k-1)}}} \qquad (8.9)$$

8.4 Introduction to Centrifugal Compressors and Basic Considerations

Centrifugal compressors are widely used in the petroleum and petrochemical industries, small gas turbines (to power road vehicles and commercial helicopters), diesel engine turbochargers, and many others. The main components of the centrifugal compressor are much the same as those for a pump; however, the compressor impeller vanes have an inducer section which extends from the inlet eye up to the region where the flow starts to turn towards the radial direction. The main function of the inducer is to move the fluid from the compressor inlet eye to the impeller vane with minimum hydraulic losses (friction and shock losses). When the compressor operates at its design point, the fluid velocity relative to the inducer vane should be aligned with the vane in order to minimize losses. As it moves through the inducer, the fluid velocity changes from the axial direction to enter the vane flow passage (channel) with a substantial component in the radial direction. Figure 8.16 shows general views of single- and double-shrouded impellers of centrifugal compressors. The figure shows a large number of vanes in comparison with a pump impeller. This is mainly to reduce the relative circulation in the channel between the vanes due to the high speed of rotation.

The axial length of the inducer section has been found to have a noticeable effect on the compressor efficiency. An inducer of large axial length having smoother curvature resulted in better performance. Figure 8.17 shows an impeller of a centrifugal compressor with a long inducer section and equipped with splitter blades. The use of splitter blades gives two advantages. The first advantage is to reduce the circulatory flow effect, which tends to decrease the slip velocity at the impeller exit. This decrease in the slip velocity allows higher work to be done by the impeller vanes and produces a larger increase in the fluid stagnation pressure through the machine. The second advantage is to increase the area of flow at the vane inlet region, thus reducing the fluid velocity at the inducer exit and allowing a higher mass flow rate without choking. The use of splitter blades is common in high speed compressors.

(a) (b)

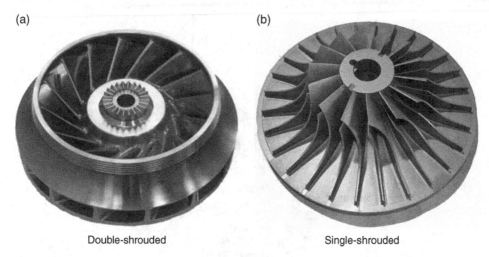

Double-shrouded Single-shrouded

Figure 8.16 General views of single-shrouded (Source: http://www.ist.rwth-aachen.de/typo3/index. php?id=950&L=1) and double-shrouded (Source: https://www.google.co.uk/search?q=Compressor +double-shrouded+impeller&source=lnms&tbm=isch&sa=X&ei=onYOU_G5KYGc0QWc1YHYCg& ved=0CAcQ_AUoAQ&biw=914&bih=430#q=centrifugal+compressor+impeller+images&tbm=isch& facrc=_&imgdii=_&imgrc=aGNUDZmPEOy0DM%253A%3BeBgfk9Tpypt4-M%3Bhttp%253A%252F% 252Fwww.merchantvesselmachinery.com%252Fimages%252Fcompressors1-large.jpg%3Bhttp%253A% 252F%252Fwww.merchantvesselmachinery.com%252Fcompressors.html%3B550%3B454) impellers of centrifugal compressors

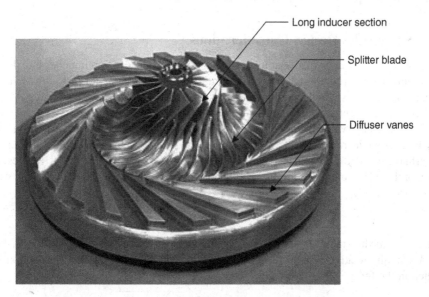

Long inducer section

Splitter blade

Diffuser vanes

Figure 8.17 A photograph of a compressor impeller having a long inducer section and splitter blades

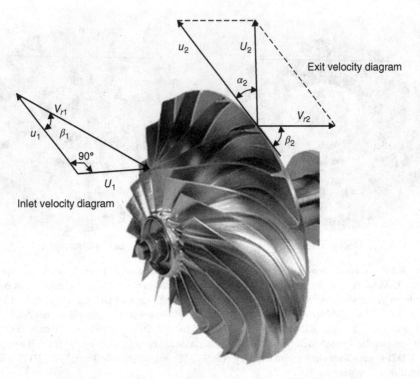

Figure 8.18 The velocity diagrams at the inducer inlet and impeller exit

The velocity diagrams at the inducer inlet and impeller exit are shown in Figure 8.18 where the impeller is equipped with backward-curved vanes.

8.5   Some Inlet Design Considerations

The height of the inducer blades at inlet is limited by the inlet eye and impeller hub diameters. These dimensions are important not only for allowing enough area for fluid flow but also for reducing friction losses and avoiding compressor choking at low flow rates. Figure 8.19 shows a schematic of a centrifugal compressor and the main dimensions.

When the relative velocity at inlet, V_{r1}, is too high the total pressure loss due to friction becomes very high. Also, increasing V_{r1} tends to increase the relative Mach number at the inducer inlet which may cause choking at small mass flow rates. Optimum performance at inlet requires suitable sizing of the inlet section that minimizes V_{r1}. For a given volume flow rate Q, we can write

$$Q = A_1 Y_1 \approx \pi \left(r_{s1}^2 - r_{h1}^2 \right) \sqrt{V_{rs1}^2 - u_{s1}^2}$$

where r_{h1} is the hub radius, r_{s1} is the inducer tip radius, u_{s1} is the inducer tip speed ($u_{s1} = \omega\, r_{s1}$), and V_{rs1} is the relative velocity at the inducer tip. Accordingly, the above equation can be written in the form,

$$Q = A_1 Y_1 \approx \pi \left(r_{s1}^2 - r_{h1}^2 \right) \sqrt{V_{rs1}^2 - \omega^2 r_{s1}^2} \qquad (8.10)$$

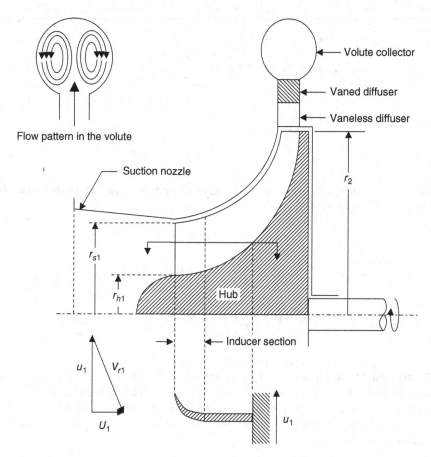

Figure 8.19 A schematic of a centrifugal compressor showing the main dimensions and the velocity diagram at the inducer inlet

If Q and r_{h1} are to be kept unchanged, there will be an optimum value of r_{s1} that results in minimum relative velocity at inlet (V_{rs1}). Accordingly, we can make the following arguments based on Eq. (8.10):

If r_{s1} is very large then Y_1 becomes very small, but u_{s1} will be very large. This tends to produce a very large V_{rs1} (Figure 8.20a).
If r_{s1} is very small then Y_1 becomes very large, which also tends produce a to very large V_{rs1} (Figure 8.20b).

The optimum value of r_{s1} is between those two extremes. If it is required to obtain maximum flow rate, Q_{max}, for given constant values of V_r and r_h, we can proceed as follows:

$$\frac{dQ}{dr_{s1}} = 0 \quad \Rightarrow \quad 2\pi r_{s1}\sqrt{V_{rs1}^2 - \omega^2 r_{s1}^2} - \pi\left(r_{s1}^2 - r_{h1}^2\right)\frac{1}{2}\frac{2\omega^2 r_{s1}}{\sqrt{V_{rs1}^2 - \omega^2 r_{s1}^2}} = 0$$

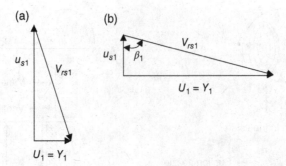

Figure 8.20 Typical velocity diagrams at the inducer inlet for the two cases of (a) large inducer tip radius and (b) small inducer tip radius

$$\text{Simplify } 2\overbrace{\left(V_{rs1}^2-\omega^2 r_{s1}^2\right)}^{Y_1^2} = \left(1-\frac{r_{h1}^2}{r_{s1}^2}\right)\overbrace{\omega^2 r_{s1}^2}^{u_{s1}^2} \Rightarrow 2Y_1^2 = k u_{s1}^2 \text{ where } k = \left(1-\frac{r_{h1}^2}{r_{s1}^2}\right)$$

$$\text{Therefore, } k = 2\left(Y_1/u_{s1}\right)^2 = 2\tan^2\beta_{s1} \ \ Or \ \ \frac{r_{h1}}{r_{s1}} = \sqrt{1-2\tan^2\beta_{s1}} \qquad (8.11)$$

The ratio r_{h1}/r_{s1} usually ranges between 0.3 and 0.6.

The maximum flow rate supplied by the compressor is related to the maximum permissible value of Y_1 which is, in turn, governed by the maximum relative Mach number at inlet M_{r1} ($=V_{rs1}/C_1$). The maximum value of M_{r1} is normally kept restricted to 0.8, in order to avoid shockwave formation. In the absence of inlet guide vanes, M_{r1} is related to the absolute inlet Mach number M_1 by

$$M_1 = M_{r1}\sin\beta_1 \qquad (8.12)$$

The optimum value of β_1 that results in maximum flow rate depends on the inlet conditions as well as the limiting value set for M_{r1}.

8.6 One-Dimensional Flow Analysis

Most of the flow processes taking place inside a typical centrifugal compressor are three-dimensional and characterized by high speeds (comparable to the speed of sound) and associated with considerable density variations. Such high speeds are essential to achieve a high stagnation pressure ratio between the compressor exit and inlet sections. However, designers always try to keep the maximum flow velocity below the speed of sound ($M < 1$) in order to avoid the formation of shockwaves. This is mainly because of the high irreversibility through these shocks that will eventually reduce the compressor efficiency. Due to such speed limitations, the stagnation pressure ratio (exit pressure/inlet pressure) per stage is normally less than

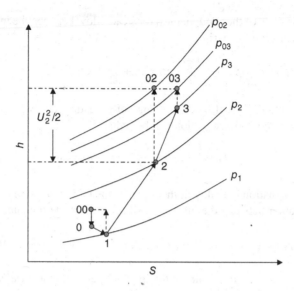

Figure 8.21 The thermodynamic processes in a centrifugal compressor

5, in most applications. Much higher pressure ratios (10:1) have been achieved per stage but at reduced efficiency. If one-dimensional analysis is used (as used in pump analysis), the flow process can be presented on the h–s diagram, as shown Figure 8.21. Points 0, 1, 2, 3, and 4 represent the following:

Point 00 – the atmospheric (or suction reservoir) condition
Point 0 – the state at inlet to the suction nozzle
Point 1 – the state at the impeller inlet eye
Point 2 – the state at the impeller exit
Point 3 – the state at the compressor discharge section

The flow processes taking place between the suction and discharge nozzles can be described as follows:

Process 00 → 0 occurs between a point far away in the atmosphere and the inlet of the suction nozzle. This process is assumed to be isentropic because there is negligible friction.
Process 0 → 1 occurs inside the suction nozzle, between the nozzle inlet and the impeller vane inlet (or inducer inlet). This process is assumed to be an adiabatic irreversible expansion process associated with friction.
Process 1 → 2 occurs inside the impeller between its inlet and exit sections. This process is assumed to be an adiabatic irreversible compression process associated with friction and with work being done by the impeller vanes.
Process 2 → 3 occurs in the diffuser flow passages between impeller exit and the exit of the compressor discharge nozzle. This process is assumed to be an adiabatic irreversible compression process associated with friction.

Applying the first law of thermodynamics for steady flow between the inlet and exit sections of the suction nozzle, we can write

$$q + h_0 + \frac{U_0^2}{2} + gz_0 = h_1 + \frac{U_1^2}{2} + gz_1 + w_{0-1}$$

Knowing that $q = 0$ (adiabatic flow), the change of potential energy $[g(z_1 - z_0)]$ is negligibly small, and $w_{0-1} = 0$, the above equation can be reduced to

$$h_0 + \frac{U_0^2}{2} = h_1 + \frac{U_1^2}{2} \quad \Rightarrow \quad h_{00} = h_{01} \tag{8.13}$$

where h_{01} is the stagnation enthalpy at the impeller inlet. Now apply the angular momentum conservation equation between the inlet and exit sections of the impeller to obtain,

$$T_d = \dot{m}\,(V_2 r_2 - V_1 r_1) \quad \Rightarrow \quad P_{input} = T_d \omega = \dot{m}\,(V_2 r_2 - V_1 r_1)\omega = \dot{m}\,(V_2 u_2 - V_1 u_1)$$

Knowing that w_{1-2} is the work done by the impeller per unit mass, we can write

$$w_{1-2} = P_{input}/\dot{m} = V_2 u_2 - V_1 u_1 \tag{8.14}$$

We can also apply the energy equation between the impeller inlet and exit sections to obtain

$$h_1 + \frac{U_1^2}{2} = h_2 + \frac{U_2^2}{2} + w_{1-2} \quad \Rightarrow \quad w_{1-2} = h_{01} - h_{02}$$

The term w_{1-2} in the above equation represents the work done on the impeller according to thermodynamic conventions. If we redefine w_{1-2} to represent the work done by the impeller, the above equation takes the form

$$w_{1-2} = h_{02} - h_{01} \tag{8.15}$$

Using Eqs (8.14) and (8.15), we get

$$h_{02} - u_2 V_2 = h_{01} - u_1 V_1 \quad \text{or} \quad h_2 + \frac{U_2^2}{2} - u_2 V_2 = h_1 + \frac{U_1^2}{2} - u_1 V_1 \tag{8.16}$$

The physical quantity on the two sides of Eq. (8.16) has the same value at the inlet and exit sections of the impeller. Accordingly, it must have the same value at any section in between. This physical quantity is labeled 'I' and is called the rotational enthalpy and therefore

$$I = h + \frac{U^2}{2} - uV = \text{Constant} \tag{8.17}$$

The above equation is important in the analysis of the flow through the impeller, especially when studying the impeller choking.

The very high velocity at impeller exit is partially transformed into pressure increase through the diffuser (which may be vaneless or equipped with vanes) and/or the volute collector.

The flow process from impeller exit (II) to the compressor discharge nozzle (III) can be considered adiabatic but associated with friction and accordingly,

$$h_{03} = h_{02} \quad and \quad p_{03} < p_{02} \tag{8.18}$$

The stagnation pressure loss $(p_{02} - p_{03})$ depends on the irreversibilities in the flow passages following the impeller which may be due to friction effects or the presence of shockwaves in the diffuser.

In normal operating condition, the whirl component (V_1) at the vane inlet is negligibly small and in this case, Eq. (8.14) and (8.15) can be reduced to

$$w_{1-2} = h_{02} - h_{01} \approx u_2 V_2 \tag{8.19}$$

8.7 Effect of Circulatory Flow (Slip)

Due to the inertia effect, the fluid inside the impeller experiences relative circulation in a direction opposite to the impeller rotation (similar to what happens in the impeller of a centrifugal pump). As a result, the actual whirl component at exit V_2' is less than V_2. The ideal (*solid lines*) and actual (*dotted lines*) velocity diagrams at the impeller exit are as shown in Figure 8.22.

The slip velocity V_s and the slip factor S_f are defined as

$$V_s = V_2 - V_2' \quad and \quad S_f = V_2'/V_2 \tag{8.20}$$

As explained in pumps, the circulatory flow effect depends on the speed of rotation, the impeller and vane shapes, the number of vanes, and the fluid properties. Stodola [5] proposed that the slip velocity is equal to the peripheral velocity resulting from the circulatory flow between any two successive vanes and so he deduced the following expression for the slip factor:

$$S_f = 1 - \frac{(\pi/Z)\sin\beta_2}{1 - (Y_2/u_2)\cot\beta_2} \tag{8.21}$$

where Z is the number of impeller vanes.

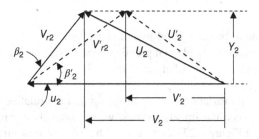

Figure 8.22 Effect of circulatory flow on the velocities at impeller exit

Stodola's equation (8.21) was found to be accurate for β_2 in the range 20–30°. Stanitz [6] obtained the following expression for S_f based on a potential flow solution:

$$S_f = 1 - \frac{0.63\,(\pi/Z)}{1-(Y_2/u_2)\cot\beta_2} \qquad (8.22)$$

Equation (8.22) was found to be accurate for nearly radial vanes ($\beta_2 = 80$–90°). Another relation for the slip factor, S_f, covering the intermediate range of β_2 (30–80°) was obtained by Busemann [7] and can be written in the form

$$S_f = \frac{A - B\,(Y_2/u_2)\cot\beta_2}{1-(Y_2/u_2)\cot\beta_2} \qquad (8.23)$$

where A and B are functions of r_2/r_1, β_2, and Z. For most pumps and compressors, $r_2/r_1 \gg 1$, which makes the value of B very close to 1.0 and A is a function of β_2 and Z only [8].

A simple approach that unifies the slip factor prediction methods adopted by Busemann, Stodola, Stanitz, and Wiesner proposed by Von Backström [9]. In his approach, an expression for the slip velocity was derived in terms of a single relative eddy, centered on the rotor axis, instead of the distributed multiple eddies (an eddy in each vane flow passage) that was initially proposed by Stodola [5]. A blade solidity parameter (blade length divided by spacing at rotor exit) was introduced and used as a prime variable for the determination of the slip. Backström recommended that solidity be used for correlating slip in centrifugal impellers as given by

$$S_f = 1 - \frac{1}{1 + \dfrac{5}{2\pi}\dfrac{(1-RR)Z}{\sqrt{\cos\beta_2}}} \qquad (8.24)$$

where RR is the vane radius ratio (vane radius at inlet/radius at exit) and Z is the number of vanes. The vane solidity, which is a geometrical characteristic, was expressed in the above equation in terms of the radius ratio, the number of vanes, and the vane exit angle. Comparisons with the slip factor predictions using Stodola, Stanitz, and Wiesner correlations and also the analytical prediction by Busemann have shown a good agreement over a wide range of impeller geometry.

A rigorous proof for the relative circulation and pressure variation in the impeller flow passages is outside the scope of this textbook, but we can simplify the problem by making the following assumptions:

- neglect viscous effects
- considering the flow to be steady and two-dimensional
- assuming adiabatic and reversible compression process (isentropic)
- considering the simple case of radial vanes ($\beta_2 = 90°$)

In order to determine the variation of the pressure p and the relative velocity V_r in the flow passage (channel) between the impeller vanes, we consider the simple case of radial vanes ($\beta_2 \approx 90°$), as shown in Figure 8.23. Starting with the 2-D momentum conservation equations (Navier–Stokes equations) in cylindrical coordinates, we can write

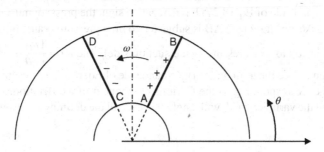

Figure 8.23   Schematic of an impeller with radial vanes

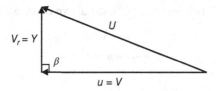

Figure 8.24   The velocity diagram at any radius r between r_1 and r_2

$$\rho\left(\frac{\partial v_r}{\partial t}+v_r\frac{\partial v_r}{\partial r}+\frac{v_\theta\,\partial v_r}{r\,\partial\theta}-\frac{v_\theta^2}{r}\right)=-\frac{\partial p}{\partial r}+F_r+\mu\left(\frac{\partial^2 v_r}{\partial r^2}+\frac{1}{r}\frac{\partial v_r}{\partial r}+\frac{1}{r^2}\frac{\partial^2 v_r}{\partial\theta^2}-\frac{v_r}{r^2}-\frac{2}{r^2}\frac{\partial v_\theta}{\partial\theta}\right)$$

$$\rho\left(\frac{\partial v_\theta}{\partial t}+v_r\frac{\partial v_\theta}{\partial r}+\frac{v_\theta\,\partial v_\theta}{r\,\partial\theta}+\frac{v_r v_\theta}{r}\right)=-\frac{1}{r}\frac{\partial p}{\partial\theta}+F_\theta+\mu\left(\frac{\partial^2 v_\theta}{\partial r^2}+\frac{1}{r}\frac{\partial v_\theta}{\partial r}+\frac{1}{r^2}\frac{\partial^2 v_\theta}{\partial\theta^2}-\frac{v_\theta}{r^2}+\frac{2}{r^2}\frac{\partial v_r}{\partial\theta}\right)$$

In this problem, there are no body forces ($F_r = F_\theta = 0$), and after applying the above simplifications ($\partial/\partial t = 0$, $\mu = 0$), the equations can be reduced to

$$v_r\frac{\partial v_r}{\partial r}+\frac{v_\theta\,\partial v_r}{r\,\partial\theta}-\frac{v_\theta^2}{r}=-\frac{1}{\rho}\frac{\partial p}{\partial r} \tag{8.25}$$

$$v_r\frac{\partial v_\theta}{\partial r}+\frac{v_\theta\,\partial v_\theta}{r\,\partial\theta}+\frac{v_r v_\theta}{r}=-\frac{1}{\rho r}\frac{\partial p}{\partial\theta} \tag{8.26}$$

For radial vanes, the radial velocity component, v_r, will be the same as the relative velocity, V_r, in our notations and v_θ becomes exactly the same as the whirl component, V, in our notations. Accordingly, we can use the velocity diagram shown in Figure 8.24 (drawn at any radius r) to write $v_\theta = V = u = \omega\, r$ and Eq. (8.26) becomes

$$V_r\omega+\omega\overbrace{\frac{\partial v_\theta}{\partial\theta}}^{=0}+V_r\omega=-\frac{1}{\rho r}\frac{\partial p}{\partial\theta} \quad\Rightarrow\quad \frac{1}{r}\frac{\partial p}{\partial\theta}=-2\rho V_r\omega \tag{8.27}$$

Since the right-hand side of Eq. (8.27) has a negative sign, the pressure must decrease in direction of rotation. Accordingly face AB is subjected to high pressure (vane impelling side) while face CD is subjected to low pressure (vane suction side). The term $\dfrac{1}{r}\dfrac{\partial p}{\partial \theta}$ is called the Coriolis pressure gradient, since the term on the right-hand side of Eq. (8.27) represents the inertia force per unit volume, corresponding to the Coriolis component of acceleration.

To determine the variation of V_r with angle θ, we recall the defining equation of the rotational enthalpy,

$$\text{Equation (8.17)} \Rightarrow I = h + \frac{U^2}{2} - uV = Constant$$

Now, since $V = u$ for radial vanes ($\beta_2 = 90°$), the above equation becomes

$$h + \frac{U^2}{2} - u^2 = Constant$$

Differentiate both sides

$$\Rightarrow dh + d\left(U^2/2\right) - d\left(u^2\right) = 0 \ \text{ or } \ dh = -d\left(U^2/2\right) + d\left(u^2\right)$$

$$\text{Rearrange} \Rightarrow dh = -d\left(\frac{U^2 - u^2}{2}\right) + d\left(\frac{u^2}{2}\right) = -\frac{1}{2}\left[d\left(V_r^2\right) - d\left(u^2\right)\right] \tag{8.28}$$

Since the flow process is assumed adiabatic and reversible (isentropic), we can make use of the first law of thermodynamics, together with the defining equation of entropy to write

$$\delta q - \delta w = du, \ \text{and} \ \delta q = Tds, \ \delta w = pdv, \ dh = du + d(pv)$$

$$\text{Therefore, } Tds = du + pdv = du + d(pv) - vdp = dh - \frac{dp}{\rho}$$

For an isentropic process ($ds = 0$), the above equation gives

$$dh = \frac{dp}{\rho} \tag{8.29}$$

Substitute in Eq. (8.28) to obtain $dp/\rho = -\frac{1}{2}\left[d\left(V_r^2\right) - d\left(u^2\right)\right]$
Divide both sides by $d\theta$

$$\Rightarrow \frac{1}{\rho}\frac{dp}{d\theta} = -\frac{1}{2}\left[\frac{d\left(V_r^2\right)}{d\theta} - \frac{d\left(u^2\right)}{d\theta}\right] \ \Rightarrow \ \frac{1}{\rho}\frac{dp}{d\theta} = -V_r\frac{dV_r}{d\theta} \tag{8.30}$$

Substitute from Eq. (8.29) in Eq. (8.30)

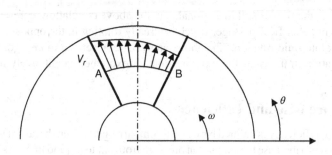

Figure 8.25 Variation of the radial velocity component in the impeller flow passage

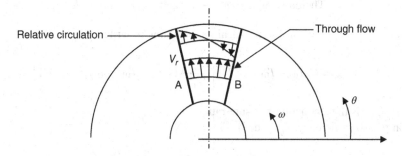

Figure 8.26 Splitting the flow to two streams, through flow and relative circulation

$$\Rightarrow \frac{1}{\rho}(-2\rho V_r \omega r) = -V_r \frac{dV_r}{d\theta} \quad \Rightarrow \quad \frac{dV_r}{d\theta} = 2\omega r \tag{8.31}$$

It is clear from Eq. (8.31) that V_r increases with θ, as shown in Figure 8.25.

The above velocity profile represents the variation of the radial velocity component across section A-B characterizing the flow in the channel between the impeller vanes. This flow can be split to two streams; the first is the through flow with uniform velocity profile and the second is pure rotation causing the relative circulation as shown schematically in Figure 8.26. This relative circulation is sometimes referred to as 'Coriolis circulation.'

In fact, the above simplified frictionless flow is still irrotational, and the relative circulation in the clockwise direction (opposite to the direction of impeller rotation) arises because of the rotation of the flow passage. This can be easily proven by calculating the vorticity in this flow regime as follows:

$$\zeta = \frac{1}{r}\frac{\partial}{\partial r}(rv_\theta) - \frac{1}{r}\frac{\partial v_r}{\partial \theta} \tag{8.32}$$

Knowing that $v_\theta = \omega r$ and using the value of $\partial v_r/\partial \theta$ obtained in Eq. (8.31), we can write

$$\zeta = \frac{1}{r}\frac{\partial}{\partial r}(r^2\omega) - \frac{1}{r}(2\omega r) = 2\omega - 2\omega = 0 \tag{8.33}$$

Now since $\zeta = 0$, the flow is still irrotational, and the above circulation represents only rotation relative to the impeller flow passage, which is already rotating in the opposite direction which makes no absolute fluid rotation. The above analysis represents a more rigorous proof of the relative circulation in the centrifugal pump impeller explained qualitatively in Section 3.4.1.

8.8   Pressure Rise and Efficiencies

The compressor efficiency is defined based on a comparison between the actual work done by the given compressor to increase the stagnation pressure from p_{01} to p_{03} and the work done by an ideal compressor to achieve the same stagnation pressure increase. The compression process in the ideal compressor is the adiabatic reversible (isentropic) process $01 \rightarrow 03i$ shown in Figure 8.27.

$$\text{Therefore,} \quad w_{ideal} = h_{03i} - h_{01} \quad \text{and} \quad w_{actual} = h_{03} - h_{01} \tag{8.34}$$

Based on Eq. (8.34), we can use the ideal gas relations to write

$$w_{ideal} = C_p T_{03i} - T_{01} = C_p T_{01} \left[\frac{T_{03i}}{T_{01}} - 1 \right] = C_p T_{01} \left[(p_{03}/p_{01})^{\frac{k-1}{k}} - 1 \right] \tag{8.35}$$

where p_{03}/p_{01} is the stagnation pressure ratio.
The compressor efficiency, η_c, is defined as

$$\eta_c = \frac{w_{ideal}}{w_{actual}} = \frac{h_{03i} - h_{01}}{h_{03} - h_{01}} = \frac{T_{03i} - T_{01}}{T_{03} - T_{01}} \tag{8.36}$$

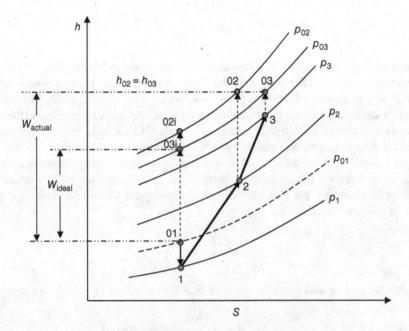

Figure 8.27   The thermodynamic processes in actual and ideal compressors

For axial inlet (no prewhirl), we also use Eq. (8.19) to write

$$w_{actual} = u_2 V_2' = S_f u_2 V_2 \tag{8.37}$$

Substituting from Eqs (8.35) and (8.37) in Eq. (8.36), we obtain,

$$\eta_c = \frac{C_p T_{01}\left[(p_{03}/p_{01})^{\frac{k-1}{k}} - 1\right]}{S_f u_2 V_2}$$

Thus, the stagnation pressure ratio (p_{03}/p_{01}) can be expressed as

$$\frac{p_{03}}{p_{01}} = \left(1 + \frac{S_f \eta_c u_2 V_2}{C_p T_{01}}\right)^{\frac{k}{k-1}} \tag{8.38}$$

The value of η_c is usually 70–80%.

For a typical impeller with 19 to 21 radial vanes, together with standard atmospheric condition at inlet, the pressure ratio, p_{03}/p_{01}, varies with the blade tip speed, u_2, as shown in Figure 8.28 (based on Eq. (8.38)). The figure shows that the pressure ratio increases with the tip speed, but material strength puts an upper limit of about 500 m/s (for light alloys), and η_c is normally less than 80%, so the highest pressure ratio per stage is normally about 5:1.

The impeller efficiency, η_I, is defined as,

$$\eta_I = \frac{w_{ideal\ impeller}}{w_{actual\ impeller}} = \frac{h_{02i} - h_{01}}{h_{02} - h_{01}} = \frac{T_{02i} - T_{01}}{T_{02} - T_{01}} \tag{8.39}$$

The flow velocities at the impeller inlet and exit are related to the impeller dimensions by the mass conservation equation

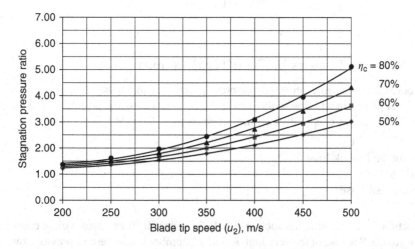

Figure 8.28 Effect of blade tip speed on the compressor stagnation pressure ratio (p_{03}/p_{01})

$$\dot{m} = \rho_1 A_1 Y_1 = \rho_2 A_2 Y_2 \tag{8.40}$$

where $A_1 = \pi\left[r_{s1}^2 - r_{h1}^2\right]$ and $A_2 = 2\pi\, r_2\, b_2$

The mechanical efficiency, $\eta_{mech.}$, has the same definition as in pumps:

$$\eta_{mech.} = \frac{Actual\ power\ input\ to\ impeller}{Shaft\ power} = \frac{\dot{m}\, w_{actual}}{Shaft\ power} \tag{8.41}$$

Example 8.8

The table below shows the stagnation pressures and temperatures measured at the suction and discharge nozzles of a centrifugal compressor. Determine the compressor efficiency, η_c.

	Stagnation pressure (kPa)	Stagnation temperature (K)
At compressor suction	100.0	316
At compressor discharge	300.0	500

Solution

$$\eta_c = \frac{T_{03i} - T_{01}}{T_{03} - T_{01}} \quad \text{but} \quad \frac{T_{03i}}{T_{01}} = \left(\frac{p_{03}}{p_{01}}\right)^{\frac{k-1}{k}} \Rightarrow \frac{T_{03i}}{316} = \left(\frac{300}{100}\right)^{0.286}$$

Therefore, $T_{03i} = 432.5\ \text{K}$

$$\eta_c = \frac{432.5 - 316}{500 - 316} \quad \Rightarrow \quad \eta_c = 63.3\%$$

8.9   Sources of Losses in Centrifugal Compressors

The sources of losses in centrifugal compressors are very much the same as those presented in centrifugal pumps in addition to the irreversibilities caused by the possible presence of shockwaves. These losses can be divided into three types:

a. friction and shock losses
b. leakage losses
c. mechanical losses

The friction losses occur in the suction nozzle, impeller, diffuser vanes, volute casing, and discharge nozzle. Because of the very high Reynolds number inside the compressor flow passages, the amount of friction loss has a strong dependence on the flow velocity as well as the surface roughness. When the compressor operates at off-design conditions (operation at reduced flow rates or at speeds different from the rated speed), flow separation may occur in the impeller (downstream of the impeller vane inlet) and also in the diffuser vanes. This flow separation causes a higher degree of turbulence and a definite increase in friction losses. The presence of shockwaves within the impeller or at the inlet to the diffuser vanes causes additional losses due to the transformation of energy from a valuable form (stagnation pressure) to a less valuable form (heat). Flow through shockwaves is irreversible and is accompanied by a loss of stagnation pressure. Both effects of friction and shock losses contribute to a decrease in the compressor efficiency, η_c.

Internal leakage occurring through various clearances separating high and low pressure compartments represents loss of energy that affects the compressor's overall efficiency. The main part of this leakage occurs in the clearance between the impeller front shroud and the casing. This clearance is exposed to the high pressure in the volute in one side and the low pressure in the suction nozzle in the other side. The amount of leakage depends on the pressure difference between the two sides, as well as the magnitude of the clearance between the impeller and casing. An experimental study for the effect of leakage on the performance of a centrifugal compressor was carried out by Mashimo et al. [10]. They reported that the reduction in compressor efficiency is strongly dependent on the size of the clearance space between the impeller and casing. A computational study on the effect of vane tip clearance on the compressor performance was reported by Eum et al. [11]. They found that the tip leakage flow strongly interacts with the main stream flow at the impeller inlet and considerably changes the secondary flow and the loss distribution inside the impeller passage.

Mechanical losses are very much the same as the mechanical losses in centrifugal pumps discussed in Chapter 3. These losses include disc friction loss, and losses in the other mechanical components such as bearings and the mechanical seal. The disc friction power loss depends on the speed of rotation, impeller geometry and size, surface roughness, fluid properties and the clearance space between the impeller and casing. A recent computational study by Cho et al. [12] indicated that the disc friction losses can be minimized by reducing surface roughness of the impeller shrouds and casing surfaces and also by optimizing the axial clearance between the impeller and the casing. The power loss in the other mechanical components depends on the size, geometry, and operating condition of these components as well as the speed of rotation. In general, the mechanical losses have negligible dependence on the mass flow rate supplied by the compressor.

8.10   Compressor Performance Characteristics

The compressor performance is normally presented in terms of three curves representing the variation of each of the stagnation pressure ratio (p_{03}/p_{01}) and the overall efficiency (η_o) with the mass flow rate ($\dot{m}$) when operating at a constant speed. To have a dimensionless parameter for the mass flow rate, the flow coefficient (Q/ND^3) is combined with the speed coefficient (ND/C) to give the following mass flow coefficient (C_m).

$$C_m = \frac{Q}{ND^3}\frac{ND}{C}\frac{\rho}{(p/RT)} = \frac{\dot{m}}{D^2\sqrt{kRT}}\frac{RT}{p} = \frac{\dot{m}}{D^2}\frac{\sqrt{RT/k}}{p} \tag{8.42}$$

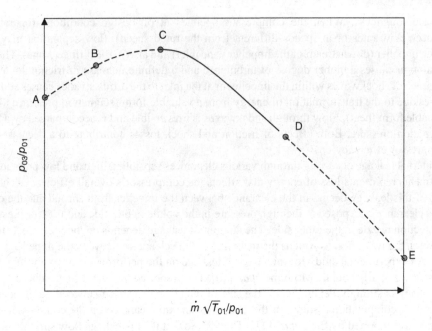

Figure 8.29 Typical compressor characteristic at a constant speed

For a given compressor handling a known gas, the values of D, k, and R are considered constants in the above equation. In defining the mass flow parameter, the constants D, k, and R are omitted in Eq. (8.42) and the parameter $\dot{m}\sqrt{T_{01}}/p_{01}$ is used with the stagnation properties p_{01} and T_{01} used as reference values. Using the same reasoning, a speed parameter $(N/\sqrt{T_{01}})$ is used instead of the speed coefficient (ND/C). Figure 8.29 shows a typical pressure ratio versus mass flow characteristic for a centrifugal compressor running at a constant speed. The variation of the mass flow rate is achieved by delivery valve throttling.

Point A in Figure 8.29 represents the shut-off condition when the delivery valve is fully closed. The region between points A and C is an unstable region due to compressor surge, and point D is a choking point at which any of the compressor flow passages is choked. So the dotted lines AC and DE represent regions of unsafe and unavailable operating conditions, respectively. The surge and choking problems are discussed in the following sections. Figure 8.30 shows typical performance characteristics of a centrifugal compressor operating at different speeds, indicating the isoefficiency lines.

8.11 Compressor Surge

Similar to pump surge explained in Section 5.5, compressor surge refers to unstable performance resulting in flow rate and pressure fluctuations that may lead to flow reversal. It occurs when the compressor fails to generate the delivery pressure required to maintain continuous flow to the downstream side. The region between points A and C on the pressure versus flow rate characteristic in Figure 8.29 has a positive slope. This region is generally unstable, since an

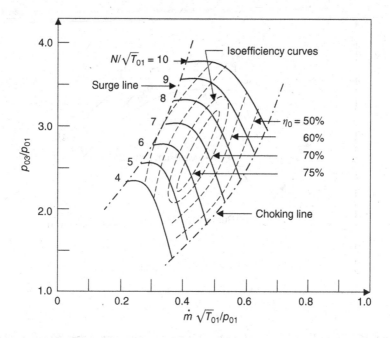

Figure 8.30 Typical performance curves for a centrifugal compressor operating at different speeds showing the isoefficiency curves

increase in the delivery pressure (due to a disturbance or blockage in the delivery side) will momentarily cause a reduction in $\dot{m}$ accompanied by a pressure increase in the delivery side. The compressor response to this reduction in $\dot{m}$ is a decrease in the delivery pressure (negative value of Δp_{03}) which will, in turn, reduce the flow rate further, and the process continues until the flow is reversed. On the other hand, the region between points C and D in the same figure is unconditionally stable since the compressor response to the same disturbance is developing higher discharge pressure (positive value of Δp_{03}). Such response is considered favorable since the process becomes self-correcting. In general, the surge problem may be initiated inside the compressor due to aerodynamic instability and may also be initiated because of pumping system unsteadiness. The aerodynamic instabilities may occur due to the rotating stall phenomenon, which is very common in centrifugal and axial-flow compressors. This phenomenon was described in detail in Section 5.9. Severe flow rate and pressure fluctuations due to compressor surge may cause failure of compressor components such as bearings and mechanical seals.

In order to protect the compressor against surge, a minimum flow rate must be maintained so that the point of operation continues to be on the right side of the surge line. Figure 8.31 shows typical performance curves of a compressor at two different speeds ($N_2 < N_1$). At speed N_1, the flow rate must be higher than $\dot{m}_A$ to avoid compressor surge. In practice, a minimum flow rate limiting value is set (say $\dot{m}_C$), below which a control system is activated in order to increase the flow rate to safeguard the compressor against surge.

The commonly used method is recycling part of the discharge through a controlled bypass, as shown in Figure 8.32. When the flow rate reaches the set value $\dot{m}_C$, the flow controller (FC)

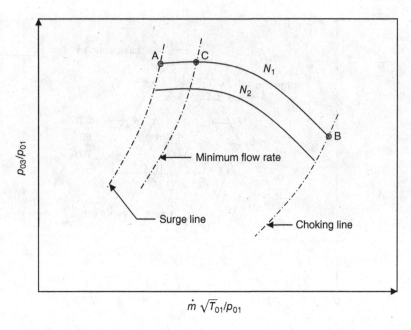

Figure 8.31 Typical compressor characteristics showing the surge and minimum flow rate lines

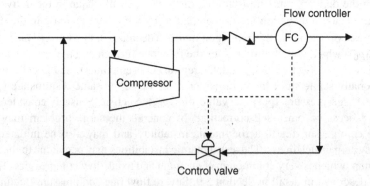

Figure 8.32 Typical Surge control system using a bypass

actuates the bypass control valve to circulate part of the discharge back to the compressor suction side. That leads to a reduction in the discharge pressure, resulting in an increase in the flow rate. The disadvantage of this method is the power loss in the bypass causing a reduction in the system efficiency. The pressing environmental issues calling for reduction in power consumption initiates a thrust towards introducing other methods of flow rate and surge controls. Among these methods is to control the flow rate using variable speed drivers and also using inlet guide vanes. These methods are discussed in Section 8.13.

8.12 Choking in Centrifugal Compressors

Similar to the case of flow in a C-D nozzle explained in Section 8.3.7, the various flow passages inside the compressor may reach choking condition. For example, if the suction nozzle is choked, the flow rate supplied by the compressor cannot be increased any further even by increasing the rotational speed or by decreasing the delivery pressure. This phenomenon puts an upper limit to the mass flow rate supplied by the compressor. Choking may occur in the suction nozzle, the impeller, or the diffuser.

8.12.1 Suction Nozzle Choking

The suction nozzle is said to be choked when the Mach number reaches unity at any section in the nozzle. From now on, all the flow properties at that section will have a star (∗) superscript. This normally takes place at the nozzle exit for a convergent only nozzle. In this case, we can use Eq. (8.9) to write

$$\dot{m}_{max} = \rho^* C^* A^* = \frac{\rho_{oo} A^* C_{oo}}{\left(\dfrac{k+1}{2}\right)^{\frac{k+1}{2(k-1)}}} \tag{8.43}$$

It is clear from this equation that if the suction nozzle is choked, the mass flow rate $\dot{m}_{max}$ becomes independent of the impeller speed as well as the delivery conditions.

8.12.2 Impeller Choking

The impeller choking occurs when the relative velocity, V_r, at any section in the impeller reaches the local speed of sound (i.e. when $M_r = 1$). Since the rotational enthalpy, I, is the same at any radius, r, as proven in Eq. (8.17), we can write

$$I = h + \frac{U^2}{2} - uV = Constant$$

By using the velocity diagrams (Figure 8.18), the above expression takes the form

$$I = h + \frac{V_r^2}{2} - \frac{u^2}{2} = Constant \tag{8.44}$$

When choking occurs at radius $r*$, then $V_{r*} = C^* = \sqrt{kRT^*}$, and by applying the above equation between the impeller inlet and the critical section (at $r*$), we can write

$$h^* + \frac{C^{*2}}{2} - \frac{u^{*2}}{2} = h_1 + \frac{U_1^2}{2} - u_1 V_1 \approx h_{o1}$$

which leads to $T^* + \dfrac{kRT^*}{2C_p} - \dfrac{u^{*2}}{2C_p} = T_{o1}$, $C_p = kR/(k-1)$

Rearrange to obtain $\dfrac{T^*}{T_{o1}}\left(1 + \dfrac{k-1}{2}\right) = 1 + \dfrac{u^{*2}}{2C_pT_{o1}}$, from which we can write

$$\frac{T^*}{T_{o1}} = \frac{C^{*2}}{C_{o1}^2} = \frac{2}{k+1}\left(1 + \frac{u^{*2}}{2C_pT_{o1}}\right) \tag{8.45}$$

If we assume isentropic flow (for simplicity) between the impeller inlet and the critical section, then

$$\rho^*/\rho_{o1} = (T^*/T_{o1})^{\frac{1}{k-1}} \tag{8.46}$$

The maximum mass flow rate (corresponding to impeller choking) can be obtained from

$$\dot{m}_{\max} = \rho^* C^* A^*$$

By using the expressions (8.45) and (8.46) for ρ^* and C^*, we can finally write

$$\dot{m}_{\max} = \rho_{o1} C_{o1} A^* \left[\frac{2}{k+1}\left(1 + \frac{u^{*2}}{2C_pT_{o1}}\right)\right]^{\frac{k+1}{2(k-1)}} \tag{8.47}$$

If the stagnation properties at the impeller inlet (ρ_{o1}, C_{o1}) are unchanged, the choking mass flow rate, $\dot{m}_{\max}$, will increase with the increase of the rotational speed, unless choking occurs in another component of the compressor.

8.12.3 Diffuser Choking

Diffuser choking is very much the same as that in the suction nozzle with the only difference being that the stagnation conditions at the diffuser inlet (ρ_{o2}, C_{o2}) are different from those at the suction nozzle inlet (ρ_{o1}, C_{o1}). The corresponding maximum flow rate can be expressed as

$$\dot{m}_{\max} = \rho^* C^* A^* = \frac{\rho_{o2} A^* C_{o2}}{\left(\dfrac{k+1}{2}\right)^{\frac{k+1}{2(k-1)}}} \tag{8.48}$$

Since the stagnation conditions at the impeller exit (ρ_{o2}, C_{o2}) depend on the impeller rotational speed, N, the choking flow rate through the diffuser will also depend on N.

Example 8.9

A centrifugal compressor draws air from the atmosphere ($p_{atm} = 101$ kPa and $T_{atm} = 25\,°C$) through a convergent nozzle as shown in Figure 8.33. The diameter at the nozzle exit is 200 mm. Determine the maximum mass flow rate that can be delivered by the compressor.

Solution

The maximum mass flow rate is obtained when the suction nozzle is choked ($M = 1$ at the throat). At this condition, the mass flow rate can be expressed as

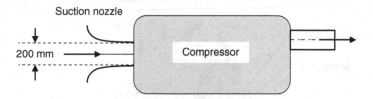

Figure 8.33 Diagram for Example 8.9

$$\dot{m}_{max} = \frac{\rho_{o1} A^* C_{o1}}{\left(\dfrac{k+1}{2}\right)^{\frac{k+1}{2(k-1)}}} \tag{I}$$

$$\rho_{01} = p_{01}/R\,T_{01} = \frac{101}{0.287 \times 298} = 1.18 \text{ kg/m}^3,\ A_2 = \pi(0.2)^2 = 0.0314 \text{ m}^2,$$

$$\dot{C}_{01} = \sqrt{k\,R\,T_{01}} = \sqrt{1.4 \times 287 \times 298} = 346 \text{ m/s}$$

Substitute in Eq. (I) $\Rightarrow \dot{m}_{max} = \dfrac{1.18 \times 0.0314 \times 346}{[1.2]^3} = 7.42 \text{ kg/s}$

8.13 Flow Rate Control in Centrifugal Compressors

The flow rate control in centrifugal compressors can be achieved using any of three methods:

a. driver speed control
b. throttling of the suction or delivery valves
c. prewhirl using inlet guide vanes

The first method is the most efficient but is only available in special cases in which the compressor is driven by a turbine or by a variable-speed motor. This method permits operation through a wide range of mass flow rates and at a reasonably high efficiency. The use of valve throttling creates irrecoverable energy loss and it can be done by throttling either the delivery or the suction valves. The suction valve throttling creates a reduction in the gas density which will result in reducing the compressor power consumption. On the other hand, throttling the delivery valve does not lower the discharge pressure and does not take advantage of the lower density as in the case of suction throttling.

In some compressors, adjustable guide vanes are used upstream of the impeller to provide some prerotation. The main reason is to reduce the mass flow rate by reducing the axial component of the fluid velocity at the impeller inlet. At the same time, the imposed prerotation at the impeller inlet will reduce the amount of work done ($w_{1\text{-}2}$), and consequently the compressor power consumption. Figure 8.34 shows a schematic sectional view of a single stage centrifugal compressor, equipped with adjustable inlet guide vanes. The effect of using the inlet guide vanes on the inlet velocity diagram is as shown in Figure 8.35.

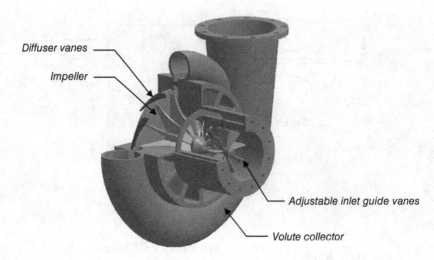

Figure 8.34 A schematic sectional-view of a centrifugal compressor showing the adjustable inlet guide vanes

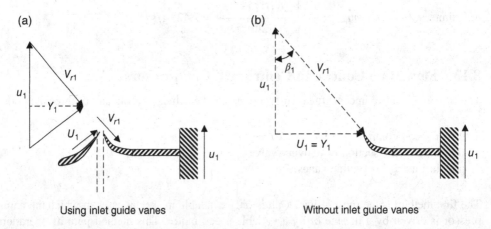

Figure 8.35 A schematic showing the effect of using inlet guide vanes for flow rate control (a) using inlet guide vanes (b) without inlet guide vanes

Based on the angular momentum conservation equation, we know that the driving torque can be expressed as $T_d = \dot{m}\,(V_2 r_2 - V_1 r_1)$. When using inlet guide vanes to impose some prewhirl, V_1 will have a positive value (instead of zero) and $\dot{m}$ becomes less because of the decrease in Y_1. Both effects tend to decrease the driving torque and so decrease the compressor power consumption. The effect of using an inlet guide for flow rate control on the compressor performance is shown in Figure 8.36.

Example 8.10
Air enters a centrifugal compressor at 101 kPa and 20 °C at a rate of 2.5 kg/s. The impeller which has an outside diameter of 50 cm rotates at 20 000 rpm. The vane angle at exit

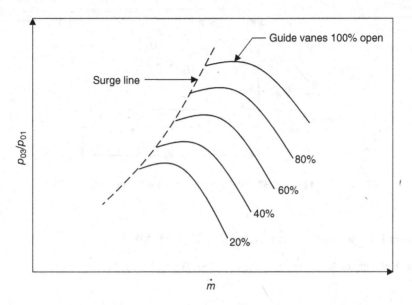

Figure 8.36 Effect of inlet guide vanes opening on the performance of a centrifugal compressor

$\beta_2 = 63°$, and the radial component of the exit velocity $Y_2 = 130$ m/s. Knowing that the slip factor $S_f = 0.9$ and the compressor efficiency is 80%, determine

a. stagnation pressure ratio
b. power required to drive the compressor
c. stagnation temperature at the compressor outlet.

Assume $\eta_{mech} = 95\%$ and $(C_p)_{air} = 1.004$ kJ/kg.K.

Solution

a. $T_{01} = 293$ K, $u_2 = \omega r_2 = 524$ m/s,
   From the velocity diagram in Figure 8.37, we can write

$$V_2 = u_2 - Y_2 \cot \beta_2 = 524 - 130 \cot 63° = 458\,\text{m/s}$$

Using Eq. (8.38)

$$\Rightarrow \frac{p_{03}}{p_{01}} = \left(1 + \frac{S_f \eta_c u_2 V_2}{C_p T_{01}}\right)^{\frac{k}{k-1}} \Rightarrow \frac{p_{03}}{p_{01}} = \left(1 + \frac{0.9 \times 0.8 \times 524 \times 458}{1004 \times 293}\right)^{\frac{1.4}{0.4}} = 5.04$$

b. Equation (8.35)

$$\Rightarrow w_{ideal} = C_p T_{01}\left[(p_{03}/p_{01})^{\frac{k-1}{k}} - 1\right] = 1.004(293)\left[(5.04)^{\frac{0.4}{1.4}} - 1\right] = 172\,\text{kJ/kg}$$

But $\eta_c = \dfrac{w_{ideal}}{w_{actual}} \Rightarrow 0.8 = \dfrac{172}{w_{actual}} \Rightarrow w_{actual} = 215\,\text{kJ/kg}$

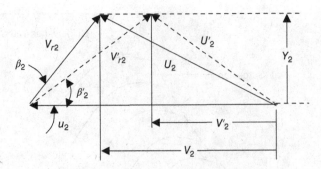

Figure 8.37   Velocity diagram for Example 8.10

Power input to impeller $= \dot{m}w_{actual} = 2.5 \times 215 = 538\,kW$

Shaft power (or BP) $= \dot{m}w_{actual}/\eta_{mech.} = 538/0.95 = \underline{567\,kW}$

or

$$\text{Shaft power} = \frac{\dot{m}w_{actual}}{\eta_{mech.}} = \frac{\dot{m}u_2 V_2'}{\eta_{mech.}} = \frac{\dot{m}S_f u_2 V_2}{\eta_{mech.}} = \frac{2.5 \times 0.9 \times 524 \times 458}{0.95 \times 10^3} = \underline{568\,kW}$$

c. Using Eq. (8.34) $\Rightarrow w_{actual} = h_{03} - h_{01} = C_p(T_{03} - T_{01})$

$$\Rightarrow 215 = 1.004(T_{03} - 293) \Rightarrow T_{03} = \underline{507K}$$

Example 8.11

A centrifugal compressor has an impeller with 35 radial vanes and is driven at a speed of 12 000 rpm. The compressor sucks air from the atmosphere where the ambient pressure and temperature are 101 kPa and 300 K respectively. Air enters the impeller axially at a speed of 72 m/s. The following data is also provided:

Air flow rate	$= 4\,kg/s$
Impeller exit area	$= 0.05\,m^2$
Static pressure at impeller exit	$= 150\,kPa$
Static temperature at impeller exit	$= 350\,K$
Impeller outer diameter	$= 0.5\,m$

a. Determine the actual flow velocity and Mach number at the impeller exit.
b. If the stagnation pressure at the compressor exit is 215 kPa, what is the compressor efficiency?
c. Determine the compressor power consumption assuming $\eta_{mech.} = 93\%$.

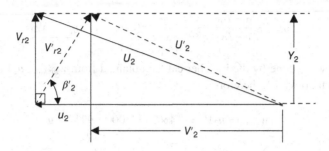

Figure 8.38 Ideal and actual velocity diagrams for Example 8.11

Solution

a. $u_2 = \omega r_2 = 314$ m/s

$$\rho_2 = \frac{p_2}{RT_2} = \frac{150}{0.287 \times 350} = 1.49 \, \text{kg/m}^3$$

$$Y_2 = \frac{\dot{m}}{\rho_2 A_2} = \frac{4}{1.49 \times 0.05} = 53.6 \, \text{m/s}$$

Since $\beta_2 = 90°$ (Figure 8.38), therefore $V_2 = u_2 = 314$ m/s

Using the Stanitz formula for the slip factor (Eq. (8.20)b), we can write

$$S_f = 1 - \frac{0.63\pi/Z}{\left(1 - \frac{Y_2}{u_2}\cot\beta_2\right)} = 1 - 0.63(\pi/Z) = 0.943$$

Therefore, $V_2' = S_f V_2 = 0.943 \times 314 = 296$ m/s

Using the velocity diagram, $U_2' = \sqrt{V_2'^2 + Y_2^2} = \sqrt{296^2 + 53.6^2} = 301$ m/s

Now, $C_2 = \sqrt{kRT_2} = \sqrt{1.4 \times 287 \times 350} = 375$ m/s

Finally $M_2 = U_2'/C_2 = 301/375 = 0.8$

b. Using Eq. (8.6) $\Rightarrow p_{02} = p_2 \left(1 + \frac{k-1}{2}M_2^2\right)^{\frac{k}{k-1}} = 150\left[1 + 0.2(0.8)^2\right]^{3.5} = 229 \, \text{kPa}$

We also know that $T_{03} = T_{02} = T_2\left(1 + \frac{k-1}{2}M_2^2\right) = 350\left[1 + 0.2(0.8)^2\right] = 395 \, \text{K}$

and $p_{03} = p_{03i} = 215$ kPa.

Now, since the process $01 \rightarrow 03i$ is isentropic (see Figure 8.27), we can write

$$T_{03i}/T_{01} = (p_{03}/p_{01})^{\frac{k-1}{k}} = (215/101)^{0.286} = 1.24 \Rightarrow T_{03i} = 395 \, \text{K}$$

To determine the compressor efficiency, we can use Eq. (8.23) to write

$$\eta_c = \frac{T_{03i}-T_{01}}{T_{03}-T_{01}} = \frac{372-300}{395-300} = 0.76 \quad \Rightarrow \quad \underline{\eta_c = 76\%}$$

c. The actual work done by the impeller can be obtained from $w_{actual} = u_2 V_2'$ since air enters axially (with no whirl), therefore

$$w_{actual} = u_2 V_2' = 314 \times 296/1000 = 93\,\text{kJ/kg}$$

The compressor power consumption = Brake power = $\dfrac{\dot{m} w_{actual}}{\eta_{mech}} = \dfrac{4 \times 93}{0.93} = \underline{400\,\text{kW}}$

Example 8.12

A centrifugal air compressor has an impeller with 19 radial vanes and is driven at a speed of 16 000 rpm. The compressor sucks air from the atmosphere where the pressure and temperature are 101 kPa and 290 K respectively. The mass flow rate supplied by the compressor is 8 kg/s. The impeller outer diameter is 55 cm and the vane width at exit is 2 cm. The measurements recorded at the impeller and compressor exit sections are as follows:

Temperature at impeller exit	$T_2 = 390$ K
Pressure at impeller exit	$p_2 = 259$ kPa abs.
Temperature at compressor exit	$T_3 = 475$ K
Pressure at compressor exit	$p_3 = 485$ kPa abs.

Knowing that the mechanical efficiency is 96% and neglecting prerotation at the impeller inlet, determine:

a. compressor power consumption and BP
b. stagnation pressure ratio, p_{03}/p_{01}.

Solution

a. $u_2 = \omega r_2 = 461$ m/s and $\rho_2 = \dfrac{p_2}{RT_2} = \dfrac{259}{0.287 \times 390} = 2.31\,\text{kg/m}^3$

Mass conservation $\Rightarrow m^{\bullet} = \rho_2\, \pi D_2 b_2 Y_2$

$$\Rightarrow 8 = 2.31\,\pi \times 0.55 \times 0.02 \times Y_2 \quad \Rightarrow \quad Y_2 = 100\,\text{m/s}$$

For radial vanes, $\beta_2 = 90°$, and we can use Eq. (8.20b) to obtain

$$S_f = 1 - \frac{0.63\pi/Z}{1-(Y_2/u_2)\cot\beta_2} = 1 - 0.63\,(\pi/19) = 0.896$$

Also, $V_2 = u_2 - Y_2 \cot \beta_2 \Rightarrow V_2 = 461$ m/s

$$w_{actual} = u_2 V_2' - u_1 V_1' = u_2 V_2' - 0 = S_f u_2 V_2 = 0.896(461)^2 \text{ J/kg}$$

$$\Rightarrow \quad w_{actual} = 190.2 \text{ kJ/kg}$$

$$\text{Shaft power (or brake power)} = \frac{\dot{m}\, w_{actual}}{\eta_{mech}} = \frac{8 \times 190.2}{0.96} = \underline{1585 \text{ kW}}$$

b. $w_{actual} = C_p(T_{03} - T_{01}) \Rightarrow 190.2 = 1.004(T_{03} - 290) \Rightarrow T_{03} = 479.5 \text{ K}$

Now, $T_{03}/T_3 = (p_{03}/p_3)^{(k-1)/k} \Rightarrow 479.5/475 = (p_{03}/485)^{0.4/1.4} \Rightarrow p_{03} = 501 \text{ kPa}$

Therefore, $p_{03}/p_{01} = 501/101 = \underline{4.96}$

References

1. Dixon, S.L. (2005) *Fluid Mechanics and Thermodynamics of Turbomachinery*, 5th edn. Elsevier Butterworth-Heinemann Publishers, Amsterdam.
2. Sharma, D.M., Vibhakar, N.N., and Channiwala, S.A. (2003) Experimental and analytical investigations of slip factor in radial tipped centrifugal fan. Proceedings of the International Conference on Mechanical Engineering 2003 (ICME2003), Dhaka, Bangladesh, 26–28 December 2003.
3. Guo, E.M. and Kim, K.Y. (2004) Three-dimensional flow analysis and improvement of slip factor model for forward-curved blades centrifugal fan. *KSME Int J*, **18** (II), 302–312.
4. Wiesner, F.J. (1967) A review of slip factors for centrifugal impellers. *ASME J Eng Power*, **89** (IV), 558–566.
5. Stodola, A. (1945) *Steam and Gas Turbines*. McGraw-Hill, New York, reprinted by Peter Smith, New York; 1927.
6. Stanitz, J.D. (1952) Some theoretical aerodynamic investigations of impellers in radial- and mixed-flow centrifugal compressors. *Trans ASME*, **74**, 473–476.
7. Busemann, A. (1928) Das Förderhöhenverhältniss Radialer Kreiselpumpen Mit Logarithisch-Spiraligen Schaufeln. *Z Angew Math Mech*, **8**, 371–384.
8. Csanady, G.T. (1930) Head correction factors for radial impellers. *Engineering, London*, **190**, 195.
9. Von Backström, T.W. (2005) A unified correlation for slip factor in centrifugal impellers. *ASME J Turbomach*, **128** (I), 1–10.
10. Mashimo, T., Watanabe, I. and Ariga, I. (1979) Effects of fluid leakage on performance of a centrifugal compressor. *J Eng Power*, **101** (III), 337–342.
11. Eum, H.J., Kang, Y.S., and Kang, S.H. (2004) Tip clearance effect on through-flow and performance of a centrifugal compressor. *KSME Int J*, **18** (6), 979–989.
12. Cho, L., Lee, S., and Cho, J. (2012) Use of CFD analyses to predict disk friction loss of centrifugal compressor impellers. *Trans Jpn Soc Aeronaut Space Sci*, **55** (III), 150–156.

Problems

8.1 A centrifugal fan is used to deliver 4 m³/s of air when running at 750 rpm. The impeller has an outer diameter of 75 cm and an exit width of 10 cm. The vanes are backward-curved with an exit angle of 70°. The volute casing converts 30% of the impeller exit velocity head into pressure. If air enters radially at a speed of 15 m/s with a density of 1.1 kg/m³, determine:

a. actual velocity at impeller exit considering a vane efficiency of 90%
b. pressure rise developed by the fan in kPa
c. power input to the impeller assuming a hydraulic efficiency of 88%.

8.2 A centrifugal fan is used to deliver air from the atmosphere to a chemical reactor. The fan is driven at 3000 rpm and has an impeller with the following dimensions:

$$D_1 = 60 \text{ cm}, \ \beta_1 = 30°$$
$$D_2 = 75 \text{ cm}, \ b_2 = 20 \text{ cm}, \ \beta_2 = 20°$$

The reactor receives air at a stagnation pressure of 4 kPa above atmospheric. Each of the fan suction and delivery nozzles has an area of 0.72 m^2. The flow enters the impeller with no whirl, and the circulatory flow effect reduces the whirl component of velocity at exit by 15%. Neglecting leakage, and assuming a hydraulic efficiency of 88% and a mechanical efficiency of 95%, determine:

a. air velocity at the impeller exit
b. static pressure at the discharge nozzle
c. fan power consumption.

Note: Neglect losses between the fan exit and the chemical reactor and assume an air density of 1.2 kg/m^3.

8.3 A centrifugal fan is used to deliver air from the atmosphere to a combustion process at a rate of $1.25 \text{ m}^3/\text{s}$ when driven at a speed of 1500 rpm. The impeller is equipped with forward-curved vanes and has the following dimensions:

$$D_1 = 40 \text{ cm}, \ b_1 = 10 \text{ cm}, \ \beta_1 = 60°$$
$$D_2 = 70 \text{ cm}, \ b_2 = b_1, \ \beta_2 = 130°$$

Each of the fan inlet and delivery sections has an area of 0.05 m^2. The circulatory flow effect increases the whirl component at inlet by 10% while reducing it at exit by 10%. Neglecting leakage and assuming a hydraulic efficiency of 88% and a mechanical efficiency of 95%, determine

a. stagnation pressure difference between the fan inlet and exit sections
b. actual air velocity at the impeller exit
c. static pressure at the fan delivery section
d. fan power consumption.

Note: Assume an average air density of 1.2 kg/m^3.

8.4 A centrifugal compressor draws air from the atmosphere ($p_{atm} = 101 \text{ kPa}$, $T_{atm} = 27°C$) through a convergent suction nozzle. The nozzle dimensions are as shown in Figure 8.39. In order to avoid choking at impeller inlet, the maximum allowable Mach number (based on the absolute velocity) at impeller inlet is restricted to 0.6. Calculate the maximum mass flow rate that can be supplied by this compressor.

8.5 The convergent–divergent nozzle shown in Figure 8.40 is supplied by compressed air from a large reservoir, as shown in the figure. The pressure and temperature in the reservoir are 300 kPa abs. and 350 K respectively. Knowing that the nozzle diameter at the throat is 50 mm and assuming isentropic flow, determine:

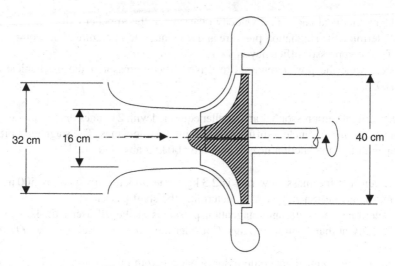

Figure 8.39   Diagram for Problem 8.4

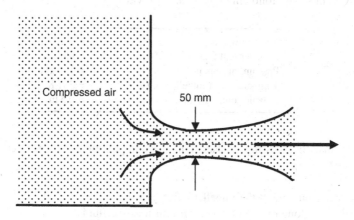

Figure 8.40   Diagram for Problem 8.5

   a. pressure at the throat if the nozzle is choked
   b. flow velocity at the throat
   c. mass flow rate supplied by the nozzle.

8.6 Air enters a centrifugal compressor at a rate of 1.5 kg/s with the following conditions at
impeller inlet:

$$p_1 = 96 \text{ kPa}, \ T_1 = 298 \text{ K}, \ U_1 = 80 \text{ m/s}$$

The impeller has an outer diameter of 50 cm and is equipped with 17 backward-curved
vanes with vane exit angle of 60°. The impeller rotates at 18 000 rpm, which results in
an exit radial velocity component, $Y_2 = 120$ m/s.

a. Draw the ideal and actual velocity diagrams at the impeller exit.
b. Determine the stagnation pressure and stagnation temperature at the compressor outlet if the compressor efficiency is 78%.
c. Determine the power required to drive the compressor if the mechanical efficiency is 95%.

8.7 A centrifugal compressor has an impeller equipped with 21 vanes that are radial at exit. The compressor has a vaneless diffuser and no inlet guide vanes. The stagnation pressure and temperature at the suction nozzle inlet are 100 kPa abs. and 300 K.

a. Given that the mass flow rate is 2.3 kg/s, the impeller tip speed is 500 m/s, and the mechanical efficiency is 96%, determine the shaft power.
b. Determine the static and stagnation pressures at the diffuser exit, knowing that the velocity at that point is 100 m/s. Consider the compressor efficiency to be 82%.

8.8 A single-stage centrifugal compressor draws air from the atmosphere, where the pressure and temperature are 100 kPa and 300 K respectively. The compressor has an impeller with 17 radial vanes and is driven at a speed of 10 000 rpm. Air enters the impeller axially at a speed of 72 m/s. The following data is also provided:

Mass flow rate	$= 6\,\text{kg/s}$
Impeller exit area	$= 0.06\,\text{m}^2$
Pressure at impeller exit	$= 130\,\text{kPa}$
Temperature at impeller exit	$= 331\,\text{K}$
Impeller outer diameter	$= 0.5\,\text{m}$

Knowing that the pressure measured at the compressor discharge section is 195 kPa determine:

a. flow Mach number at the impeller exit
b. actual work done by the impeller per unit mass of fluid
c. power required to drive the compressor, assuming a mechanical efficiency of 96%.

8.9 A centrifugal air compressor has an impeller with 13 backward-curved vanes and is driven at a speed of 12 000 rpm. The compressor draws air from the atmosphere where the pressure and temperature are 100 kPa and 300 K respectively. The impeller dimensions are:

Impeller outer diameter	$D_2 = 480\,\text{mm}$
Impeller width at exit	$b_2 = 28\,\text{mm}$
Vane angle at exit	$\beta_2 = 70°$

The pressure and temperature measurements recorded at the impeller and compressor exit sections are:

Temperature at impeller exit	$T_2 = 343$ K
Pressure at impeller exit	$P_2 = 135$ kPa abs.
Stagnation temperature at compressor exit	$T_{03} = 375$ K
Stagnation pressure at compressor exit	$P_{03} = 180$ kPa abs.

The flow in the suction nozzle has negligible amount of friction (can be assumed isentropic). Knowing that air enters the impeller axially and the compressor mechanical efficiency is 96%, determine:

a. Mach number at the impeller exit
b. mass flow rate supplied by the compressor
c. compressor power consumption
d. compressor efficiency.

8.10 It is required to design a centrifugal compressor for delivering air at a rate of 4.0 kg/s with a stagnation pressure ratio, $p_{03}/p_{01} = 2.4$. The impeller is assumed to have 17 radial vanes ($\beta_2 = 90°$) and to be driven at a speed of 10 000 rpm. The impeller width is designed to limit the radial velocity component at impeller exit (Y_2) to 85 m/s. The compressor will suck air from the atmosphere where the pressure and temperature are 100 kPa and 288 K, respectively. The flow is assumed to enter the impeller with no prerotation, and the compressor and impeller efficiencies are assumed $\eta_c = 75\%$ and $\eta_I = 85\%$. Determine:

a. impeller outer diameter, D_2
b. flow Mach number at impeller exit, M_2
c. vane width at impeller exit, b_2
d. expected compressor brake power, assuming $\eta_{mech.} = 95\%$.

8.11 A single-stage centrifugal air compressor is driven at a speed of 16 000 rpm and delivers a mass flow rate of 15 kg/s. The stagnation pressure and temperature at the impeller inlet are 101 kPa and 288 K respectively. The impeller is equipped with radial vanes and has an outer diameter of 0.55 m and the vane width at exit is 37 mm. Assuming that air enters the impeller axially, that the impeller efficiency is 90%, and that the slip factor is 0.9, determine:

a. stagnation pressure and temperature at impeller exit
b. power required to derive the compressor assuming $\eta_{mech.} = 95\%$.
c. Prove that the radial velocity component at impeller exit is approximately 105 m/s.
d. What is the Mach number at impeller exit?

9

Multiphase Flow Pumping

Production systems in the oil and gas industries often require transportation of a mixture of water, oil, and gas for long distances from production well to the processing facility. The conventional production operation, in which fluids are separated before being pumped and compressed through separate lines, costs about 30% more when compared to a multiphase pumping facility. Moreover, by eliminating the separation equipment, even greater savings can be achieved in offshore production sites. Although, multiphase pumps operate less efficiently than conventional pumps by 30–50%, there are a number of advantages to using multiphase pumps. These include the increase in production capacity by lowering the backpressure on wells, the reduction in equipment capital cost, and the reduction in operational complexity. In this chapter, an introduction to multiphase flow basics and examples of pumps used in multiphase flow and their theory of operation will be discussed. The effect of multicomponent flow on the operation of conventional centrifugal pump performance is also discussed.

9.1 Introduction

Multiphase flow is encountered in an increasing number of applications and industries where different types of pumps are used. This can be found in pumping liquid and solid in water sewage treatment plants, pumping gas–liquid–solid multiphase mixtures in different separations, and transportation processes in the oil and gas industry. Also, multiphase flow pumping occurs in heat exchangers, steam generators, chemical reactors, and many other petrochemical processes. Furthermore, artificial gas lift pumping facilities and cavitation phenomena involve pumps under multiphase flow conditions. Therefore, understanding multiphase flow behavior and evaluating the performance characteristics of different pumps under multiphase flow conditions is essential to the engineer.

Pumping Machinery Theory and Practice, First Edition. Hassan M. Badr and Wael H. Ahmed.
© 2015 John Wiley & Sons, Ltd. Published 2015 by John Wiley & Sons, Ltd.

One of the important classifications of multiphase flows is based on the presence or absence of heat transfer. Therefore, the adiabatic two-phase flow is differentiated from the diabatic flow where heat transfer is involved. In the latter case, the flow and heat transfer processes are coupled as a thermo-hydrodynamic problem. In this case, heat transfer my cause phase change and hence a change of phase distribution and consequently flow pattern. This will create changes in the hydrodynamics parameters, such as a pressure drop throughout the system, which also changes the heat transfer characteristics. Moreover, it will create additional complexities in the analysis of multiphase flow where hydrodynamic instabilities are introduced, and there is the occasional departure from thermodynamic equilibrium between the phases. In order to avoid such complexities, integral analysis and experiments have so far been conducted, based on the assumptions of fully developed flow patterns and without heat addition to the flow. Therefore, extensive information has become available on flow patterns, phase distribution, and pressure drop in adiabatic flows, often for two-component, gas–liquid mixtures such as in the oil and gas industries. This information provides basic understanding of the pattern transition mechanism and pressure drop characteristics. Then, in order to apply adiabatic conditions knowledge to the systems that experience diabatic conditions with mass transfer, certain modifications are required.

9.1.1   Two-Phase Flow Parameters

Before proceeding with the subject of two-phase flow pumping, it is necessary to define relevant two-phase flow parameters and their terminology. Usually, the void fraction (α) is one the most important parameters, when dealing with gas–liquid flows. The void fraction is generally defined as the volume occupied by the gas phase over the total volume of the pipe over a specific distance along the piping system. This is can be understood as if a pipe of length L and cross-section A is suddenly isolated by closing valves at both ends, the contents can be analyzed, and the total volume ($\forall_g$) of the gas in the pipe can be determined. The volumetric average value of void fraction (α) can be written as:

$$\langle \alpha \rangle = \frac{\forall_g}{A \cdot L} \tag{9.1}$$

The void fraction can be also defined at a certain cross-section as the area average void fraction ($\bar{\alpha}$). In this case, it is defined as the area occupied by gas (A_g) over the pipe cross-sectional area (A) and can be written as:

$$\bar{\alpha} = \frac{A_g}{A} \tag{9.2}$$

When there are a large number of dynamic variations of two-phase with time, the void fraction measurement is obtained at a local infinitesimal volume of the pipe, and the integral void fraction over the measuring volume along the time is defined. The definition for void fraction is introduced as a function of the instantaneous reading and the average void fraction over the time and space is written as

$$\langle \alpha \rangle = \frac{\iint \alpha(v,t) dv \cdot dt}{\int dv \cdot \int dt} \tag{9.3}$$

In many practical applications, the symbol (α) is used to refer to the average volumetric concentration without bothering to define exactly the type of average is to be taken.

Another important ratio for two-phase flow is defined based on the mass and is known as mass quality (x). It is defined as the mass of the gas phase over the total mass of both phases (m) and can be written as:

$$x = \frac{m_g}{m} \tag{9.4}$$

For unsteady or non-uniform two-phase flow, the mass quality is defined as the average taken over a specific surface and for a period of time:

$$\langle x \rangle = \frac{\iint G_g dA \cdot dt}{\int GdAdt} \tag{9.5}$$

where (G_g) is the mass flux of gas and (G) is the total mass flux. Note that flux is a vector quantity. On the other hand, volumetric flux (J) is defined as the volume flow rate (Q) per unit area. In the two-phase flow subject, fluxes are commonly used to represent the scalar component in the direction of motion along a pipe or duct.

$$j = \frac{Q}{A} \tag{9.6}$$

The relations between volumetric fluxes, void fraction, volume flow rate, and specific volumes are listed below. If the liquid volumetric flux (j_f) is written as a function of void fraction and specific volume of liquid:

$$j_f = (1-\alpha)v_f \tag{9.7}$$

The gas volumetric flux (j_g) can be written as

$$j_g = \alpha v_g \tag{9.8}$$

where

$$j_f + j_g = j \tag{9.9}$$

and the volume flow rate of both gas (Q_g) and liquid (Q_f) are written as

$$Q_g = \int j_g \, dA \tag{9.10}$$

$$Q_f = \int j_f \, dA \tag{9.11}$$

The volumetric and mass fluxes can be correlated as

$$j_f \cdot \rho_f = G_f \tag{9.12}$$

$$j_g \cdot \rho_g = G_g \tag{9.13}$$

$$G = G_f + G_g \tag{9.14}$$

Using Eqs (9.8)–(9.14), the following relations can be obtained.

$$\frac{j_f}{j_g} = \frac{(1-\alpha)v_f}{\alpha v_g} \tag{9.15}$$

$$\frac{G_f}{G_g} = \frac{m_f}{m_g} = \frac{1-x}{x} \tag{9.16}$$

Therefore, the relation between void fraction and mass quality can be expressed as

$$\frac{1-x}{x} = \frac{v_f}{v_g} \frac{(1-\alpha)}{\alpha} \frac{\rho_f}{\rho_g} \tag{9.17}$$

Another important parameter used in many two-phase flow analyses, is the drift velocity (v_{fj}). Drift velocities for both gas and liquid in a two-phase flow mixture are defined as the difference between the actual phase velocity (vv) and the total volumetric flux, and can be written as

$$v_{fj} = v_f - j \tag{9.18a}$$

$$v_{gj} = v_g - j \tag{9.18b}$$

Then, the drift flux represents the volumetric flux of the component relative to a surface moving at the average velocity. Two drift fluxes of gas with respect to liquid phase and vice versa are defined as

$$j_{fg} = (1-\alpha) \cdot (v_f - j) \tag{9.19a}$$

$$j_{gf} = \alpha \cdot (v_g - j) \tag{9.19b}$$

where

$$j_{fg} = -j_{gf} \tag{9.20}$$

Equation (9.20) represents a symmetry which is an important and useful property of the drift flux. Also, it can be proved that the drift velocity is proportional to the relative velocity between phases, and Eq. (9.19) can be rewritten as

$$j_{fg} = \alpha(1-\alpha) \cdot \left(v_{fg}\right) \tag{9.21}$$

9.1.2   Flow Patterns

It is very important to consider the flow pattern effect when dealing with two-phase flow. In the case of a single-phase flow, the behavior only depends on whether the flow is in the laminar or turbulent regime. Similarly, the hydrodynamic behavior of two-phase flow, such as pressure drop and velocity distribution, varies with the observed flow pattern (or regime). However, in contrast to single-phase flow, it is complex to generalize one principle for gas–liquid flows that can be used as a framework for solving all practical problems. For example, for two-phase flow, phenomenological principles – such as Prandtl's mixing-length theory, the methods of analogies such as Colburn's j factors, or boundary-layer theory simplifications – cannot be considered in this case. Alternatively, flow patterns are used to provide a picture of the phase boundaries, which allows for various order-of-magnitude calculations using integrated forms of the momentum and continuity equations. These calculations determine which variables should be investigated and what kind of flow behavior is expected. Several techniques are used to identify the flow pattern in piping systems, including the measurements of hot wire probes, conductance probes, and other sampling probes. However, the visual or photographic representation when available in transparent piping is considered the best. In this case, both still pictures and movies are used to construct the flow pattern maps.

Several flow patterns are observed in varieties of pipe orientations. In co-current upward flow of gas and liquid in vertical pipe, liquid and gas phases typically distribute themselves into a distinct flow structure which depends on the mass flow rate of each phase. The sketch of these flow patterns are presented in Figure 9.1. These flow patterns can be classified into: bubbly, slug, churn, and annular flow patterns. In bubbly flow, the gas phase is dispersed in the form of discrete bubbles in the continuous liquid phase. The shapes of these bubbles are approximately spherical and their sizes widely varies depending on the pipe size and the magnitude of the gas flow rate. As the gas volume flow rate increases, the proximity of the bubbles decreases which causes bubbles to collide and coalesce to form larger bubbles. If the size of these big bubbles is approximately similar to the pipe diameter, the flow is called a slug flow. These bubbles are characterized by a bullet shape with hemispherical nose and a blunt tail end. They are commonly referred to as Taylor bubbles. Taylor bubbles are separated from one another by slugs of liquid, which include several small bubbles and are surrounded by a thin film of liquid close to the pipe wall. Within this film some liquid may flow downward due to gravity, while the total net flow is in the upward direction.

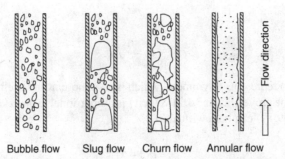

Figure 9.1   Flow patterns in co-current upward vertical gas/liquid flow

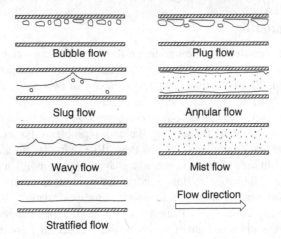

Figure 9.2   Flow patterns for horizontal gas-liquid flow

In horizontal pipes, flow patterns for fully developed flow have been reported in numerous studies. The typical flow patterns of two-phase flow in horizontal pipe are sketched in Figure 9.2. It should be noted that the transition between one flow pattern and another in horizontal pipes is gradual. This classification is, however, is subjective to the visual interpretation of individual investigators. In some cases, statistical analysis of pressure fluctuations is used to distinguish different flow patterns.

The prediction of two-phase flow patterns has been a very crucial subject and has received a great deal of attention from many researchers. It has been concluded that the approximate prediction of flow pattern may be done quickly using what is called flow pattern maps. Since the pipe orientation plays an important role in distributing two-phase flow in pipes, specified flow pattern maps are developed for each piping orientation. Examples of the most cited maps available in the literature for fully developed gas-liquid flows for vertical pipes are shown in Figures 9.3 and 9.4. In addition to the flow pattern maps, flow pattern prediction based on mechanistic models has also been used in critical applications such as the nuclear industry. This is mainly due to their greater accuracy, especially for large pipe diameters and for fluids with physical properties different from air/water at atmospheric pressure. On the other hand, maps such as

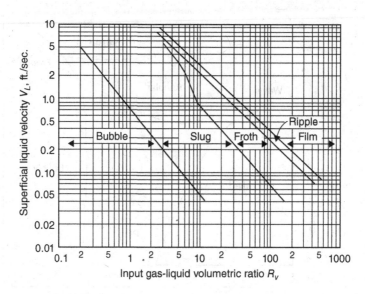

Figure 9.3 Flow-pattern regions in concurrent liquid–gas flow in up-flow through vertical pipes (Govier *et al.* 1957)

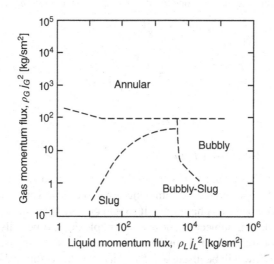

Figure 9.4 Flow regime map for vertical gas–liquid flow by Hewitt and Roberts (1969) for flow in a 3.2 cm diameter tube

that developed by Baker (1954) provide flow pattern prediction for horizontal two-phase flow as shown in Figure 9.5.

Where

$$\lambda = \left(\rho'_G \rho'_L \right)^{1/2} \qquad (9.22)$$

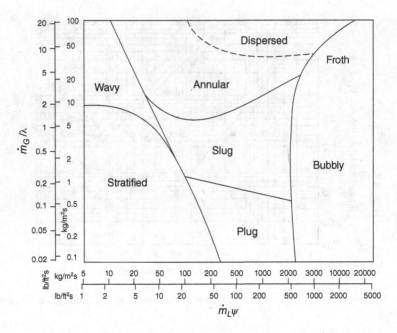

Figure 9.5 Flow-pattern regions in concurrent liquid–gas flow through horizontal pipe (Baker 1954)

$$\psi = \frac{1}{\sigma'}\left[\frac{\mu_L'}{(\rho_L')^2}\right]^{1/3} \tag{9.23}$$

9.1.3 Flow Pattern Transition

The flow pattern can depend on the way the two phases are introduced into the channel and also on the existence of bends, surface contaminants, and so on. However, discussion here will be mainly for fully developed two-phase flow. It should be noted that the definition of fully developed in two-phase flow is subject to additional complexities, especially if heat transfer takes place. Therefore, for the sake of focusing on the subject of two-phase flow pumping, only adiabatic two-phase flow will be discussed. Note also that the additional effects of flow pattern transition and changes in two-phase flow parameters exist as the flow passes through the pumping equipment. For example, in most pumping processes, bubble coalescence occurs and formation of larger bubbles or slugs can take place, but higher pressures lead to suppression of large gas bubbles.

For example, the key feature in determining the onset of flow pattern transition from bubbly to slug flow is the frequency of bubble coalescence. One theory that can be adapted here is based on the work by Radovcich and Mossis (1962). They considered the bubbles to be cubic frames having motion relative to one another at the average velocity of $\bar{C}$, and they expressed the collision frequency as a function of bubble diameter and void fraction as follows:

$$f \propto \frac{C}{D_b\left[(0.74/\alpha)^{1/3}-1\right]^5} \tag{9.24}$$

For low void fraction, this frequency is low and becomes very high for void fractions greater than 0.3. In the case of high void fraction, bubbly flow is unlikely to persist, and flow pattern transition to plug or slug may occur.

Several situations for flow pattern transitions are found such as slug to churn, churn to annular, or annular to mist flow. These transitions are also affected by phenomena such as flooding and/or flow reversal. Considering a tube in which liquid is fed in at a higher level than the gas, liquid can fall in the form of liquid-film downward. However, if the gas velocity increases excessively, the liquid can be carried upward and a climbing film is formed. On the other hand, if the gas velocity is reduced, flow reversal may occur and liquid may fall again below the liquid injection point. For example, the transition to churn flow is characterized by the flooding breakdown of the gas bubbles in slug flow. In other cases, the two-phase in which there is a continuous flow of liquid-film in the upward direction is characterized by annular flow pattern. This phenomenon of flooding has been investigated by many researchers and one of the most widely used correlations to describe flooding is that developed by Wallis (1961) and Hewitt and Wallis (1963):

$$V_G^* = V_G \rho_G^{1/2}\left[gd\left(\rho_f-\rho_G\right)\right]^{-1/2} \tag{9.25}$$

$$V_f^* = V_f \rho_f^{1/2}\left[gd\left(\rho_f-\rho_G\right)\right]^{-1/2} \tag{9.26}$$

The reversal flow transition is represented quite well by the gas flow rate equivalent to $V_G^* = 1$, and the flooding transition is represented by

$$V_G^{*1/2} + V_f^{*1/2} = 1 \tag{9.27}$$

9.1.4   Modeling of Multiphase Flow

There are several approaches to modeling two-phase flow. In general, these approaches can be classified into two main categories: mixture models and separated flow models. In the mixture model, definitions for two-phase flow parameters such as density and viscosity are established based on the single-phase properties. In this case, the two phases are assumed to move with the same velocity and the interaction between the phases is neglected. On the other hand, separated flow models take into account the difference in velocity between phases, the interfacial forces, and sometimes specific parameters related to the distribution of the two phases within the piping system which is known as the flow pattern. In this section, a sample model of each category will be discussed in order to give the reader a basic understanding of the challenges involved in the two approaches. The homogenous model will be introduced as an example of the mixture models and the one-dimensional two-fluid model will represent the separated flow models.

9.1.4.1   Homogenous Flow Model

The homogenous flow model provides the simplest technique for analyzing two-phase flows. In this model, suitable average properties are evaluated and the mixture is treated as a pseudo-fluid that obeys the governing equations of single-phase flow. Therefore, all the standard methodologies of fluid mechanics can be applied in this case. The average velocity, temperature, and chemical potential are assumed to be the same for both phases and therefore homogenous equilibrium can be assumed. However, for rapid fluid acceleration or large pressure changes, equilibrium theory can be inaccurate for describing the flow in this case.

The basic equations for steady one-dimensional homogenous equilibrium flow in a duct are as follows:

Continuity equation

$$\dot{m} = \rho_m v A = const \tag{9.28}$$

Momentum equation

$$\int_A \left[p - \left(p + \frac{dp}{dz} \right) \partial z \right] dA = \int_S \tau \partial z \partial s + \int_A \frac{d}{dz} \left(G_f v_f + G_G v_G \right) dA + \int_A \rho' g dz dA \tag{9.29}$$

where the local density is defined as

$$\rho' = \alpha' \rho_G + (1 - \alpha') \rho_f \tag{9.30}$$

The momentum equation can also be written as

$$\dot{m} \frac{dv}{dz} = -A \frac{dp}{dz} - P \tau_w - A \rho_m g \cos \theta \tag{9.31}$$

where P is the perimeter.

Energy equation

$$\frac{dq_e}{dz} - \frac{dw}{dz} = \dot{m} \frac{d}{dz} \left(h + \frac{v^2}{2} + g z_g \right) \tag{9.32}$$

where q_e is the heat transfer per unit length, w is the net work done, and z_g is the vertical coordinate. We also often write the momentum equation as the summation of three terms:

$$\frac{dp}{dz} = \left(\frac{dp}{dz} \right)_{Accelaration} + \left(\frac{dp}{dz} \right)_{Frictional} + \left(\frac{dp}{dz} \right)_{Gravitational} \tag{9.33}$$

Equation (9.29) can be written as

$$\frac{dp}{dz} = -\frac{\dot{m}\,dv}{A\,dz} - \frac{P}{A}\tau_w - \rho_m g\cos\theta \tag{9.34}$$

In Eq. (9.34), the mixture density can be expressed in various ways, as a function of either the volume fraction or the mass fraction:

$$\rho_m = \alpha\rho_G + (1-\alpha)\rho_f \tag{9.35}$$

$$\frac{1}{\rho_m} = \frac{x}{\rho_G} + \frac{(1-x)}{\rho_f} \tag{9.36}$$

In the momentum equation, the wall friction can be expressed in terms of friction coefficient and hydraulic mean diameter D as

$$\tau_w = \frac{1}{2}C_f\rho_m v^2 \tag{9.37}$$

Therefore, the frictional pressure drop in Eq. (9.33) can be written as

$$\left(\frac{dp}{dz}\right)_F = 2C_f\rho_m\frac{v^2}{D} \tag{9.38}$$

Substituting by volumetric mass flow rates:

$$\left(\frac{dp}{dz}\right)_F = \frac{2C_f Gj}{D} \tag{9.39}$$

Using the definition of the average velocity:

$$v = \frac{G}{\rho_m} = G\left(\nu_G x + (1-x)\nu_f\right) \tag{9.40}$$

$$\left(\frac{dp}{dz}\right)_F = \frac{2C_f G^2}{D}\left(\nu_G x + (1-x)\nu_f\right) \tag{9.41}$$

Using similar manipulations, the gravitational and acceleration pressure drop can be expressed as

$$-\left(\frac{dp}{dz}\right)_A = G^2\left\{\nu_{fG}\frac{dx}{dz} + \frac{dp}{dz}\left[x\frac{d\nu_G}{dp} + (1-x)\frac{d\nu_f}{dp}\right] - (\nu_f + x\nu_{fg})\frac{1}{A}\frac{dA}{dz}\right\} \tag{9.42}$$

$$-\left(\frac{dp}{dz}\right)_G = \frac{1}{v_f + v_{fg}} g\cos\theta \qquad (9.43)$$

9.1.4.2 One-Dimensional Two-Fluid Model

The two-fluid model is formulated by considering each phase separately, so the conservation of mass, momentum, and energy is considered for each phase. However, the averaged-parameters of one phase are dependent on the other, so interaction terms are included in these equations. These terms account for the mass, momentum, and energy transfer between the two phases across the interface. These terms should obey the balance laws at the interface which are commonly known as the interfacial jump conditions.

Local instantaneous conservation equations are derived by writing the general integral balance equations for mass, momentum, and energy for a small differential control volume in the flow field. The selected control volume is divided into two sub-volumes representing the two phases and separated by moving the boundary (interfacial area). Using Leibniz's rule and the Gauss theorem, the equation is transformed from integral form to local instantaneous phase equations similar to the derivation of local instantaneous equations in single-phase flow. Then the conservation equations can be written as

Mass conservation

$$\frac{\partial \rho_k}{\partial t} + \nabla \cdot \left(\rho_k \vec{u}_k\right) = 0 \quad k = 1, 2 \qquad (9.44)$$

and at the interface

$$\sum_{k=1,2} \dot{m}_k = 0 \qquad (9.45)$$

Momentum conservation

$$\frac{\partial (\rho_k u_k)}{\partial t} + \nabla \cdot \left(\rho_k \vec{u}_k \vec{u}_k\right) + \nabla p_k - \nabla \cdot \bar{\bar{\tau}}_k - \rho_k \vec{g}_k = 0 \quad k = 1, 2 \qquad (9.46)$$

and at the interface

$$\dot{m}_1 \left(\vec{u}_1 - \vec{u}_2\right) + (P_1 - P_2) \cdot \vec{n}_1 + (\bar{\bar{\tau}}_2 - \bar{\bar{\tau}}_1) \cdot \vec{n}_1 = 0 \qquad (9.47)$$

Energy conservation

$$\frac{\partial}{\partial t}\left[\rho_k\left(e_k + \frac{1}{2}u_k^2\right)\right] + \nabla \cdot \left[\rho_k\left(e_k + \frac{1}{2}u_k^2\right)\vec{u}_k\right] - \nabla \cdot \left(\bar{\bar{\sigma}} \cdot \vec{u}_k\right) + \nabla \cdot \vec{q}_k - \rho_k \vec{g}_1 \cdot \vec{u}_k = 0 \qquad (9.48)$$

where

$$\bar{\bar{\sigma}} = -P\delta_{ij} + \bar{\bar{\tau}} \tag{9.49}$$

and the interfacial condition is

$$\dot{m}_1\left(e_1 + \frac{1}{2}u_1^2\right) + \dot{m}_2\left(e_2^2 + \frac{1}{2}u_2^2\right) + \left(\vec{q}_1 - \vec{q}_2\right)\cdot\vec{n}_1 - \left(\vec{n}_1\cdot\bar{\bar{\sigma}}_1\right)\cdot\vec{u}_1 - \left(\vec{n}_2\cdot\bar{\bar{\sigma}}_2\right)\cdot\vec{u}_2 = 0 \tag{9.50}$$

It is worth mentioning that, in most practical flow systems, the local instantaneous equations cannot be applied directly because of two main physical and mathematical difficulties:

1. Existence of moving discontinuities in the flow field which is represented by the moving deformable interfaces.
2. Fluctuations of the flow variables caused by turbulence and moving interfaces.

Although the local instantaneous equations represent the exact mathematical model for such a problem, they cannot be solved. Therefore, in order to derive a solvable set of equations for two-phase flow applications, averaging procedures are required as explained earlier in details by Banerjee and Chan (1980). In summary, a volume-averaging procedure can be applied by integrating each term over the control volume ($\forall_k$) for any variable $F = F(\bar{x}, t)$.

The volumetric average of the variable (F) can be written as

$$\langle F \rangle = \frac{1}{v_k}\int_{v_k} F\,dv \tag{9.51}$$

where the volumetric concentration for phase k is given by:

$$\langle \alpha_k \rangle = \frac{v_k}{v}\langle \alpha_k \rangle = \frac{\forall_k}{\forall} \tag{9.52}$$

and the time averaging over a sample period defined by:

$$\bar{F} = \frac{1}{T}\int_0^T F\,dt \tag{9.53}$$

Using the general volume and time averaging (composite averaging), the conservation of mass equation becomes

$$\frac{\partial}{\partial t}\overline{\alpha_k\langle\rho_k\rangle} + \nabla\cdot\overline{\alpha_k\langle\rho_k\vec{u}_k\rangle} = -\overline{\langle\dot{m}_k\rangle}_i \tag{9.54}$$

where

$$-\overline{\langle \dot{m}_k \rangle}_i = \overline{\rho_k \left(\vec{u}_i - \vec{u}_k \right) . \vec{n}_k} \tag{9.55}$$

and the conservation of momentum

$$\frac{\partial}{\partial t} \overline{\alpha_k \langle \rho_k u_k \rangle} + \nabla . \overline{\left(\alpha_k \left(\left\langle \rho_k \vec{u}_k \vec{u}_k \right\rangle \right) \right)} + \nabla \overline{\alpha_k p_k} - \nabla . \overline{\alpha_k \langle \bar{\bar{\tau}}_k \rangle}$$

$$= \overline{\alpha_k \langle \rho_k g \rangle} - \overline{\left\langle \dot{m}_k \vec{u}_{ki} \right\rangle} + \overline{\langle \vec{n}_k . \bar{\bar{\tau}}_k \rangle}_i + \overline{\langle \vec{n}_k . \tau_k \rangle}_w \tag{9.56}$$

The conservation of energy is written as

$$\frac{\partial}{\partial t} \overline{\alpha_k \left\langle \rho_k \left(h_k + \frac{\vec{u}_k . \vec{u}_k}{2} \right) \right\rangle} + \nabla . \overline{\alpha_k \left\langle \rho_k \vec{u}_k . \left(h_k + \frac{\vec{u}_k . \vec{u}_k}{2} \right) \right\rangle}$$

$$- \alpha_k \frac{\partial}{\partial t} \overline{\langle p_k \rangle} + \nabla . \overline{\alpha_k \langle \vec{q}_k \rangle} - \nabla . \overline{\alpha_k \langle \vec{n}_k . (\bar{\bar{\tau}}_k . u_k) \rangle} \tag{9.57}$$

$$= - \overline{\left\langle \left[\dot{m}_k \left(h_k + \frac{\vec{u}_k . \vec{u}_k}{2} \right) + \vec{n}_k . \vec{q}_k - \vec{n}_k . \vec{u}_k . \bar{\bar{\tau}}_k \right] \right\rangle}_i - \overline{\langle \vec{n}_k . \vec{q}_k \rangle}_w + \overline{\alpha_k \langle \rho_k \vec{u}_k . \vec{g} \rangle}$$

In addition to the basic governing equations, the average jump conditions are defined as follows:

Interfacial mass

$$\sum_{k=1,2} \langle \bar{\dot{m}} \rangle_{ki} = 0$$

$$\text{or } \overline{\langle \dot{m}_1 \rangle} i = - \overline{\langle \dot{m}_2 \rangle} i \tag{9.58}$$

Interfacial momentum

$$\sum_{k=1,2} \overline{\langle \dot{m}_k \bar{u}_k - n_k . \bar{\bar{\sigma}}_k \rangle} = 0 \tag{9.59}$$

And this can also be expressed as

$$\sum_{k=1,2} \frac{1}{\forall} \int_{a_i} [\dot{m}_k u_k + \bar{n}_z . \bar{n}_k p_k - \bar{n}_z . (\bar{n}_k . \bar{\bar{\tau}}_k)] da = 0 \tag{9.60}$$

Interfacial energy

$$\sum_{k=1,2} \frac{1}{\forall} \int_{a_i} \left[\dot{m}_k \left(h_k + \frac{u_k^2}{2} \right) + p_k (\bar{n}_k \cdot \bar{u}_k) + \bar{n}_k \cdot (\bar{q}_k - \bar{\bar{\tau}}_k \cdot \bar{u}_k) \right] da = 0 \qquad (9.61)$$

This means that the net mass, momentum, and energy at the interface are zeroes. In order to obtain a solvable model, the average of products should be expressed in terms of the averaged variables. Therefore, distribution parameters need to be introduced and can be defined as

$$C_1 = \frac{\overline{\alpha_k \langle u_k \rangle}}{\bar{\alpha}_k \langle u_k \rangle} \qquad (9.62)$$

$$C_2 = \frac{\overline{\langle \rho_k u_k \rangle}}{\langle \bar{\rho}_k \rangle \langle \bar{u}_k \rangle}, \text{etc.} \qquad (9.63)$$

Then, the two-fluid model equations can be derived from the local instantaneous equations, considering each phase separately. Two sets of conservation equations including mass, momentum, and energy are developed for each phase. As discussed earlier, the volume and time average of the above equations results in interfacial terms. Then the time-volumetric average of the conservation equations can be simplified to a one-dimensional equation assuming the following:

1. Space and time distribution parameters (C_1, C_2, etc.) are equal to unity.
2. The equation of state is valid for average quantities.
3. We can neglect the phase viscous stress derivatives and the associated energy dissipation.
4. We can neglect the longitudinal heat transfer.
5. Pressure is assumed to be uniform up to the interface, that is, the pressure difference between vapor and the liquid exists discontinuously at the interface (Banerjee and Chan 1980).

Now, if we consider

$$\alpha_G = \alpha \text{ and } \alpha_L = (1 - \alpha) \qquad (9.64)$$

the conservation of mass is reduced to

$$\text{vapor} \frac{\partial}{\partial t} (\alpha \rho_G) + \frac{\partial}{\partial z} (\alpha \rho_G u_G) = \langle \dot{m}_{LG} \rangle = \Gamma \qquad (9.65)$$

$$\text{liquid} \frac{\partial}{\partial t} ((1 - \alpha) \rho_L) + \frac{\partial}{\partial z} ((1 - \alpha) \rho_L u_L) = -\Gamma \qquad (9.66)$$

where $\langle \dot{m}_{LG} \rangle_i = \Gamma$ is the vapor generation rate per unit volume if mass transfer takes place. The interfacial jump condition is expressed as

$$\langle \dot{m}_{LG} \rangle_i + \langle \dot{m}_{GL} \rangle_i = 0 \qquad (9.67)$$

The conservation of momentum for vapor is written as

$$\frac{\partial}{\partial t}[\alpha\rho_G u_G] + \frac{\partial}{\partial z}[\alpha\rho_G u_G^2] + \alpha\frac{\partial P_G}{\partial z} - \alpha\rho_G g_z = -F_{Gi} - F_i - F_{wG} \qquad (9.68)$$

where
F_{Gi} = momentum exchange due to mass transfer per unit volume
F_i = interfacial force restraining the vapor flow
F_{wG} = wall to vapor frictional force per unit volume.
Also, the conservation of momentum for liquid is written as

$$\frac{\partial}{\partial t}[(1-\alpha)\rho_L u_L] + \frac{\partial}{\partial z}[(!-\alpha)\rho_L u_L^2] + (1-\alpha)\frac{\partial P_L}{\partial z} - (1-\alpha)\rho_L g_z = -F_{Li} - F_i - F_{wL} \qquad (9.69)$$

where
F_{Li} = momentum exchange due to mass transfer per unit volume
F_{LG} = the interfacial force restraining the liquid flow.
Utilizing the above assumption, the conservation of energy for vapor is written as

$$\frac{\partial}{\partial t}\left[\alpha\rho_G\left(h_G + \frac{u_G^2}{2}\right)\right] + \frac{\partial}{\partial z}\left[\alpha\rho_G\left(h_G + \frac{u_G^2}{2}\right)\right] - \alpha\frac{\partial p_G}{\partial t} - \alpha\rho_G u_G g_z = q_i + q_{wG} + \alpha q_G''' \qquad (9.70)$$

where
q_i = interfacial heat transfer rate to vapor phase per unit volume
q_G''' = internal heat generation rate
q_{wG} = wall to vapor heat transfer per unit volume.
Also, the conservation of energy for liquid is written as

$$\frac{\partial}{\partial t}\left[(1-\alpha)\rho_L\left(h_L + \frac{u_L^2}{2}\right)\right] + \frac{\partial}{\partial z}\left[(1-\alpha)\rho_L\left(h_L + \frac{u_L^2}{2}\right)\right]$$

$$-(1-\alpha)\frac{\partial p_L}{\partial t} - (1-\alpha)\rho_L u_L g_z = -q_i + q_{wL} + (1-\alpha)q_L''' \qquad (9.71)$$

In the above equations, there are seven unknown variables ($\alpha, u_G, u_L, h_L, h_G, P_L, P_G$) and there are only six conservation equations. Therefore a seventh equation is required; it can be obtained using the constitutive relations. This closure requirement which includes constitutive relations for both liquid and vapor densities can be obtained, for example, from the equation of state. Also, the saturation properties and both phases' temperatures can be obtained from the equation of state. Moreover, closure requirements can be obtained using the transport terms at the interface ($\Gamma, \Gamma_{LG}, F_{Gi}, F_{Li}, q_i$) or at the wall ($q_{wG}, q_{wL}, F_{wG}, F_{wL}$). To summarize, the following can be written for the two-fluid model:

No. of conservation equations = 6
Thermal non-equilibrium $T_G \neq T_L$
No restriction for slip $u_G \neq u_L$

Boundary conditions at the wall have four variables $(q_{wG}, q_{wL}, F_{wG}, F_{wL})$
Boundary conditions at the interface have three variables (Γ, q_i, F_i)
Mechanical equilibrium $p_L = p_g$ (not valid for many applications)

9.1.4.3 Constitutive Relations for Interfacial Area Concentration

In order to find the constitutive relations, interfacial terms for mass, momentum, and energy need to be identified. These interfacial terms are a function of the interfacial area per unit volume, which is known as the interfacial area concentration (a_i), and of the driving force, as explained by Ishii and Mishima (1984).

For mass transfer

$$\Gamma = a_i \dot{m}_k \tag{9.72}$$

where $\dot{m}_k$ is the mass exchange rate per unit interfacial area.

For momentum drag

$$F_i = a_i C_D (u_G - u_L) \tag{9.73}$$

For heat transfer

The interfacial heat transfer rate is

$$q_i = a_i h_i (T_L - T_G) \tag{9.74}$$

These constitutive relations are strongly dependent on flow patterns. A comprehensive review of available constitutive relations may be found in Kelly and Kazimi (1982), and Ishii and Mishima (1984). In general, it should be noted that two-phase flow structure is characterized by two geometrical properties:

1. Volumetric concentration of vapor phase, also known as void fraction, which identifies the phase distribution.
2. Interfacial area concentration, which describes the area for mass momentum and energy transport between the two phases.

For example, the interfacial area concentration for smooth stratified flow shown in Figure 9.6a can be obtained from geometrical parameters per unit length of tube with diameter (D) as follows:

$$a_i = \frac{x \times L}{volume} \tag{9.75}$$

which can be written as

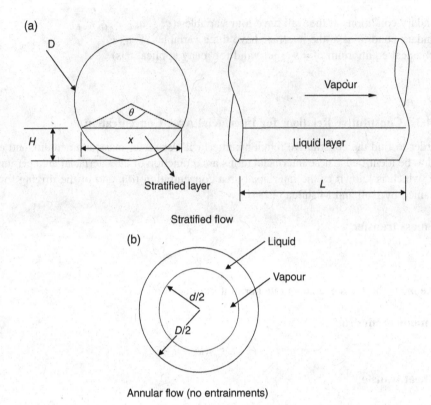

Figure 9.6 Geometrical parameters for stratified flow pattern: (a) stratified flow (b) annular flow (no entrainments)

$$a_i = \frac{4\sin(\theta/2)}{\pi D} \qquad (9.76)$$

Also, the void fraction for stratified flow is related to the angle (θ) and can be expressed as

$$\alpha = 1 - \frac{1}{2\pi}(\theta - \sin\theta) \qquad (9.77)$$

Note that the above relations are not valid for stratified wavy flow. Performing similar analysis for annular flow with no entrainment, a_i can be obtained from geometrical parameters shown in Figure 9.6b as

$$a_i = \frac{4\sqrt{\alpha}}{D} \qquad (9.78)$$

The value of interfacial area concentration has been the interested of many researchers in the past. For annular mist flow, Ishii and Mishima (1984) introduced an expression for interfacial area. The concentration can be expressed as a function of the entrained droplets ratio as

$$a_i = \frac{4C_{an}}{D}\sqrt{\frac{\alpha}{1-\alpha_{fd}}} + \frac{\alpha}{1-\alpha_{fd}} \cdot \frac{3\alpha_{fd}}{r_{sm}} \tag{9.79}$$

where
α_{fd} = liquid droplet fraction in the gas core alone
C_{an} = roughness parameter due to waves in the film ($C_{an} \geq 1$).
The first term in the right-hand side represents the interfacial area from the wavy interface, while the second term represents the entrained droplet. Then the droplet void fraction in the core approximated by

$$\alpha_{fd} \cong E\frac{J_f}{J_g} \tag{9.80}$$

where E is the entrainment fraction of annular flow for quasi-equilibrium conditions given by:

$$E = \tanh\left(7.25 * 10^{-7} W_e^{1.25} R_{ef}^{0.25}\right) \tag{9.81}$$

Several definitions developed for the entrainment are given by Rohatgi (1982) and Katatoka and Ishii (1984). For example, the value of E is determined experimentally as shown by Ishii and Mishima (1984) in Figure 9.7. The values of Weber number (W_e) and liquid Reynolds number (R_{ef}) are determined as

$$W_e = \frac{\rho_g J_g^2 D}{\sigma}\left(\frac{\Delta\rho}{\rho_g}\right)^{\frac{1}{3}} \tag{9.82}$$

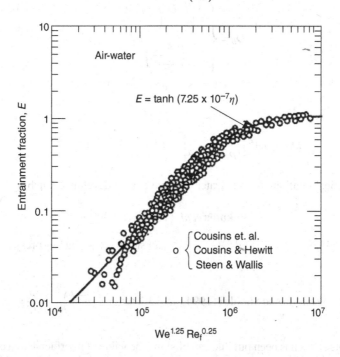

Figure 9.7 Correlation for entrainment factor and comparison with data (Ishii and Mishima 1984)

$$R_{ef} = \frac{\rho_f J_f D}{\mu_f} \tag{9.83}$$

In other cases, the interfacial area concentration for bubbly flow assuming spherical bubbles can be expressed as a function of the number of bubbles and surface area of each bubbles as

$$a_i = \frac{N 4\pi D_b^2}{\frac{\pi}{4} D^2 l} \tag{9.84}$$

where D_b is the bubble diameter, l and D is the channel length and diameter.
The void fraction of bubbly flow can be written as

$$\alpha = \frac{V_G}{V_T} = \frac{N \times \frac{4}{3}\pi D_b^3 \times \frac{1}{8}}{\frac{\pi}{4} D^2 l} \tag{9.85}$$

Then,

$$a_i = \frac{6\alpha}{D_b} \tag{9.86}$$

Kelly and Kazimi (1982) introduced a relationship for D_b:

$$D_b = \begin{cases} D_{b0} & \alpha < 0.1 \\ \\ D_{b0}\left(\dfrac{a\alpha}{1-\alpha}\right)^{1/3} & \alpha \geq 0.1 \end{cases} \tag{9.87}$$

where

$$D_{b0} = 0.45\left[\frac{\sigma}{(\rho_L - \rho_G)g}\right]^{1/2}\left(1 + 1.34[(1-\alpha)U_L]^{1/3}\right)^{-1} \tag{9.88}$$

Several empirical relations for air–water data have been developed, such as

$$\text{Fukuma } et\ al.\,(1987): a_i = 300\alpha \tag{9.89}$$

$$\text{Tabei } et\ al.\,(1989): a_i = 2100\alpha^{1.25}(1-\alpha)^{0.75} \ \text{(for steam} - \text{water at low pressure } (P < 3\,\text{bar}))$$
$$\tag{9.90}$$

$$\text{Zeitoun and Shoukri (1994): } a_i = 3.24\alpha^{0.757}\left(g\frac{\Delta\rho}{\sigma}\right)^{0.55}\left(\frac{\mu}{G}\right)^{0.1} \tag{9.91}$$

Comparable effort has been put into establishing the value of interfacial area concentration in the case of slug flow. Ishii and Mishima (1984) introduced a model for a_i for slug or churn turbulent flow, with a_i approximated by considering a mixture of Taylor bubbles separated by liquid slug containing small droplets as

$$a_i = \frac{4.5}{D} \frac{\alpha - \alpha_{GS}}{1 - \alpha_{GS}} + \frac{3\alpha_{GS}}{r_{Sm}} \frac{1-\alpha}{1-\alpha_{GS}}$$

(9.92)

where
α_{GS} = average void fraction in the liquid slug section
D = hydraulic diameter
r_{Sm} = Sauter bubble diameter.
Many correlations based on air–water data are also available in the literature. For example:

$$\text{De Jesus and Kawaji (1990)} a_i = 1.535 \left(\frac{dP}{dz}\right)_f^{0.12} J_G^{1.2} J_L^{-0.14} (1-\alpha)^{1.6}$$

(9.93)

$$\text{Tomidan et al. (1978)} a_i = 0.22 \left(\frac{dP}{dz}\right)_t \alpha$$

(9.94)

where
$\left(\frac{dP}{dz}\right)_f$ = friction pressure gradient

$\left(\frac{dP}{dz}\right)_t$ = total pressure gradient.

In power generation applications, the exchange of mass across a liquid–vapor interface usually takes the form of vapor generation (Γ) per unit volume. Ishii (1975) derived the expression for (Γ) neglecting the surface tension and the energy dissipation due to interfacial viscous shear. Then, the vapor generation can be written as a function of interfacial thermal conditions:

$$\Gamma h_{LG} = q_{Li} + q_{Gi}$$

(9.95)

$$\text{or } \Gamma h_{LG} = a_i [h_{Li}(T_L - T_i) + h_{Gi}(T_G - T_i)]^i$$

(9.96)

where
q_{Li}, q_{Gi} = heat transfer rate from liquid and vapor to the interface, respectively
h_{lG} = latent heat of evaporation.
Other relations can be used to calculate the rate of evaporation or condensation.

9.1.4.5 Constitutive Relations of Interfacial Momentum Transfer

The momentum exchange between the two phases (F_i) is dependent on the flow conditions and the interfacial structure. The liquid to gas interfacial momentum transfer is modeled as a summation of two terms:

$$F_i = F_{id} + F_{vm}$$

(9.97)

where

F_{id} = interfacial drag force

F_{vm} = the virtual mass force due to relative acceleration between phases.

Ishii and Mishima (1984) developed a correlation for the interfacial drag force based on extensive experimental work, and given as

$$F_{id} = \frac{-1}{8}C_D\rho_C a_i U_r |U_r| \tag{9.98}$$

where

U_r = relative velocity between phases

ρ_c = continuous phase density

C_D = drag coefficient for bubbly or droplet flow

C_D = interfacial friction factor F_i (stratified and annular).

Values for the drag coefficient (C_D) are represented by Ishii and Mishima (1984) for different flow patterns.

The second term in Eq. (9.97) is the virtual mass (F_{vm}) effect, or the interference effect. When the vapor bubble or liquid droplet is accelerated relative to the surrounding fluid, an increase in the momentum transport occurs. This term is obtained by Ishii and Chawala (1979) as

$$F_{vm} = C_{vm}\rho_L\alpha_{vm} \tag{9.99}$$

Another relation was also developed by Lahey (1979) for bubbly flow as

$$F_{vm} = \alpha\rho_L C_{vm}\left\{ \left[\frac{\partial}{\partial t}(u_G - u_L) + u_G\frac{\partial}{\partial z}(u_r)\right] + \left[(u_G - u_L)\left[(\lambda-2)\frac{\partial u_G}{\partial z} + (1+\lambda)\frac{\partial u_L}{\partial z}\right]\right] \right\} \tag{9.100}$$

where

$$\lambda = 2(1-\alpha) \tag{9.101}$$

$$C_{vm} = \frac{1}{2}\frac{(1+2\alpha)}{(1-\alpha)} \tag{9.102}$$

Moreover, he adopted the above equation for slug flow after expressing C_{vm} as a function of Taylor bubble size. He also found that the value of F_{vm} is minimal for droplet flow.

9.1.4.6 Constitutive Relations for Interfacial Momentum Transport

Interfacial momentum transport is important in phase change processes. In this case, one phase receives the momentum when converted to the other phase. For example, in the evaporation process, gas velocity increases, taking some momentum from the vaporized liquid (Schwellnus and Shoukri, 1991). The momentum changes and the proportion taken from each phase are related to the ratio (η). Then, the liquid and gas interfacial momentum forces can be related as

$$F_{Li} = -G_{uL}\frac{dx}{dz} - (1-\eta)(u_G - u_L)G\frac{dx}{dz} \tag{9.103}$$

$$F_{Gi} = -G_{uL}\frac{dx}{dz} - \eta(u_G - u_L)G\frac{dx}{dz} \tag{9.104}$$

$\eta = 0.5$, as recommended by Wallis (1969).

9.1.4.7 Constitutive Relations for Interfacial Heat Transfer

Interfacial heat transfer exists as a result of thermal non-equilibrium in the heat transfer rate per unit volume. It is modeled as a summation of sensible and latent heat transfer components as

$$q_i = a_i h_i(T_L - T_G) + G\frac{dx}{dz}h_{LG} \tag{9.105}$$

The interfacial convective heat transfer coefficient (h_i) for bubbly flow is obtained from the rate of growth or collapse of bubbles as

$$h_i a_b(T_L - T_i) = \rho_G h_{LG}\frac{dV_b}{dt} \tag{9.106}$$

A non-dimensional convective heat transfer coefficient for bubbly flow was developed by Schwellnus and Shoukri (1991) as

$$N_u = \frac{h_i d}{K_L} = 2 + 0.15R_{eb}^{1/2}P_{rL}^{1/3} \tag{9.107}$$

Also, for stratified flow, Kim *et al.* (1985) used

$$N_u = 0.966 \times 10^{-3}\text{Re}_L^{0.98}P_r^{0.95}F_r^{0.8} \tag{9.108}$$

where
F_r = Froude number
P_r = Prandtl number.

9.1.4.8 Constitutive Relations for Interface Properties

At the interface, several approximations may be considered. For example, if thermodynamic equilibrium at the interface is assumed, the interface temperature is given as

$$T_i = T_{sat}(P_i) \tag{9.109}$$

Another approximation is assuming the velocity of the interface to be equal to the mean velocity of the two phases to calculate the momentum exchange as

$$u_i = \frac{1}{2}(u_L + u_G) \tag{9.110}$$

Moreover, some correlations were developed for specific target industries. This is mainly depended on how much accuracy is expected and the whether the consequences of inaccurate prediction can be tolerated in the operation of the two-phase system. For example, in nuclear safety analysis codes, such as CATHENA, the velocity at the interface for bubbly or slug flows is assumed to be

$$u_{Gi} = u_{Li} = u_i = \frac{\alpha \rho_G u_G + (1 - \alpha) \rho_L u_L}{\alpha \rho_G + (1 - \alpha) \rho_L} \tag{9.111}$$

and for annular or stratified flows is given by

$$u_i = u_L$$

Example 9.1

Starting from the basic governing equation for two-phase flow, develop an expression for the change in the liquid film height in a stratified flow pattern along a straight pipe as a function of two-phase flow parameters. Provide the necessary assumptions and the appropriate two-phase flow terms to describe the flow.

Solution

In this model the governing equations are written for each phase (k) and an averaging technique is used to convert the information of three-dimensional into one-dimensional analysis. The equations are written at a node (i) along the selected axis.

Continuity equation

$$\frac{\partial}{\partial t} \overline{\alpha_k \langle \rho_k \rangle} + \frac{\partial}{\partial z} \overline{\alpha_k \langle \rho_k u_k \rangle} = -\left\langle \dot{m}_k \right\rangle_i \tag{9.112}$$

Momentum equation

$$\frac{\partial}{\partial t} \overline{\alpha_k \langle \rho_k u_k \rangle} + \frac{\partial}{\partial z} \overline{\alpha_k \langle \rho_k u_k^2 \rangle} + \bar{\alpha}_k \frac{\partial}{\partial z} \overline{\langle p_k \rangle} - \frac{\partial}{\partial z} \overline{\alpha_k \langle \tau_{zz,k} \rangle} - \Delta \overline{p}_{ki} \frac{\partial \overline{\alpha_k}}{\partial z} = \overline{\alpha_k \langle \rho_k F_{zik} \rangle} - \left\langle \dot{m}_k u_k \right\rangle_i$$

$$- \overline{\langle \Delta p_{ki} \rangle}_i + \overline{\langle \vec{n} \cdot \tau_z \rangle}_i + \overline{\langle \vec{n}_{kw} \cdot \tau_z \rangle}_w \tag{9.113}$$

For smooth stratified flow, as shown in Figure 9.6a, we can write

$$\alpha_L = \frac{A_l}{A} \tag{9.114}$$

$$\theta = \cos^{-1}\frac{R-h_L}{R} \tag{9.115}$$

$$l = \sqrt{2Rh_L - h_L^2} \tag{9.116}$$

$$A_L = R^2\theta - (R-h_L)\sqrt{2Rh_L - h_L^2} \tag{9.117}$$

$$p_i = p_g \tag{9.118}$$

$$T_g = T_{sat} \tag{9.119}$$

The liquid pressure can be written as

$$p_L = \frac{1}{A_l}\int \rho_L g(h_L - y)\dot{d}y \tag{9.120}$$

and

$$\Delta p_{Li} = p_L - p_i \tag{9.121}$$

$$\alpha' = \frac{\partial\alpha_L}{\partial h_L} \tag{9.122}$$

$$\Delta P'_{Li} = \frac{\partial\Delta p_{Li}}{\partial h_L} \tag{9.123}$$

In deriving a simplified set of the governing equations, the momentum and continuity equations of both phases are considered first. The liquid continuity equation (at $k = L$) can be written as

$$\frac{\partial\alpha_L\rho_L}{\partial t} + \frac{\partial\alpha_L\rho_L u_L}{\partial z} = -\dot{m}_L \tag{9.124}$$

Since ρ_L is constant, Eq. (9.56) can be rewritten as

$$\rho_L\frac{\partial\alpha_L}{\partial t} + \rho_L\frac{\partial\alpha_L u_L}{\partial z} = -\dot{m}_L \tag{9.125}$$

The void fraction changes can be expressed as a function of liquid height (h_L)

$$\frac{\partial\alpha_L}{\partial t} = \frac{\partial\alpha_L}{\partial h_L}\cdot\frac{\partial h_L}{\partial t} = \alpha'\cdot\frac{\partial h_L}{\partial t} \tag{9.126}$$

also

$$\frac{\partial \alpha_L u_L}{\partial z} = \alpha_L \cdot \frac{\partial u_L}{\partial z} + u_L \cdot \frac{\partial \alpha_L}{\partial z} \tag{9.127}$$

and

$$\alpha_L \cdot \frac{\partial u_L}{\partial z} + u_L \cdot \frac{\partial \alpha_L}{\partial h_L} \cdot \frac{\partial h_L}{\partial z} = \alpha_L \cdot \frac{\partial u_L}{\partial z} + u_L \cdot \alpha' \cdot \frac{\partial h_L}{\partial z} \tag{9.128}$$

then

$$\rho_L \cdot \frac{\partial \alpha_L u_L}{\partial z} = \rho_L \cdot \alpha_L \cdot \frac{\partial u_L}{\partial z} + \rho_L \cdot u_L \cdot \alpha' \cdot \frac{\partial h_L}{\partial z} \tag{9.129}$$

Substituting from Eqs (9.59) and (9.61) into Eq. (9.51), the following expressions can be obtained for liquid continuity as

$$\frac{\partial h_L}{\partial t} + \frac{\alpha_L}{\alpha'} \cdot \frac{\partial u_L}{\partial z} + u_L \cdot \frac{\partial h_L}{\partial z} = \frac{-\dot{m}_L}{\rho_L \alpha'} \tag{9.130}$$

and the liquid momentum as

$$\rho_L \frac{\partial \alpha_L u_L}{\partial t} + \rho_L \cdot \frac{\partial \alpha_L u_L^2}{\partial z} + \alpha_L \cdot \frac{\partial p_L}{\partial z} - \frac{\partial \alpha_L \tau_{zz,L}}{\partial z} - \Delta p_{Li} \cdot \frac{\partial \alpha_L}{\partial z} = \alpha_L \rho_L F_{ziL} - \dot{m}_L u_i - \Delta p_{Li} + \tau_i a_i - \tau_L a_L \tag{9.131}$$

If gas density is assumed to be constant, the continuity and momentum equation can be derived similarly.

The continuity equation for gas is written as

$$\rho_g \frac{\partial \alpha_g}{\partial t} + \rho_g \frac{\partial \alpha_g u_g}{\partial z} = -\dot{m}_g \tag{9.132}$$

and the gas momentum equation is:

$$\rho_g \frac{\partial \alpha_g u_g}{\partial t} + \rho_g \cdot \frac{\partial \alpha_g u_g^2}{\partial z} + \alpha_g \cdot \frac{\partial p_g}{\partial z} - \frac{\partial \alpha_g \tau_{zz,g}}{\partial z} - \Delta p_{gi} \cdot \frac{\partial \alpha_g}{\partial z} = \alpha_g \rho_g F_{zig} - \dot{m}_g u_i - \Delta p_{gi} - \tau_i a_i - \tau_g a_g \tag{9.133}$$

Since the interfacial term is highly dependent on the flow pattern, smooth stratified flow is assumed and all interface pressure fluctuation terms are neglected. With extensive manipulations, which can be found in many multiphase flows, the momentum equation for the one-dimensional two-fluid model can be written for liquid as

$$\frac{\partial u_L}{\partial t} + u_L \cdot \frac{\partial u_L}{\partial z} + \frac{1}{\rho_L} \cdot \frac{\partial p_i}{\partial z} + \frac{1}{\rho_L \alpha_L} \cdot [\alpha_L \partial \Delta P'_{Li} + \Delta P_{Li} \alpha'] \cdot \frac{\partial h_L}{\partial z} = \frac{\tau_i a_i}{\rho_L \alpha_L} - \frac{\tau_L a_L}{\rho_L \alpha_L} + \frac{\dot{m}_L}{\rho_L \alpha_L}(u_L - u_i)$$

$$(9.134)$$

and for gas as

$$\rho_g \left[u_g \cdot \frac{\partial \alpha_g}{\partial t} + u_g \cdot \frac{\partial \alpha_g u_g}{\partial z} \right] + \rho_g \left[\alpha_g \cdot \frac{\partial u_g}{\partial t} + \alpha_g u_g \cdot \frac{\partial u_g}{\partial z} \right] + \alpha_g \cdot \frac{\partial p_g}{\partial z} = \dot{m}_L u_i - \tau_i a_i - \tau_g a_g \quad (9.135)$$

It should be noted that $\dot{m}_L = -\dot{m}_g$ and the interfacial pressure $p_i = p_g$. Therefore, Eqs (9.66) and (9.67) can be rearranged as

$$-\frac{\partial p_g}{\partial z} = \rho_g \left[\frac{\partial u_g}{\partial t} + u_g \cdot \frac{\partial u_g}{\partial z} \right] + \frac{\dot{m}_L}{\alpha_g}(u_g - u_i) + \frac{\tau_i a_i}{\alpha_g} + \frac{\tau_g a_g}{\alpha_g} \quad (9.136)$$

Recalling $\frac{\partial P_i}{\partial z}$ from Eq. (9.66) and equating it with Eq. (9.68), the following can be obtained, using $\alpha = \alpha_g$:

$$\frac{\partial h_L}{\partial t} + u_L \frac{\partial u_L}{\partial z} + \frac{1}{\rho_L \alpha_L} [\alpha(\Delta p'_{Li}) + \Delta p_{Li} \alpha'_L] \frac{\partial h_L}{\partial z} = -\frac{\tau_L a_L}{\rho_L \alpha_L} + \frac{\tau_i a_i}{\rho_L} \left(\frac{1}{\alpha_L} - \frac{1}{\alpha} \right) + \frac{\dot{m}_L}{\rho_L} \left(\frac{u_L - u_i}{\alpha_L} + \frac{u_g - u_i}{\alpha} \right)$$

$$+ \frac{\rho_g}{\rho_L} \left(\frac{\partial u_g}{\partial t} + u_g \frac{\partial u_g}{\partial z} \right) + \frac{\tau_g a_g}{\rho_l \alpha} \quad (9.137)$$

Note that similar analysis can be performed for different flow patterns, and this can be found in many textbooks such as Wallis (1969).

Example 9.2

A differential pressure transducer in a particular experiment was used to measure the average void in a test pipe, as shown in Figure 9.8. The transducer is installed on a piping configuration

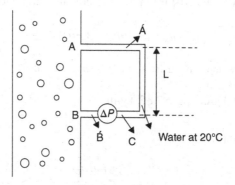

Figure 9.8   Schematic of experimental setup

as shown, with a vertical length (L) of 25.4 cm. The instrumentation line is filled with water at room temperature of 20°C. Calculate the pressure drop (ΔP) across the transducer for the following cases:

a. The pipe is empty.
b. The pipe is filled with water at room temperature.
c. The test pipe is filled with steam/water mixture at 200°C with $\bar{\alpha} = 0, 0.25, 0.5, 1$

Solution
Assume static fluid between point A and A',

$$P_A = P_{A'}$$

and between B and B',

$$P_B = P_{B'}$$

At point C,

$$P_C = P_{A'} + \rho_w g L = P_A + \rho_w g L$$
$$\Delta P = P_C - P_B = (P_A + \rho_w g L) - P_B \qquad \text{(I)}$$
$$= (P_A - P_B) + \rho_w g L$$

In the test pipe

$$P_A - P_B = \bar{\rho} g L \qquad \text{(II)}$$

Using the homogenous model

$$\bar{\rho} = (\alpha \rho_g + (1 - \alpha)\rho_f) \qquad \text{(III)}$$

Substituting from Eqs (II) and (III) in Eq. (I), the pressure difference can be calculated as

$$\Delta P = \rho_w g \cdot L - (\alpha \rho_g + (1 - \alpha)\rho_f) \cdot g \cdot L$$

The following can be determined from the property tables:
At 20°C $\rho_w \cong 1000$ kg/m^3
At 200°C $\rho_f \cong 864.3$ kg/m^3, $\rho_g \cong 7.85$ kg/m^3

a. Where the pipe is empty

$$P_A - P_B = \bar{\rho} g L = 0$$

$$\Delta P = \rho_w g \cdot L = 2.491 \text{ kPa}$$

b. Where the pipe is filled with water at room temperature

$$P_A - P_B = \rho_w g L$$

$$\Delta P = \rho_w g \cdot L - \rho_w g \cdot L = 0$$

c. Where the pipe is filled with steam/water mixture

$$\bar{\alpha} = 0$$

$$\Delta P = \rho_w g \cdot L - \left(\alpha \rho_g + (1-\alpha)\rho_f\right) \cdot g \cdot L$$

$$= 1000 \times 9.81 \times 0.254 - (0 \times 7.85 + (1-0) \times 864.3) \times 9.81 \times 0.254$$

$$= 0.337 \text{kPa}$$

We can compute the other value at
$\bar{\alpha} = 0.25$
$\bar{\alpha} = 0.5$
$\bar{\alpha} = 1$
Finally, the answers are tabulated as follows:

$\bar{\alpha}$	ΔP (kPa)
Pipe empty	2.491
Pipe filled with water at room temp.	0
$\bar{\alpha} = 0$ at 200°C	0.337
$\bar{\alpha} = 0.25$	0.871
$\bar{\alpha} = 0.5$	1.404
$\bar{\alpha} = 1$	2.471

9.2 Multiphase Flow through Centrifugal Pumps

In centrifugal pumps, centrifugal force is considered the dominant force compared with gravity. Therefore, Sachdeva et al. (1992) have shown that the diffuser performance can be neglected, especially for multiphase flow; they considered the impeller behavior only to determine the centrifugal pump performance. They developed a simple model for gas–liquid flow through centrifugal pumps based on an analytical solution. Several numerical models are available, especially in the nuclear industry, such as Zakem (1980) and Furuya (1985), but the analytical solution presented by Sachdeva et al. (1992) is the most suitable approach for the purposes of this book. But note that the development of such a model for handling multiphase through pumps is a quite challenging task given the complexity of multiphase flow behavior and the dependence of the analysis on the flow patterns and the interaction between phases. Furthermore, this model becomes more challenging given the complexity of pump geometry and multiphase flow over a rotating impeller.

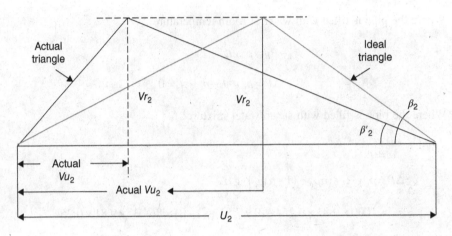

Figure 9.9   Effect of slip on output velocity triangle

For pumping a single-phase liquid, the actual velocity triangle deviates from the ideal case due to the impeller slip, as shown in Figure 9.9. This results in a reduction in useful head produced, as explained by Stepanoff (1957). These mechanical and hydraulic pump losses for single-phase flow pumping have not yet been adequately quantified. In multiphase flow pumping, the effects of impeller slip and the slip between phases make it more difficult to analyze. Therefore, given the above-mentioned challenges, simplified formulation of a multiphase pumping model is considered the most appropriate approach to be presented.

The performance of a centrifugal pump under two-phase flow condition will be determined through three main steps:

1. develop a one-dimensional pump model under liquid phase condition only, or use the pump curve published by the manufacturer
2. develop a one-dimensional two-phase flow model
3. evaluate the difference in the pump performance for cases (1) and (2) in order to determine the additional degradation in the pump head due to the free gas in the pump.

The performance curve (Figure 9.10) of the centrifugal pump operating under two-phase flow (curve D) is expected to deviate from the actual pump curve published by the pump manufacturer (curve C). In this section, a two-phase flow model will be developed in order to predict the new pump curve. Assuming ideal flow (no recirculation and no impeller slip), the two-fluid approach is used for the two-phase flow governing equations along the streamline (z) which is parallel to the impeller blades, as shown in Figure 9.11.

Continuity equation

The continuity equation along the streamline (z) for gas and liquid flows can be written as for gas

$$\dot{m}_G = \rho_G V_G \alpha A_G \tag{9.138}$$

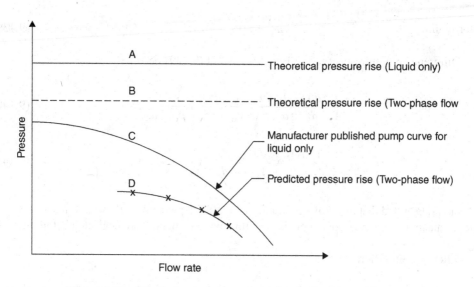

Figure 9.10   Effect of two-phase flow on the pump performance curve

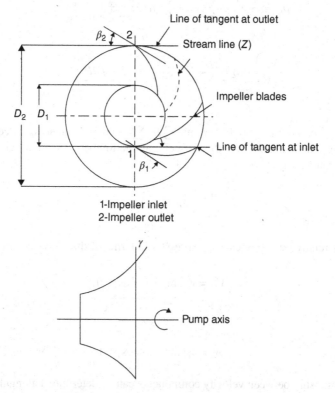

Figure 9.11   Pump geometrical parameters

for liquid $$m_L = \rho_L V_L (1 - \alpha) A_L$$ (9.139)

Differentiate with respect to (z), Eqs (9.138) and (9.139) become:

$$\frac{1}{A_z} \cdot \frac{dA_z}{dz} + \frac{1}{\alpha} \cdot \frac{d\alpha}{dz} + \frac{1}{V_G} \cdot \frac{dV_G}{dz} + \frac{1}{\rho_G} \cdot \frac{d\rho_G}{dz} = 0$$ (9.140)

$$\frac{1}{A_z} \cdot \frac{dA_z}{dz} + \frac{1}{(1 - \alpha)} \cdot \frac{d\alpha}{dz} + \frac{1}{V_L} \cdot \frac{dV_L}{dz} = 0$$ (9.141)

It should be noted that the above equations are highly dependent on flow pattern and flow pattern transition. However, for simplicity, the flow pattern changes are neglected in this analysis.

Momentum equation

The one-dimensional momentum equation can be written for each phase, assuming steady state as explained by Wallis (1969):

for gas $$\rho_G V_{rG} \frac{\partial V_{rG}}{\partial r} = \sum b_G + \sum f_G - \frac{\partial P}{\partial r} + \left(\frac{dP}{dz} \right)_{G, friction}$$ (9.142)

for liquid $$\rho_L V_{rL} \frac{\partial V_{rL}}{\partial r} = \sum b_L + \sum f_L - \frac{\partial P}{\partial r} + \left(\frac{dP}{dz} \right)_{L, friction}$$ (9.143)

where $\sum b_L$, $\sum b_G$ are the body forces and $\sum f_L$, $\sum f_G$ are the balancing forces for liquid and gas respectively. In an axial pump, the geometrical relation between angles and changes in the elevation (z) can be written as

$$\frac{dr}{dz} = \sin\beta(r) \cdot \cos\gamma(r)$$ (9.144)

Therefore the relative velocity can be expressed in terms of the above equation as

$$V_{rL} = V_L \sin\beta(r) \cdot \cos\gamma(r)$$ (9.145)

and

$$A_z = A_r \sin\beta(r) \cdot \cos\gamma(r)$$ (9.146)

Then the relationship between velocity components can be determined along the impeller path for different angles (γ) along the impeller.

The body forces due to the impeller rotation are also expressed as a function of angular speed:

$$\sum b_G = \rho_G \omega^2 r \tag{9.147}$$

and

$$\sum b_L = \rho_L \omega^2 r \tag{9.148}$$

In order to calculate the balancing forces, flow pattern is considered the key parameter that needs to be considered. For example, if bubbly flow is assumed to exist everywhere through the impeller, the drag forces for gas and liquid are given by Wallis (1969) as

$$f_{L,drag} = -C_D \frac{\alpha}{(1-\alpha)^{2.78}} \rho_L |V_{rL} - V_{rG}| (V_{rL} - V_{rG}) \frac{3}{8 r_b} \tag{9.149}$$

and

$$f_{G,drag} = C_D \frac{1}{(1-\alpha)^{1.78}} \rho_L |V_{rL} - V_{rG}| (V_{rL} - V_{rG}) \frac{3}{8 r_b} \tag{9.150}$$

The drag forces are responsible for reducing the useful head produced in the pump. It is expected that the change of two-phase flow pattern can affect the value of drag forces, but bubbly flow is considered a common flow pattern that can be observed through the pump impeller. This is mainly due to the strong mixing effect generated by the impeller rotation. Also, it should be noted that the above equations for the drag forces account for the effect of bubble swarm (bubble interaction forces). Another flow pattern of interest is churn-turbulent flow. Wallis (1969) and Craver (1984) found that the term C_D/r_b in the above equations could be replaced by a function of the void fraction $(1 - \alpha)$. As the value of C_D/r_b reduced, the gas–liquid lag increased, and consequently the liquid phase accelerated more, causing the useful pump energy to decrease. This energy reduction is responsible for the increase in liquid kinetic head. In the present analysis, the flow pattern in the pump casing is assumed to be bubbly flow in all cases and the transition to churn is neglected, as recommended by Sachdeva et al. (1992).

The other important forces in two-phase flow are the apparent mass forces (virtual mass forces). These virtual mass forces cannot be ignored for bubbly flow as they tend to reduce the gas–liquid velocity lag. They can be expressed as

$$f_{vm,G} = -C \rho_L V_{rG} \frac{d}{dr} (V_{rG} - V_{rL}) \tag{9.151}$$

and

$$f_{vm,L} = -C \left(\frac{\alpha}{1-\alpha} \right) \rho_L V_{rG} \frac{d}{dr} (V_{rG} - V_{rL}) \tag{9.152}$$

The constant (C) in the above equations can be taken as 0.5 for a spherical bubble shape. Also, the friction forces for each phase are calculated as suggested by Wallis (1969) and Craver (1984) and given in Sachdeva et al. (1992).

Using the above set of equations for gas and liquid phases, considering the geometrical relations given by Eq. (9.146), the momentum equation is reduced to

$$\frac{g}{\rho_L} \cdot \frac{dP}{dr} = \omega^2 r - \frac{1}{2} \frac{d}{dr} \left(V_{rL}^2 \right) \tag{9.153}$$

By integrating Eq. (9.152) along the impeller streamline, assuming linear relation between the inlet and outlet angle (β), the new pumping curve for two-phase will be obtained. It should be noted that the angle (γ) is assumed constant for an axial flow pump and zero for a radial flow pump. Therefore, the relation between the inlet and outlet angle can be written as

$$\beta(r) = \frac{\beta_2 - \beta_1}{r_2 - r_1}(r - r_1) + \beta_2 \tag{9.154}$$

Equation (9.143) can be rewritten as

$$\cos(\gamma) \int_0^z dz = \int_{r_1}^r \frac{dr}{\sin\beta(r)} \tag{9.155}$$

and the integration can be obtained as

$$z = \frac{\sec(\gamma)}{M} \left[\log \tan \frac{M(r - r_2) + \beta_2}{2} - \log \tan \frac{M(r_1 - r_2) + \beta_2}{2} \right] \tag{9.156}$$

In order to determine the pump curve, a value for the term C_D/r_b is required. Normally this term can be obtained experimentally. Based on 326 data points for diesel–CO_2 of Sachdeva (1988) and Lea and Bearden (1982), the following correlation is obtained:

$$\frac{C_D}{r_b} = k \frac{P_{in}^{E1}}{\alpha_{in}^{E2} Q_L^{E3}} \tag{9.157}$$

The solution vectors ($d\dot{m}_L/dz$, $d\dot{m}_L/dz$, $d\dot{m}_g/dz$, $d\rho_g/dz$, $d\alpha/dz$, dP/dz) are solved along each point of the impeller and at different flow rates for the following constants obtained for Eq. (9.157), as listed in Table 9.1. The performance curve of the pump, operating under two-phase flow condition, is

Table 9.1 Correlation constants of Eq. (9.157)

Pump type	k	E1	E2	E3
Axial K-70 pump	9.53×10^{-4}	3.33	2.83	5.92
Radial C-72 pump	6.65×10^{-4}	5.21	5.22	8.94
Axial I-42 pump	5.7×10^{10}	2.36	6.64	5.87

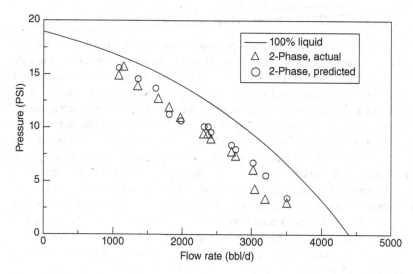

Figure 9.12   Performance of the centrifugal pump under two-phase flow condition

given in Figure 9.12. As shown in the figure, the dynamic pressure generated by the pump is lower in the case of two-phase flow and this value is expected to decrease as the void fraction increases.

9.2.1   Air Injection for Minimizing Cavitation Erosion in Hydraulic Pumps

In a centrifugal pump, one of the techniques used for minimizing cavitation erosion is injection of air inside the pump casing. By this method, the pump is artificially protected against cavitation damage by injecting small amounts of air into the cavitating region as discussed many years ago by Mousson (1942) and Anon (1945). In this method, the permanent gas in the cavitation bubble would greatly reduce the pressure originating from a collapsing bubble. Several studies have shown that the air injection technique provides an efficient method for solving cavitation erosion problems in pumps. Also, many older spillways are being retrofitted with air injection systems. However, there are differing opinions of the effects of entrained air on the performance of centrifugal pumps and it is often considered as a damaging factor in pumps. This, however, can be true if a large amount of air in the order of 5–6% exists in the impeller's eye, which consequently causes a loss in the total pump flow and an increase in the level of noise in the pumps. On the other hand, a small amount of entrained air (below about 1%) has been found to cushion the pulsation effects of cavitation and consequently reduces pump noise and minimizes erosive damage. Moreover, the addition of just 0.5% has been proved to reduce the suction pressure pulsation level in a 6-inch inlet end suction pump by 82%, as discussed by Allan and Phillip (1998). This can explain why some centrifugal pumps get quieter as the *NPSH* margin is reduced and more cavitation is observed in the pump. In this case, the normally dissolved gases in liquid quieten and cushion the cavitation implosions that occur in these pumps.

9.2.2 Centrifugal Pump Conveying Slurries

In many applications such as water treatment and sewage plants, the flow of solid–liquid mixtures is pumped through centrifugal pumps. This flow pattern is known as slurry flow. In this case, the flow is characterized based on the size of the solid particles entrained in the liquids. When the solid particles are small so that their settling velocity is much less than the turbulent mixing velocities, the flow will be well mixed, given that the volume concentration of the particles is low or moderate. In order to analyze the flow in this case, a homogeneous flow regime can be considered. The typical slurry pipelines in many practical applications have all the particle sizes of the order of tens of microns or less. However, if larger particles are present, vertical gradients will occur in the concentration, and the regime is termed heterogeneous. In this, the large particles tend to sediment faster and so a vertical size gradient will also occur. The limit of the heterogeneous flow regime occurs when the particles form a packed bed at the bottom of the pipeline. Furthermore, when a packed bed develops, the flow regime is known as a saltation flow. The solid particles in this case may be transported either due to the bed movement or by the suspending fluid. Further analyses of these flow regimes, their transitions, and pressure gradients can be found in Stepanoff (1965).

Centrifugal pumps are extensively used for slurry pipeline transportation systems. The effect of solid particles on a centrifugal pump performance is a major consideration in pump selection and slurry system design. For this purpose, the accuracy of predicted head and efficiency reduction factors for centrifugal pumps operating in slurry flow regime are essential for the design and optimization of the piping system. As discussed by Khalil et al. (2013), the performance characteristics of centrifugal pumps operating for slurry flow are affected by the size and concentration of the solid particles, abrasive property of the slurry, pumping pressures, pipe diameter, reactivity between the solids, the liquid, and the surfaces, viscosity of the liquid, critical velocity, and the slurry properties.

To evaluate the pump performance, many empirical correction factors are obtained from experimental results for single-stage centrifugal pumps. When a pump at a given speed operates under slurry flow condition instead of single-phase water, the head decreases, while the power drawn increases. The flow of solid particles through the centrifugal pump creates hydraulic losses due to the relative motion of slurry particles, which have greater inertia and cannot accelerate as rapidly as the carrier liquid, as discussed by Engin and Gur (2001). Other studies that have investigated the effect of solid concentration on the performance of a centrifugal slurry pump are discussed by Khalil et al. (2013). They mainly conclude that high solid concentration has a strong influence on the pump head, efficiency, and power consumption.

9.3 Multiphase Pumping for the Oil and Gas Industry

As many oil fields produce mixtures of oil and gas with different ratios, multiphase flow pumping without separation will greatly reduce the machinery and platform space required. Figure 9.13 shows the difference in complexity between using conventional pumps with a satellite platform and multiphase pump with a much simpler arrangement. As shown in Figure 9.13b, the multiphase production systems have significant advantages over conventional operations. Using multiphase pumping eliminates the need for an offshore structure and other process facilities. This technology offers great savings due to the reduced footprint

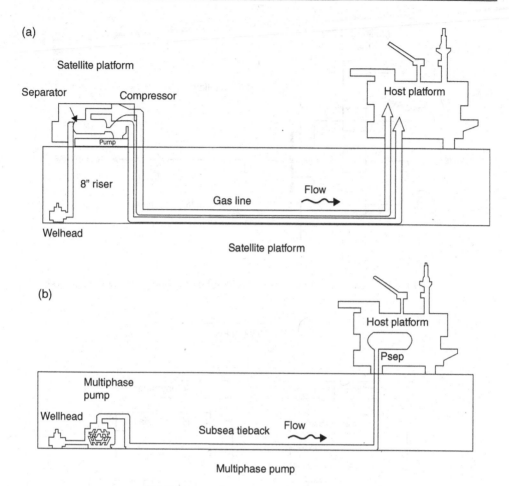

Figure 9.13   Schematic of production system

detailed in Figure 9.14 by elimination of conventional separation system described in details by Nikhar (2006). It is clear that multiphase pumping technology allows marginal oil fields to become more economic, and the field life to be extended. However, the decision to implement multiphase pumping continues to be complex and depends on many parameters. In this case, the expected boost in production should be compared against the total cost of the pumping system, maintenance, and the power requirement.

Two multiphase pumping categories are utilized for multiphase flow pumping as shown in Figure 9.15. These categories come under multistage rotodynamic pumps and positive displacement pumps. In the rotodynamic category, the mixture is progressively compressed through many stages such as in the helicon-axial or multistage centrifugal pump. In this category, higher speed is required at higher gas volume fractions (GVF) in order to accommodate the full range of GVF. In the positive displacement category – such as rotary screw pumps – two screws are timed by external gears, and the delivery pressure depends on the flow resistance, and finally the pump delivers whatever mixture is intended at the exit pressure. Several pump manufacturers

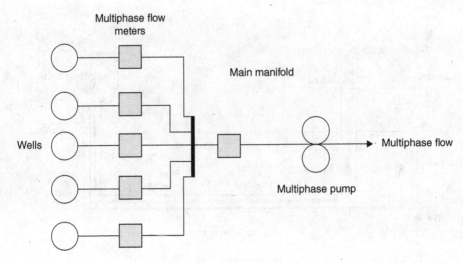

Figure 9.14   Reduced footprint and by application of multiphase pumping (Nikhar 2006)

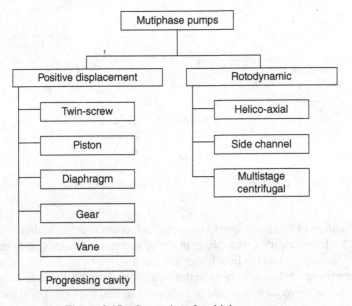

Figure 9.15   Categories of multiphase pumps

have produced a variety of pumps that handle crude oil with large amounts of entrained gas. The use of these pump installations has increased rapidly over the past years, as indicated by Scott (2002) and shown in Figure 9.16. Scott indicated that the helicon-axial pumps only represent a small number of the total multiphase pump installation and they are mainly used for off-shore and subsea applications. On the other hand, twin-screw pumps are by far the most popular multiphase pumps in use. These pumps are designed to handle high GVF and fluctuating inlet conditions. A schematic of a twin-screw pump is shown in Figure 9.17.

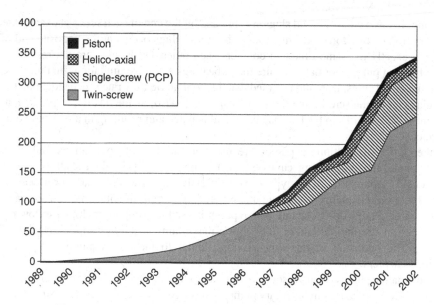

Figure 9.16 Usage of multiphase pumps (adapted from Scott 2002)

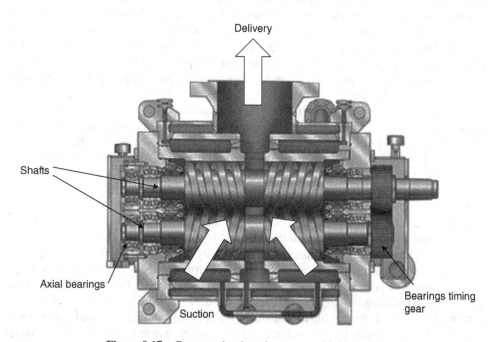

Figure 9.17 Cross-section in twin-screw multiphase pumps

Figure 9.17 shows the operation of a twin-screw pump. As shown in the figure, the multi-phase flow mixture enters at one end of the pump and splits into two flow streams which feed into the inlets located on opposite sides of the pump and then delivered along the length of the screw to the outlet end. The pump is designed so as to equalize the stresses associated with flow slugging. In these pumps, liquid slugs are split and hit the end of each screw at exactly the same time. Therefore, any force or thrust caused by liquid slugs occurs at the opposite end of each screw at exactly the same time and counters each other, producing zero net resultant force. Other types of pumps normally require thrust bearings and have limitations on their capability to handle liquid slugs. It should be noted that the volumetric flow rate is dependent on the pitch and diameter of the screws and the rotational speed. Also, as the gas is compressed, a small amount of liquid will slip back, causing internal leakage and resulting in a reduced volumetric efficiency.

Other types of multiphase pumps are used in varieties of applications and industries. For example, the progressing cavity pump has been widely used in shallow wells in oil production (Scott 2002). This type is very effective for low flow rates (<30 000 bbl/day) and it has a unique ability to tolerate a considerable amount of solids. For a moderate range of flow rates, a large double-acting piston pump is used in pumping multiphase flow mixture. This type can handle multiphase flow with a maximum capacity of approximately 110 000 bbl/day. The operation of this type is very similar to the piston pump discussed in the previous chapters.

Rotodynamic pumps are also used in multiphase pumping. The principle of operation of these pumps is similar to that of many rotary pumps. The angular momentum is created as the fluid is subjected to centrifugal forces caused by the impeller rotation. This momentum is converted into a pressure as the flow slows down and is directed through a stationary diffuser. Several models of this pump type are used in multiphase pumping such as helico-axial and multistage centrifugal designed by several manufacturers. It is important to note that helico-axial multiphase pumps have a high degree of inherent flexibility, as the speed can be varied for a required performance. Also, the number of pumps installed can be phased in as required, and the number of pump stages can be changed to match pressure change requirements if different capacity is needed at some point in the life of the field.

Note that a generalized model for multiphase pump performance does not exist, as explained in the previous section. This is mainly due to the complex and highly proprietary internal pump geometry in addition to the complex modeling of multiphase flows. Scott (2002) presented a schematic diagram (Figure 9.18) showing the typical performance of a multiphase pump. The pump curve shown in Figure 9.18 is constructed on the basis of specific GVF, suction pressure, and liquid density and viscosity. However, if this conditions change, the curve should be changed to a different one in these conditions. Scott (2002) used a steady-state multiphase flow simulator and rigorous interpolation routines to model the twin-screw pump and to determine the effect of different flow parameters on the pump performance.

While multiphase pumping eliminates the complexity encountered with a satellite platform, the complex nature of multiphase increases the difficulty of analysis of such systems. Moreover, the slug flow pattern, which is commonly found in multiphase flow systems, adds more complexity to the design of such systems, due to the flow instabilities characterizing this flow pattern. In this case, a slug catcher located on the host platform should be carefully designed in order to overcome severe slugging and to avoid the resulting dynamic forces that may damage the flow lines. Also, the high viscosity of the multiphase mixture results in a higher pressure drop in the production line. Therefore detailed analysis of a system carrying a multiphase flow mixture is essential to ensure efficient operation. Furthermore, a detailed flow assurance study should be performed in order to ensure that the temperature in the flow lines is higher than the cloud point for wax deposition and the hydrate formation temperature. Finally, the variation in

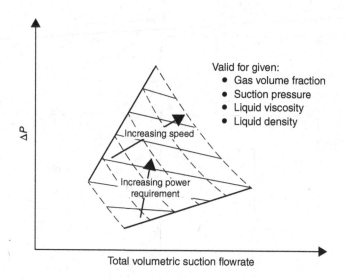

Figure 9.18   Typical multiphase pump performance curve (adapted from Scott 2002)

the performance of the multiphase pump should be investigated in order to ensure pump operability for the desired flow conditions.

9.4   Airlift Pump: an Example of Non-Conventional Pumping

The operational principle of airlift pumps has been understood since 188; however, practical use of airlift did not start until around the beginning of the twentieth century. An airlift pump is a device for raising liquids or mixtures of liquids and solids through a vertical pipe with length (L), partially submerged in the liquid (H_s), by means of compressed air introduced into the pipe near the lower end. The air then moves up in the pipe riser, carrying the liquid for a distance of (L_s), as shown in Figure 9.19. The pump operates by 'aerating' the liquid in the discharge pipe as the injected air lowers the overall specific gravity of the liquid mixture column. In comparison with other type of pumps, the manufacturing of airlift pumps is very simple. Moreover, airlift pumps do not have any moving parts and do not suffer any lubrication or wear problems. Therefore, theoretically, the maintenance of such pumps has a lower cost, and these pumps operate with higher reliability. In general, airlift pumps are used for lifting corrosive and/or toxic substances in chemical industries, conveying slurries in mining, lifting manganese nodules from deep-sea bed (Hatta *et al.* 1998), sludge removal in sewage treatment plants (Storch 1975), and lifting radioactive materials in nuclear industries, and so on. Moreover, they are easy to use in irregularly shaped wells where other deep well pumps do not fit. Airlift pumps are not available from suppliers, but they are designed and built according to the need and for specific operating conditions. Airlift pumps are generally efficient when the static liquid level is high, but the efficiency drops significantly as the available static head is reduced. In general, the main disadvantage of airlift pumps is their low efficiencies at low submergence, compared to other pumps.

Many studies have been performed to investigate the performance of airlift pumps operating in two-phase flow (Nicklin 1963; Clark and Dabolt 1986; Kassab *et al.* 2009). For airlift pumps

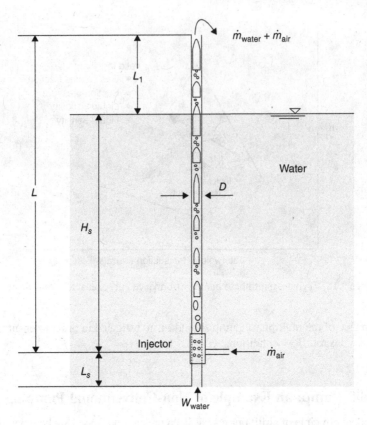

Figure 9.19   Theory of operation of airlift pump

conveying solid particles, several experimental studies are reported in the literature, but only a few studies have been carried out to analyze their performance theoretically, such as Kassab *et al.* (2007). Moreover, only uniform solid particles have been used to investigate pump performance. For example, Kato *et al.* (1975a) studied the performance of airlift pumps lifting uniform solid particles at low submergence ratios. They analyzed the pump performance based on the available two-phase flow theories. The model was developed by coupling the momentum equation of two-phase flow and the equation of motion of a single solid particle. The performance of the airlift pump was calculated, and its fundamental characteristics were obtained, neglecting the compressibility of air. They validated their model using the experimental data obtained for a 19 mm diameter pipe as a riser and small glass balls (density = 2600 kg/m^3) of 3.75 and 7.57 mm diameter as solid particles.

The study performed by Stenning and Martin (1968) is considered to be one of the most important studies to evaluate airlift pump performance. Using the continuity and momentum equations, they assumed one-dimensional flow in the pump riser and used the two-phase flow parameters to solve the governing equations. They found that a one-dimensional flow model provides a good basis for evaluating the performance of airlift pumps. Moreover, Clark and Dabolt (1986) introduced a general design equation for airlift pumps operating in slug flow regime by integrating the differential momentum equation over the pump riser length. Their model was validated experimentally. They indicated that the analysis presented by Nicklin (1963) was accurate only in the design of short pumps as the gas density changes over the pipe riser length.

Although the geometry of the pump is very simple, the theoretical study of its performance is very complicated due to the complex nature of two-phase flow encountered. Among the early classical theories used to model the airlift pump performance are those proposed by Harris, Lorenz, Gibson, and Swindin, as mentioned by Stepanoff (1929). Harris considered the force of buoyancy of the air bubbles as the motive force of the pump. He analyzed the motion of the bubble, and obtained the relation between the bubble size, two-phase slip, and the head produced. On the other hand, Lorenz developed his model by writing Bernoulli's equation for the differential head corresponding to a given flow in the pipe riser. In his analysis, he assumed variable pressure, total density of the mixture, and integrated over the head limits to obtain the relation between the variables involved. Stepanoff (1929) used thermodynamics theory in studying the effect on the efficiency of the airlift pump of the submergence, riser diameter, air-to-water ratio, climate, and the pressure at the water surface. He found that treating the airlift pump thermodynamically has a definite advantage in explaining its operation.

Yoshinaga and Sato (1996) developed a theoretical model based on the momentum equation combined with some empirical correlations available for multiphase flow. They studied the effects of pipe diameter, the submergence ratio, and the size and density of the solid particles on pump performance. In their experimental work, they used ceramic spheres of diameters 6.1 and 9.9 mm (density = 3630 kg/m^3). Two pipes of 26 and 40 mm diameters were used with the uniform ceramic balls. Several combinations of the ceramic balls were lifted using the 40 mm diameter pipe. Submergence ratios of 0.6, 0.7, and 0.8 were tested. The theoretical model was validated by comparison with the experimental results.

Another theoretical analysis of three-phase flow in a vertical pipe was performed by Hatta *et al.* (1998). They used the basic governing equations for a one-dimensional multifluid model to develop the pump model. They consider the flow pattern transitions under different flow conditions in solving the mass and momentum equations. Their analysis was later extended to include the effects of air compressibility, Hatta *et al.* (1999). They used a sudden change in the pipe diameter to account for the compressibility of air. They found that the motion of the solid particles depends strongly on the gas volumetric flux as well as the submergence ratio.

Airlift pump operating under multiphase flow conditions was also modeled by Margaris and Papanikas (1997) using a system of differential equations derived from the fundamental conservation equations of continuity and momentum using a separated flow model. They developed a general mathematical model that is applicable to a wide range of pump installations. They were able to program their equations and developed a computational tool that can be used to optimize airlift pump design. Their model used to obtain important parameters such as drag coefficients of both solid and liquid, pump efficiency, and optimum values of pipe riser diameter, length, and the location of the air injection point.

9.4.1 Modeling Airlift Pump Performance

The model of the airlift pump for two-phase flow aims to predict the liquid mass flow rate and pump efficiency as a function of the air mass flow injected at the bottom part of the pipe riser. As shown in Figure 9.19, the geometric parameters (L, H_s, L_s, and D), the pressure conditions (P_a, P_{in}), and the fluid properties are considered as the input data to the theoretical model equations. Where L is the

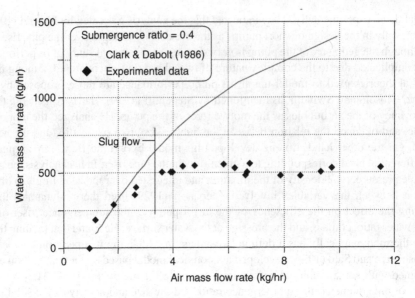

Figure 9.20   Comparison between the results of Clark and Dabolt (1986) and the experimental results of Kabbab *et al.* (2009)

riser-pipe length, H_s is the static head of water, L_s is the length of the suction part of the pipe, D is the pipe diameter, P_a is the atmospheric pressure, and P_{in} is the air injection pressure.

The theoretical predictions using the model of Clark and Dabolt (1986), together with the corresponding experimental results of Kassab *et al.* (2001), are presented in Figure 9.20 for a submergence ratio of 0.4. The agreement between these two results is quite reasonable up to an air mass flow rate of 3.4 kg/hr. However, the theoretical model does not predict the experimental data for air mass flow rates higher than 3.4 kg/hr. This comparison showed that the Clark and Dabolt (1986) model is suitable only for the first region of pump performance, where the flow regime is slug, and it is essential to develop a model which is more general to cover wider range of operating flow conditions. Therefore, a modified model for the performance of the airlift pumps is proposed by Kassab *et al.* (2009) for airlift pumps operating under two-phase flow conditions. Their approach in determining airlift pump performance is based on driving both the continuity and momentum equations, assuming one-dimensional two-fluid flow and considering the vertical pipe riser is partly full of liquid.

Since the pipe riser is filled with liquid to a static head of H_s, then the static pressure, P_o, at the base of the pipe is given by Bernoulli's equation as follows:

$$P_o = P_a + \rho_L g H_s - \frac{1}{2}\rho_L V_1^2 \tag{9.158}$$

where ρ_L is the liquid density, and V_1 is the water velocity at the inlet section. Neglecting the density changes of the air, the continuity equation can be written as

$$A V_2 = Q_g + Q_L = Q_g + A V_1 \tag{9.159}$$

where V_2 is the mixture velocity of air and water leaving the injector.

Dividing all terms of Eq. (9.159) by $Q_L = AV_1$, gives:

$$V_2 = V_1 \left(1 + \frac{Q_g}{Q_L}\right) \tag{9.160}$$

Neglecting the air mass flow rate compared to the liquid mass flow rate, the continuity equation can be written as

$$\rho_2 AV_2 = \rho_L AV_1 \tag{9.161}$$

So

$$\rho_2 = \rho_L \frac{V_1}{V_2} \tag{9.162}$$

Substituting Eq. (9.160) in Eq. (9.162), we obtain

$$\rho_2 = \frac{\rho_L}{\left(1 + \frac{Q_g}{Q_L}\right)} \tag{9.163}$$

The momentum equation applied to the injector as a control volume, neglecting the wall friction, is given by

$$P_2 = P_o - \rho_L V_1 (V_2 - V_1) \tag{9.164}$$

Substituting for V_2 from Eq (9.160) into Eq.(9.164), then

$$P_2 = P_o - \frac{\rho_L V_1 Q_g}{A} \tag{9.165}$$

Hence, combining Eq. (9.158) and Eq. (9.165), gives

$$P_2 = P_a + \rho_L g H_s - \frac{1}{2}\rho_L V_1^2 - \frac{\rho_L V_1 Q_g}{A} \tag{9.166}$$

Neglecting momentum changes caused by the flow adjustment after the mixer, the momentum equation for the upper portion of the pump can be written as suggested by Stenning and Martin (1968) in the form:

$$P_2 - P_a = \tau \frac{Lb}{A} + \frac{W}{A} \tag{9.167}$$

where τ is the average wall shear stress, b is the wetted perimeter of the pipe, and W is the total weight of the gas and liquid in the pipe.

An expression for the average shear stress, τ, was suggested by Griffith and Wallis (1961) as

$$\tau = f\rho_L \left(\frac{Q_L}{A}\right)^2 \left(1 + \frac{Q_g}{Q_L}\right) \tag{9.168}$$

where f is the friction factor assuming that the water alone flows through the pipe.

The weight of the fluid in the pipe equals the total weight of liquid plus gas, which can be obtained from

$$W = L\left(\rho_L A_L + \rho_g A_g\right) \tag{9.169}$$

where A_L is the area for the liquid phase, and A_g is the area for the gas phase.

$$A = A_L + A_g \tag{9.170}$$

$$Q_g = A_g V_g \tag{9.171}$$

$$Q_L = A_1 V_L = A V_1 \tag{9.172}$$

Substituting in Eq. (9.169) and neglecting the density of gas with respect to the liquid density, we obtain

$$W = L\frac{\rho_L A}{\left(1 + \frac{Q_g}{sQ_L}\right)} \tag{9.173}$$

where s is the slip ratio which equals

$$s = \frac{V_g}{V_L} \tag{9.174}$$

where V_g and V_L are the actual velocities of gas and liquid, respectively.

Substituting from Eqs (9.168) and (9.173) into Eq. (9.165), we get

$$P_2 = P_a + \frac{4fL}{D}\rho_L V_1^2 \left(1 + \frac{Q_g}{Q_L}\right) + \rho_L \frac{L}{\left(1 + \frac{Q_g}{sQ_L}\right)} \tag{9.175}$$

This equation, obtained by Stenning and Martin (1968), can be rewritten as

$$\frac{H_s}{L} - \frac{1}{\left(1 + \frac{Q_g}{sQ_L}\right)} = \frac{V_1^2}{2gL}\left[(K+1) + (K+2)\frac{Q_g}{Q_L}\right] \tag{9.176}$$

where K is the friction factor, which is given by

$$K = \frac{4fL}{D} \tag{9.177}$$

Stenning and Martin (1968) used the above equations in their analytical model for a fixed value of the slip ratio (s), and the friction factor (K). Physically, the slip ratio changes as the water and air mass flow rates change. Also, the friction factor changes with changes in the flow conditions. Therefore, Kassab et al. (2009) considered the variation in the slip ratio as expressed by Griffith and Wallis (1961) for slug flow in the form

$$s = 1.2 + 0.2\frac{Q_g}{Q_L} + \frac{0.35\sqrt{gD}}{V_1} \tag{9.178}$$

Also, the friction factor is obtained using the Colebrook equation, as listed by Haaland (1983), where the friction factor, f, may be obtained by solving the following equation:

$$\frac{1}{\sqrt{f}} = -2.0\log\left(\frac{\varepsilon/D}{3.7} + \frac{2.51}{Re\sqrt{f}}\right) \tag{9.179}$$

where ε is the pipe roughness and Re is the Reynolds number.

Also, the pump efficiency is calculated using the definition given by Nicklin (1963) as

$$\eta = \frac{\rho g Q_L (L - H_s)}{P_a Q_a \ln\frac{p_{in}}{P_a}} \tag{9.180}$$

The modified model by Kassab et al. (2009) was able to predict the airlift pump performance over an extended range of pump operation, as shown in Figures 9.21 and 9.22. The calculation procedure of the pump performance can be summarized as follows:

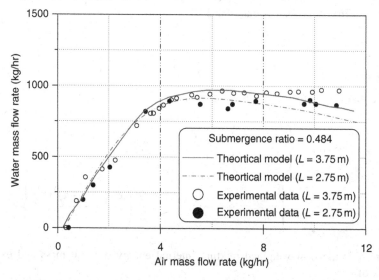

Figure 9.21   Prediction of airlift-pump performance using Kassab et al. (2009) model

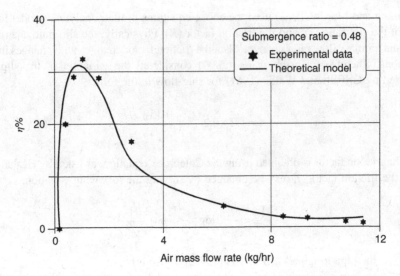

Figure 9.22 Comparison between the calculated efficiency and experiments (Kassab *et al.* 2009)

1. The geometrical parameters of L, D, pipe roughness ε, and the water density ρ and viscosity μ are known. Then for a known air inlet pressure the inlet air mass flow rate is assigned.
2. Select a static head H_s for a certain submergence ratio.
3. Assume a value of water mass flow rate.
4. Compute the coefficient of the friction f from Colebrook Eq. (9.180), and calculate the slip ratio s from Eq. (9.178).
5. Calculate the value of friction factor (K) from Eq. (9.177).
6. Calculate the value of the left-hand side and the right-hand side of Eq. (9.176).
7. Repeat steps 3–6 until the total difference between the left-hand side and the right-hand side of Eq. (9.176) becomes less than 0.001.

Problems

9.1 Prove that the energy equation can be written as

$$\frac{dp}{dz}\left[xv_g + (1-x)v_f\right] = \left\{\frac{dE}{dz} - \frac{Q_{wf}}{W}\right\}$$

$$+ \left\{p\frac{d}{dz}\left[xv_g + (1-x)v_f\right] + \frac{G^2}{2}\frac{d}{dz}\left(\frac{x^3v_g^2}{\alpha^2} + \frac{(1-x)^3v_f^2}{(1-\alpha)^2}\right)\right\} + g\sin(\theta)$$

where E is the flow-weighted mixture internal energy per unit mass and expressed in terms of

$$E = x\varepsilon_g + (1-x)\varepsilon_f$$

and W is the total mass flow rate.

Explain how the total pressure gradient can be obtained using the two-phase flow parameters discussed.

9.2 Saturated water at 7.5 MPa enters the bottom of a vertical steam generator (no preheat) in a power plant. The steam generator has a flow area of 4 m^2 and a wetted perimeter of 1000 m. Heat is transferred to the steam generator through 95% of the wetted perimeter at a constant heat flux of 50 kW/m^2. If the steam generator mass flow rate is 2000 kg/s and the flow length is 5 m. Assuming a constant two-phase flow friction factor of 0.005, what is the pressure drop for homogenous, equilibrium two-phase flow. List all the assumptions used in making the pressure drop calculations.

9.3 In a mining process, a mixture of water, air, and sand is pumped over a distance, L, in an inclined pipe of diameter D. The angle of inclination is θ.

    a. Write down a set of kinematic equations for the three-phase flow, assuming the flow to be one-dimensional.

    b. Discuss in general the boundary conditions and interfacial relationships required to solve the set of equations.

9.4 Consider a horizontal channel in a steam power generator of constant cross-sectional flow area. The inlet density, pressure, and average velocity of the water are 670 kg/m^3, 11 MPa, and 10 m/s respectively. The channel outlet pressure and average velocity are found to be 10 MPa and 15 m/s respectively. Assume that flow is homogenous and at equilibrium, and two-phase flow exists at the outlet.

    a. Determine the void fraction and quality at the outlet.

    b. Determine the size of the pump required.

9.5 Derive the energy equations for each phase separately for the case of an airlift pump. Assume that the expansion of the gas is isothermal and that there is no heat or mass transfer between the phases. Assume that the losses in the foot piece are negligible and that the liquid is incompressible. Derive an expression for the mechanical efficiency of the pump.

9.6 Consider the following data from a gas well in Brazil. The well is vertical with an internal diameter of $d = 0.062$ m and roughness of $\varepsilon = 0.00018$ m, the flow pattern is an annular. The production rates are $q_G = 188{,}000$ m^3/day and $q_o = 8.5$ m^3/day, expressed at standard conditions (10^5 Pa and 25°C, $\rho_{GSC} = 0.78$ kg/m^3). At one location in the wellbore, the following physical properties are measured. Assume incompressible liquid phase and no mass transfer between the phases and the liquid film thickness is 0.00023 m. Determine the total pressure gradient at the given location and the pumping power required for production.

References

1. Allan, R.B. and Phillip, A.M. (1998) Effects of entrained air, NPSH margin, and suction piping on cavitation in centrifugal pumps. Proceedings of the 15th International Pump Users Symposium, March 1998.

2. Anon. (1945) Cavitation in hydraulic structures: a symposium. *Proc ASCE*, **71** (7).

3. Baker, O. (1954) Design of pipelines for simultaneous flow of oil and gas. *Oil Gas J*, **26**.

4. Banerjee, S. and Chan, A.M.C. (1980) Separated flow models I analysis of the average local instantaneous formulations. *Int J Multiphase Flow*, **6**, 1–24.
5. Clark, N.N. and Dabolt, R.J. (1986) A general design equation for airlift pumps operating in slug flow. *AICHE J*, **32**, 56–64.
6. Engin, T. and Gur, M. (2001) Performance characteristics of a centrifugal pump impeller with running tip clearance pumping solid-liquid mixtures. *J Fluid Eng*, **123**(III), 532–538.
7. Furuya, O. (1985) An analytical model for prediction of two-phase (non-condensable) flow pump performance. *J Fluid Eng*, **107**, 139–147.
8. Govier, G.W., Radford, B.A. and Dunn, J.S.C. (1957) The upwards vertical flow of air-water mixtures. *Can J Chem Eng*, **58–70**.
9. Hatta, N., Fujimoto, H., Isobe, M. and Kang, J. (1998) Theoretical analysis of flow characteristics of multiphase mixtures in a vertical pipe. *Int J Multiphase Flow*, **24** (IV), 539–561.
10. Hatta, N., Omodaka, M., Nakajima, F., Takatsu, T., Fujimoto, H. and Takuda, H. (1999) Predictable model for characteristics of one-dimensional solid-gas-liquid three-phase mixtures flow along a vertical pipeline with an abrupt enlargement in diameter. *Trans ASME J Fluid Eng*, **121**, 330–342.
11. Hewitt, G.F. and Roberts, D.N. (1969) Studies of two-phase flow patterns by simultaneous X-ray and flash photography. *Report AERE-M 21 59*. HMSO, London (quoted by Collier *et al.* 1994).
12. Hewitt, G. F., and Wallis, G. B. (1963). Flooding and associated phenomena in falling film flow in a vertical tube. Multiphase Flow Symposium, Winter Annual Meeting of ASME, Philadelphia, pp. 62–74.
13. Ishii, M. and Mishima, K. (1984) Two-fluid model and hydrodynamic constitutive relations. *Nucl Eng Des*, **82**, 107–126.
14. Kassab, S.Z., Kandil, H.A., Warda, H.A. and Ahmed, W.H. (2007) Experimental and analytical investigations of airlift pumps operating in three-phase flow. *Chem Eng J*, **131**, 273–281.
15. Kassab, S.Z., Kandil, H.A., Warda, H.A., Ahmed, W.H. (2009) Airlift pumps characteristics under two-phase flow conditions. *Int J Heat Fluid Flow*, **30**, 88–98.
16. Kelly, J.E. and Kazimi, M.S. (1982) Interfacial exchange relations for two-fluid vapour-liquid flow: a simplified regime-map approach. *Nucl Sci Eng*, **81**, 305–318.
17. Khalil, M.F., Kassab, S.Z., Abdel Naby, A.A. and Azouz, A.. (2013) Performance Characteristics of centrifugal pump conveying soft slurry. *Am J Mech Eng*, **1** (5), 103–111.
18. Lea, J.F. and Bearden, J. (1982) Effect of gaseous fluid on submersible pump performance. JPT, SPE 9218. December 1982.
19. Margaris, D.P. and Papanikas, D.G. (1997) A generalized gas-liquid-solid three-phase flow analysis for airlift pump design. *J Fluid Eng ASME*, **119**:995–1002.
20. Mousson, J.M. (1942) Cavitation Problems and their Effects Upon the Design of Hydraulic Turbines. Proceedings of the Second Hydraulics Conference, State University of Iowa, Bull; 27 June 1942.
21. Nikhar, H.G. (2006) *Flow assurance and multiphase pumping*. M.Sc. thesis. Texas A&M, College Station, TX.
22. Nicklin, D.J. (1963) The air lift pump theory and optimization. *Trans Inst Chem Eng*, **41**, 29–39.
23. Radovcich, N.A. and Mossis, R. (1962), The transition from two phase bubble flow to slug flow, *Report No. 7-7673-22*, The Office of Naval Research, Department of Mechanical Engineering, MIT.
24. Rohatgi, U.S. (1982) Assesment of TRAC codes with Dartmouth College counter-current flow tests. Nuclear Technology, Vol. **69**, pp. 100–106.
25. Sachdeva, A.R. (1988) Two-phase flow through electric submersible pumps. Ph.D. thesis. University of Tulsa, Tulsa, OK.
26. Sachdeva, A.R. (1992) Multiphase flow through centrifugal pump. University of Tulsa, Tulsa, OK.
27. Schwellnus, C.F. and Shoukri, M. (1991) A two-fluid model for non-equilibrium two-phase critical discharge. *Can J Chem Eng*, **69**, 188–197.
28. Scott, S.L. (2002) Multiphase pump survey. 4th Annual Texas A&M Multiphase Pump User Roundtable; Houston; May 2002.
29. Shoukri, M. (1995a) *Two-Phase Flow and Heat Transfer Fundamental and Application to CANDU Reactor Thermal Hydraulic, Short Course*. McMaster University.
30. Shoukri, M. (1995b) *Two-phase flow and heat transfer course*. McMaster University.
31. Stenning, A.H. and Martin, C.B. (1968) An analytical and experimental study of air lift pump performance. *J Eng Power Trans ASME*, **90**, 106–110.
32. Stepanoff, A.J. (1929) Thermodynamic theory of airlift pump. *ASME*, **51**, 49–55.
33. Stepanoff, A.J. (1957) *Centrifugal and Axial Flow Pumps*. John Wiley & Sons, Inc., New York.

34. Stepanoff, A.J. (1965) *Pumps and Blowers, Selected Advanced Topics*. John Wiley & Sons, Inc., New York.
35. Storch, B. (1975) Extraction of sludges by pneumatic pumping. *2nd Symposium on Jet Pumps, Ejectors and Gas Lift Techniques*, Churchill College, Cambridge, England, G4.51–60.
36. Wallis, G.B. (1969) *One-dimensional Two-Phase Flow*. McGraw Hill Books, New York.
37. Yoshinaga, T. and Sato, Y. (1996) Performance of airlift pump for conveying coarse particles. *Int J Multiphase Flow*, **22** (II), 223–238.
38. Zakem, S. (1980) Determination of gas accumulation and two-phase slip velocity in a rotating impeller. *J Fluid Eng*, **102**, 446–455.

10

Pump Selection Guidelines

10.1 Introduction

Designing a new pumping system is one of the important engineering tasks that require prior knowledge of various system components including pumps. We need to keep in mind that the problem is mainly economic, which requires system optimization in order to achieve minimum cost of operation. Pumping system reliability is also considered as one of the important issues in many industrial applications. In fact, pump reliability becomes more important than all other factors in some special applications such as fire pumps, fuel pumps in airplane, and rocket systems and electric submersible pumps (ESP) used in oil production. In most cases, the pump selection process cannot be isolated from the piping system specifications and operating condition. For example, the total head required may change after years of operation, due to erosion and fouling in pipes and pipe fittings. Of course, the total cost of pumping depends on the characteristics of the fluid mover as well as the piping system components. If we set aside the pumping system and focus on the fluid mover, the cost of pumping depends mainly on the following:

a. Initial cost of the fluid mover (including installation and commissioning)
b. Estimated lifetime of the fluid mover
c. Cost of power consumption
d. Cost of maintenance
e. Cost of operation (labor, supervising, monitoring, etc.)

Careful design of the pumping system layout is of direct relevance to proper pump operation. For example, the location of the suction reservoir relative to pump location has a direct effect on the available net positive suction head ($NPSHA$) which influences the occurrence of cavitation, and the same applies to the design of the pump suction pipe (pipe diameter, number and types of pipe fittings, etc.). Erosion and vibration resulting from cavitation will definitely affect the pump

Pumping Machinery Theory and Practice, First Edition. Hassan M. Badr and Wael H. Ahmed.
© 2015 John Wiley & Sons, Ltd. Published 2015 by John Wiley & Sons, Ltd.

performance and may lead to pump failure. In general, pump selection necessitates determination of the pump type and size, materials of pump components, and pump design features. Several factors have a direct effect on pump selection including type and purity of the pumped fluid, system layout, system requirements, and pump reliability and maintainability. Detailed analyses of the factors of direct relevance to pump selection are presented in the following sections.

10.2   Bases of Pump Selection

A careful study of the proposed pumping system is essential for the proper selection of the type and size of the pump to be used. While most of the pump users purchase off-the-shelf units, some of them require pumps to be designed according to their specifications. One of the major problems facing the engineer in designing a new pumping system or modifying an existing one is the presence of more than one scenario, and the designer is supposed to identify the optimum. For example, in cases when the cost of power is a major part of the total operating cost, the designer will definitely give priority to pumps with topmost efficiency. That will not be the case when selecting a pump to be used in aerospace applications (aircrafts, rocket systems, etc.) where the size and weight of the pump become the most important. In other applications, the reliability of the pumping system become the deciding factor in pump selection, as in the case of ESPs used in oil production facilities. In such applications, the loss of production due to downtime of equipment is very costly. Accordingly, the bases of pump selection are numerous and the following are the common ones:

a. type of pumped fluid
b. operating condition
c. reliability and maintainability
d. initial and operating cost
e. other factors and/or special requirements

A brief discussion of each of the above factors is given in the following sections.

10.3   Selection Based on Type of Pumped Fluid

The effect of fluid properties on the hydraulic performance of dynamic and displacement pumps was discussed in detail in Chapters 5 and 7. In general, the fluid properties of direct relevance to pump selection are the following:

a. fluid viscosity and density
b. fluid chemical activity (corrosiveness)
c. flammability or toxicity of the pumped fluid
d. presence of solid particles in the fluid (e.g. sea water, crude oil, etc.)
e. presence of suspended materials (e.g. sewage pumping)
f. presence of gas contents (e.g. natural gas in oil production facilities)
g. the fluid vapor pressure and its variation during normal operation

The fluid viscosity has a significant effect on the performance of centrifugal pumps. For fluids of low viscosity (e.g. water, gasoline, and kerosene) the performance of centrifugal pumps will be almost the same as that provided by the manufacturer. However, for fluids of higher viscosity, the pump performance can be predicted [1]. In general, fluids of viscosities much higher than water cause a considerable increase of hydraulic losses that results in a substantial reduction in the overall pump efficiency. For this reason, centrifugal pumps are avoided when pumping liquids of high viscosity and displacement pumps become a better choice.

Pumping fluids of a corrosive nature requires the use of special materials for manufacturing pump components (e.g. pump casing, impeller, and shaft) in order to avoid periodic failure of such components because of excessive corrosion [2]. For example, pump casings made from reinforced fiberglass with large thickness are being used to ensure optimal mechanical strength and chemical resistance. The presence of solid particles in the pumped fluid (such as sand particles in sea water or in crude oil) causes erosion in pump components, especially the impeller vanes, due to the large velocity of particles impacting the vanes. The rate of erosion depends on several factors such as type and size of solid particles, particle concentration, flow rate, pump speed of rotation, material of pump components, and pump geometry. As a result of erosion, the inner surfaces of the impeller and casing will develop higher roughness that will eventually increase hydraulic losses causing a reduction in the overall pump efficiency. Sometimes, special coatings are used for protecting the pump components against erosion and corrosion. Such coatings, if properly applied, may reduce the roughness of the pump inner surfaces leading to an improvement in the hydraulic efficiency.

The temperature of the pumped fluid should be considered in the calculations of the pump critical flow rate. Since the vapor pressure depends mainly on the fluid temperature, the design of the pump suction side should be based on the maximum possible temperature (or the minimum possible value of $NPSHA$). In some hot countries, the high fluid temperature during the summer may be the cause of cavitation. Another problem may arise from the presence of foreign material in the pumped fluid (as in sewage pumps) which may cause pump malfunctioning due to impeller clogging. The impeller used in such pumps may be of open-type (single-shrouded or with no shrouds) or has a wider flow passage in order to avoid clogging.

The type of pump seal is strongly influenced by the type of pumped fluid. Simple seals (such as stuffing box/gland packing seal) are used for preventing fluid leakage in pumps handling water. Flammable and/or toxic liquids (e.g. liquid fuels, liquid ammonia, liquid CO_2) require special design of the pump sealing system since a small amount of leakage can cause catastrophic incidents. Special sealing systems (such as mechanical seals) are used in pumps handling toxic or flammable fluids in accordance with standards. Also, canned motor pumps (or seal-less pumps) are sometimes used to prevent fluid leakage when pumping such liquids. In conclusion, the type of fluid pumped has a strong influence on the selection of the type of pump to be used in a specific application.

10.4   Selection Based on Operating Condition

Pumps are used for the delivery of a required flow rate of a certain fluid when operating in a given pumping system. The details of the pumping system should be completely known including the following:

a. location of the pump/pump station relative to the fluid in the suction reservoir
b. diameter, length, and surface roughness of all pipes
c. type and location of all pipe fittings (valves, bends, filters, flowmeters, etc.)
d. normal operating temperature of the pumped fluid and expected range of temperature changes
e. availability of pump priming system
f. type and characteristics of the prime mover (simple induction motor, variable speed motor, diesel engine, steam turbine, etc.)
g. presence of a booster pump
h. methods of flow rate control (valve throttling, inlet guide vanes, speed of prime mover, etc.)
i. mode of pumping system operation (continuous, intermittent, etc.)
j. possible future expansions of the system

In large systems (water and oil pipelines, cooling water systems in large power stations and oil refineries, etc.), the designer may consider operating a number of pumps in parallel or in series. This may also be very useful in cases where the changes in the flow rate are considerable (e.g. city water systems). In such systems, the number of pumps operating in a specific time period can be optimized in order to achieve topmost system efficiency. In addition, operating a number of pumps in a given system has the advantage of adding flexibility for possible modifications, in order to meet future system expansions.

Based on the given details of the pumping system and possible expansions, the system $H–Q$ curve can be determined as described in Section 2.3 and the required pump total head can be obtained for the selected range of flow rate as shown in Figure 10.1. In the calculations of the system major and minor losses, the designer should allow for additional losses to take care of changes in the pipe roughness due to corrosion effects and also possible change of pipe diameter due to scale deposits. Figure 10.1 shows the system operating window that should match as much as possible the pump operating window.

The pump operating window may be determined by a number of limiting boundaries. The first is the range of flow rate that guarantees avoiding low and high flow rate cavitation (due to suction and discharge recirculation as well as low *NPSHA*) in addition to maintaining

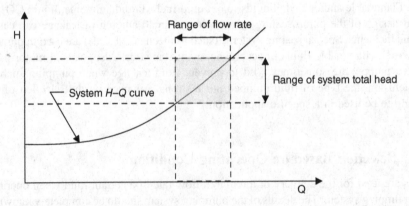

Figure 10.1   Range of system flow rate and corresponding range of the pump total head

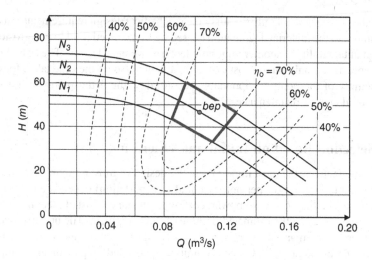

Figure 10.2 Typical operating window satisfying a minimum efficiency of 70% when operating between the speed limits N_1 and N_3

high-enough (acceptable minimum) overall efficiency. The second set of limiting boundaries is determined by the pump H–Q characteristics at the minimum and maximum operating speed (for a variable speed drive) or the minimum and maximum impeller diameters that can be used (for a constant speed drive). Figure 10.2 shows a typical pump operating window for a variable speed drive and satisfying a minimum overall efficiency. Ideally, the two operating windows should coincide for optimum operation. However, for practical considerations, the pump operating window should be slightly above the system requirements. This is mainly to absorb any miscalculations in the system H–Q curve and also to allow for future changes due to aging factors for both pump and piping system.

10.5   Selection Based on Reliability and Maintainability

In some applications, the pump reliability becomes the most important factor in the pump selection process. For example, multistage submersible pumps used in crude oil production need to have trouble-free operation for a long time period in order to avoid interruption of oil production. In this case, the downtime of equipment is very costly and should be minimized. Such pumps operate in a very harsh environment because of handling a mixture of oil, water, and sometimes gas contaminated by sand particles. Most of these pumps are equipped with journal bearings lubricated by the pumped fluid. The pumps used in aircrafts, rockets, and spacecraft represent another example in which pumps are designed to have high reliability and also meet a number of design constraints, including weight and size limitations. Pumps used in the auxiliary cooling system in nuclear power plants represent a third example in which pump reliability is extremely important. Fire pumps used in stationary or mobile firefighting systems must also have higher reliability. In general, the pump reliability depends on the pump design, manufacturing, and type of materials used. The materials used in manufacturing various pump components are selected not only to satisfy the mechanical design requirements but also to have high resistance to the material degradation caused by the pumped fluid (e.g. corrosion, erosion).

Pump maintainability is another important factor in pump selection. The cost of pump maintenance depends not only on the cost of spare parts but also on the time taken for dismantling and reassembly of the pump components. Accordingly, the better pump design should have minimum total cost of maintenance.

10.6   Selection Based on Initial and Operating Cost

The economic factor is the most important in the design of most pumping systems. The direct cost of operation per year (or per unit of liquid transportation) depends on the initial cost of equipment and installation, cost of power, cost of maintenance, cost of labor including periodic technical inspection, and the estimated lifetime of the equipment. The unscheduled shut down of the pumping system and the corresponding loss of production due to pump failure can be very costly. To avoid such loss, standby units are installed as part of the pumping system and also diesel/gasoline driven units are installed for use in case of power failure.

The initial cost of the pump is greatly affected by the pump design and materials of its components which are again influenced by the type of pumped fluid [3]. The changes in the pump lifetime and cost of maintenance resulting from using expensive materials should be assessed. The cost of unscheduled shut down of the pumping system because of pump failure should also be considered.

In some applications, the cost of power is negligibly small, due to low total head, low flow rate, intermittent operation, or the combined effect of these factors. In such applications, the designer may opt to select a cheaper pump even if it has lower efficiency. In other applications, the cost of power constitutes the main operating cost and the use of a highly efficient pump becomes a necessity. It is also important to keep in mind the matching between the best efficiency point of the pump and that of the prime mover. A considerable mismatch between these two points will result in an unnecessarily high cost of power. The calculations of the cost of power may be straightforward in simple systems, but it can be quite involved in others, especially when the flow rate variation dictates the use of a different number of pumps in parallel or in series at different times. With a variable speed prime mover, the cost of power should be based on operation at the pump rated speed.

10.7   Other Factors Affecting Pump Selection

There are many other factors relevant to pump operation having a strong influence on pump selection. The following represent some examples:

a. Submersible multistage pumps used in oil production have size and shape constraints imposed by the borehole. The outer diameter of the pump casing is normally in the range of 4–7 inches and may reach 10 inches in some cases. Due to the very high head requirements, the number of pump stages may reach 30 or more. Also, due to size limitations, the impeller is normally of mixed-flow type.

b. Weight and size limitations are also found in pumps used in aircraft, rockets, and spacecraft. Because of these limitations, the pump may be driven at a very high speed in order to develop the required pressure head. In some cases, the impeller of these pumps is equipped with an inducer for reducing cavitation pitting.

c. Environmental and economic factors may dictate the use of seal-less pumps for eliminating fluid leakage in cases of handling toxic, hazardous, or expensive fluids. In this case, the impeller is powered by a magnetic drive, and both are housed in a self-contained casing.

d. In some applications, the pump is used not only as a liquid mover but also as a fluid meter. In such cases, displacement pumps are used since the number of shaft revolutions gives a reasonably accurate measure of the volume of supplied fluid.

e. Handling a mixture of liquid and suspended solids or liquid and gas require special considerations in pump design and materials used.

f. Pumps used in the food industries may require a very low level of mixing or shear that makes lobe or screw pumps the most suitable.

g. In some cases, there are difficulties of using shaft power (mechanical work) for driving the pump. The designer may opt in these cases to use airlift or jet pumps.

h. Pumps used in medical applications may have special design to meet certain operating conditions (handling non-Newtonian fluids, precise flow rate requirements, minimum pressure pulsation, etc.). Examples are gear pumps, peristaltic pumps, and the artificial heart.

In conclusion, the process of pump selection is multidimensional since it involves the piping system characteristics, current and future flow rate requirements, type of pumped fluid, operating conditions, environmental constraints, economic factors, and other special requirements. In each case, the designer should clearly identify the set of objectives relevant to pump selection and also list the targeted priorities (e.g. economical optimization, reliability, safety and environmental constraints, or maintainability) in order to come up with the right decision.

References

1. Karassik, I.J. (1986) *Pump Handbook*, 2nd edn. McGraw-Hill, New York.
2. Anderson, H.H. (1980) *Centrifugal Pumps*. 3rd edn. Trade & Technical Press, Morden Surrey.
3. Warring, R.H. (1979) *Pumps Selection, Systems, and Applications*, 2nd edn. Trade & Technical Press, Morden Surrey.

Index

Pumping Machinery Theory and Practice, First Edition. Hassan M. Badr and Wael H. Ahmed.
© 2015 John Wiley & Sons, Ltd. Published 2015 by John Wiley & Sons, Ltd.

Printed in the United States
By Bookmasters